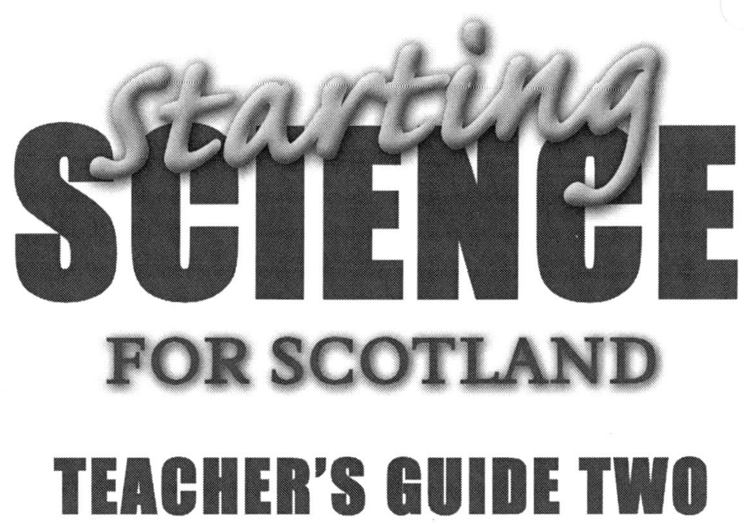

TEACHER'S GUIDE TWO

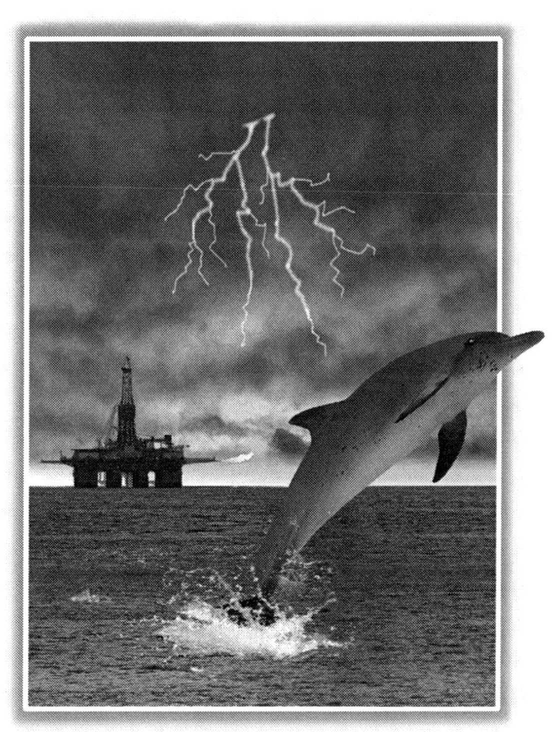

OXFORD
UNIVERSITY PRESS

OXFORD
UNIVERSITY PRESS

Great Clarendon Street, Oxford OX2 6DP

Oxford University Press is a department of the University of Oxford. It furthers the University's objective of excellence in research, scholarship, and education by publishing worldwide in

Oxford New York

Auckland Bangkok Buenos Aires Cape Town Chennai Dar es Salaam Delhi Hong Kong Istanbul Karachi Kolkata Kuala Lumpur Madrid Melbourne Mexico City Mumbai Nairobi São Paulo Shanghai Taipei Tokyo Toronto

Oxford is a registered trade mark of Oxford University Press in the UK and in certain other countries

First published 2003

British Library Cataloguing in Publication Data
Data available

ISBN 0 19 914895 3

10 9 8 7 6 5 4 3 2

Typeset in ITC Franklin Gothic
by IFA Design, Plymouth, UK

Printed in Great Britain
by The Basingstoke Press

Introduction

Starting Science for Scotland is a completely new course prepared to support the science component of Environmental Studies under the 5–14 National Guidelines.

The material has been arranged into two volumes:
Book 1/Teacher's Guide 1 concentrate on Levels D and E (with a sprinkling from Levels C and F); Book 2/Teacher's Guide 2 concentrate on Levels E and F (with a sprinkling from Levels C and D), thereby ensuring progression throughout the course. The topic titles have been chosen to reflect those used typically in Scottish schools. Differentiation is integral throughout (the students' books use the proven format of *Starting off, Going further*, and *For the enthusiast*). A further level of differentiation is available through the additional volume: *SEN Copymasters*, that accompanies the course.

The materials presented in the *Teacher's Guide* allow for assessment to be carried out with ease – all items have been allocated a 'guide level' to assist you in determining where students are at. They are not prescriptive and should be used alongside your professional judgement. You are fully in charge of how much or how little assessment you wish to carry out.

There are two new features in this volume: chapter 20 is a selection of extension materials based outside the Guidelines; the *Glossary of terms* covers chapters 2–20 and can be used during the course or a revision aid prior to a test or examination.

David Coppock
Summer 2003

Contents

Key: *W/S* = Work sheet, *H/W* = Homework, *P/T* = Practical notes/Technician's notes, *I* = Investigation, *F/S* = Fact sheet, *T* = Test, *E* = Examination, *M/S* = Mark scheme, *W* = White, *B* = Blue

Chapter 14: Metals

Chapter 15: Light and sound

Chapter 16: Microorganisms/biotechnology

How to use this guide

The following gives a brief introduction to what is provided within the *Teacher's Guide*:

Chapter overview: this sets the context for the topic and includes listings of the book pages, worksheets, and homework sheets and their guide level. It allows the teacher to more carefully plan appropriate approaches to the topics and their suitability to individual students.

Worksheets: a selection of worksheets has been provided for chapters 11–19. However, they have been prepared to fit in with the topic under discussion rather than any arrangement constraints of the students' books. The numbering system used is therefore an indicator of where a worksheet might firstly/most appropriately be introduced. (It can of course be introduced at any point thereafter.) For some worksheets there may be more than one appropriate point of introduction.

Teacher/technician notes: These are provided for all worksheets. The facility is available for the inclusion of safety/risk information and CLEAPPS/SSERC references. Note: for some worksheets, practical notes have been omitted. All sheets are listed according to the level from which the content/context is taken.

Homework sheets: homework sheets are provided for each chapter (chapters 11–19). Like the activity worksheets, they are numbered according to their point of inclusion within the course. All sheets are listed according to the level from which the content/context is taken.

Assessment tests: these are provided for chapters 11–19. For chapters covering a narrow range of materials/levels the tests are presented singly: and therefore as appropriate for the whole ability range. These are labelled Blue/White. Where chapters cover a broad range of materials and levels the test has been presented twice: Blue (for higher abilities) and White. A Blue test offers a higher range of assessment levels. A White test offers a lower range of assessment levels. It is left to the teacher to decide which test is most suitable for an individual student according to professional judgement. Each test comes with a guide for the assignment of a level according to the marks obtained. But again, confirmation of this is for professional judgement.

Investigations: a series of graded investigations has been provided for each chapter. It is therefore possible to give the same basic context for an investigation to all students but presented differently and at a different level of operation. Each Investigation is presented at levels C, D, E, F. It is left to professional judgement as to which is given to an individual student. The investigations utilize writing frames to assist students in the organization and presentation of their work.

Skills' level guide: in order to assist with the interpretation of 'practical' skills a grid has been prepared to help with showing how progression may be achieved through investigative work. This can be used when using the *Investigations* and when assessing practical work in class. These have been used when grading the Investigations.

Internet web links: a list of Internet web links has been provided. This gives a number of sites chosen for each chapter as offering starting points for Internet research on a given topic. These are presented without recommendation and no responsibility can be accepted by Oxford University Press or the authors for the content or accuracy of any site. It should be noted that sites will change their location without notice.

Students' books' note: the students' books are presented in a three-level differentiated format *Starting off, Going further,* and *For the enthusiast.* (A further level of differentiation is provided in the *SEN Copymasters* pack.)

Glossaries/(revision lists): An additional facility is provided by the glossary of terms. A separate sheet is provided for each of the course chapters (chapters 2–20). These can be used as they are or as revision lists prior to a test or examination.

Chapter 20: This is a special chapter of extension material for use outside the prescribed area of study.

Skills level guide

We have used the following interpretations of the National Guidelines' skills levels in order to show progression more clearly.

Level	Preparing for a task	Carrying out a task	Reviewing and reporting on a task
C	Select resources/apparatus appropriate to task. Answer questions about planning and suggest outcomes. Suggest how to carry out a fair test.	Use simple apparatus (provided) to collect information. Measure length, weight, time, area, and volume using appropriate apparatus. Record (with assistance) in sketches, simple tables, graphs (given axes), and annotated diagrams.	Answer questions about results and their meaning. Answer questions about possible explanations. Give a short report of findings in response to questions.
D	Select resources/apparatus appropriate to task. Answer questions about planning and suggest outcomes, giving reasons. Able to plan a fair test. Aware of things that could be changed (variables).	Use apparatus (provided) to measure distance, time, weight, area, volume, temperature in large and small units. Select from given options appropriate way of recording findings.	Give explanations for outcomes. Give an organized report about what they did including conclusions. Identify shortcomings in their approach.
E	Select resources/apparatus appropriate to task. Decide on appropriate strategies for a fair test and predict outcomes, giving reasons. Plan a valid and reliable test for a given hypothesis. Aware of variables.	Use a range of apparatus to estimate and measure accurately distance, time, weight, area, volume, temperature in large and small units. Select from a range of options and use without assistance an appropriate way of recording results e.g. choose own scales for graphs.	Link results with hypothesis. Write illustrated, systematic, and structured report. Criticize their approach, identify shortcomings, and suggest improvements.
F	Formulate own hypothesis. Decide on appropriate strategies to investigate hypothesis. Planning includes finding out how one variable is dependent on another.	Use wide range of apparatus. Estimate and measure accurately. Select from a range of options and use without assistance an appropriate way of recording results, e.g. choose own scale for graphs.	Link results with hypothesis. Write illustrated, systematic, and structured report. Evaluate fully the relevance and reliability of evidence and identify shortcomings/ limitations.

Internet web links

Please note: These internet web links are used entirely at the reader's risk/discretion. Oxford University Press cannot be held responsible for the contents of any site listed here. The listing of a site does not imply any recommendation of any kind. As with lists of all web links, the following selection is liable to change.

General:

Planet Science
www.planet-science.com
ASE **www.ase.org.uk** includes the 'Science year CD ROMS'
National Institutes of Health in USA
www.nih.gov
BBC
www.bbc.co.uk
Channel 4 **channel4.com** 'Science in focus'

Ch 11: Earth in space

European Space Agency: human space flight
www.esa.int www.esa.int/export/esaHS/
Earth observation
www.esa.int/export/esaSA/earth.html
NASA
www.nasa.gov
Royal Observatory Edinburgh
www.roe.ac.uk
Just for fun: 'powers of 10'
www.micromagnet.fsu.edu/primer/java/scienceoptics/powersof10/index.html

Ch 12: Forces

Franklin Institute, Philadelphia
www.fi.edu especially **sln.fi.edu wind/wind energy/windmills**
BBC revision materials for 11-14: look out forces
www.bbc.co.uk/schools/11_16/subjectsr_2.shtml#sc.
K'nex
www.knex.com this is kit materials and pupils may have used this in technology, there is a design competiton each year, the site has links to classroom work
ESA's materials on space flight and weightlessness see Ch11

Ch 13 : Keeping the body working

Learning connections
www.learning-connections.co.uk /
www.learning-connections.co.uk/curric/cur_pri/h_body/links.html
For pictures and descriptions
www.innerbody.com
FAQs, information, 'National smile week'
www.dentalhealth.org.uk
BBC human body
www.bbc.co.uk/science/humanbody
Madscience
www.madsci.org for teachers; lots of information and an ask a scientist question line

Ch 14: Metals

Access either
www.webelements.com or periodic table at **www.chemsoc.org** then follow the information for the metal you are interested in.
General chemistry information
www.chemdex.org (Sheffield University)
www.chem.ed.ac.uk 'Bunsen learner' (Edinburgh University)

Ch 15: Light and sound

Florida State University
www.micro.magnet.fsu.edu/optics/index.html for microscopy/ living things topics
(lots of other things of interest in this site)
Optics for kids
www.opticalres.com/kidoptx.html#StartKidOptx
How stuff works: lenses, telescopes, musical instruments, hair colouring
www.howstuffworks.com

Ch 16: Microbes/biotechnology

Edinburgh University has great micro site
www.bto.ed.ac
Microbial world
helios.bto.ed.ac.uk/bto/microbes/microbes.html
Molecular expressions has great pictures, (of) microscopes
www.micro.magnet.fsu.edu '
NHS helpline lots of disease information /encyclopaedia
www.nhs.direct.nhs.uk
National Centre for Biotechnology
www.ncbe.reading.ac.uk

Ch 17 :Changing materials

Weather forecasting and much more, webcams, rainfall animations
www.metoffice.com
Look at the US weather at
www.nws.noaa.gov
Science Museum exhibition the 'Challenge of materials' offers information quizzes
www.sciencemuseum.org.uk/on-line/challenge/
Crystals (ice-cream making), catalysts, acids
www.chemsoc.org

Ch 18: Electromagnetism/electronics

Microprocessors: for something different
www.micro.magnet.fsu.edu
Channel 4 **channel4.com** 'Science in focus' series
How stuff works: mobile phones, computers
www.howstuffworks.com
Exploratorium in San Francisco
www.erie.net
Institute of Physics (for teachers)
www.iop.org

Ch 19: Our environment

Natural History Museum: biodiversity, 'Quest challenge on-line' **investigations www.nhm.ac.uk**
Desert ecology from Edinburgh University
www.helios.bto.ed.ac.uk/bto/desertecology
Global climate change
www.exploratorium.edu/climate/index.html
Evolution
www.ucmp.berkeley.edu/history/eveolution.html
National Environment Research Council
www.nerc.ac.uk

11 Earth in space

This chapter looks at the Solar system. Many pupils find this a difficult topic to grasp conceptually, but it is hoped that by providing a varied selection of laboratory and homework activities a better understanding will be achieved. It is recommended that this topic is studied during the winter months, between October and March. This enables many of the 'dark' activities to be carried out by pupils at the end of the school day

Assessment opportunities

Formative assessment opportunities are provided by worksheets, homework sheets, and an investigation.

The **worksheets** cover material at levels C, D, and E for attainment targets for knowledge and understanding. Teachers may wish to use these worksheets not only as part of practical activities but also to provide evidence of pupil achievement.

Worksheet	Level
11.1a	C
11.1b	D
11.2a	C
11.2b	D
11.3	E

The **homework sheets** cover material at levels C and E for attainment targets for knowledge and understanding. These homework sheets can be used individually as a follow-up to work done in class or assembled into a homework booklet allied closely to schemes of work.

Homework sheet	Level
11.1a	C
11.1b	E
11.1c	E
11.3	E

The **investigation** covers all three skill areas at levels C, D, E, and F. It is written in a way that allows for pupils to be assessed in all three skill areas at one level. Alternatively, customised assessments can be constructed enabling pupils to be assessed at different levels in all three skills. The latter approach is more time consuming, but it does provide the opportunity for pupils to show evidence of achievement at different levels in different skills in the same investigation. Teachers will need to use their professional judgement when deciding which level is appropriate to individual pupils. It is envisaged that pupils will show progression through the levels as they work through their science course.

Summative tests are provided at two levels, white and blue. The white test contains questions covering attainment target levels C, D, and E. The blue test contains questions covering attainment target levels D, E, and F. Each test has a total of 30 marks and will take about 30 minutes for pupils to complete, although this can be varied depending on pupil ability. Mark schemes are provided together with suggested grade/level boundaries. It is envisaged that these tests will be given to pupils on completion of the material covered in Chapter 11.

ICT opportunities

The use of data loggers/remote sensors can extend the range, speed, and sensitivity of measurements in many of the worksheets for this chapter. Once downloaded onto a PC, data-handling programs can be used to analyse information gathered, data can be manipulated, and appropriate graphs etc. presented. The Internet provides pupils with access to a huge range of scientific information. A list of suitable websites is included in this Teacher's Guide.

Students' book chapter 11 contents and guide levels

11.1	Earth in space	*Starting off*	C
	The changing seasons	*Going further*	E
	When the light goes out	*For the enthusiast*	E
11.2	A snapshot of our Solar System	*Starting off*	C
	Data from space (1)	*Going further 1*	D
	Data from space (2)	*Going further 2*	D
	People in space	*For the enthusiast*	D
11.3	Amazing space	*Starting off*	E
	Mysteries of space	*Going further*	E/F

11.1a The Solar System W/S

Name: Date: Group:

What you need:

Scissors, glue.

What to do:

1 Cut out this diagram and stick it into your book.

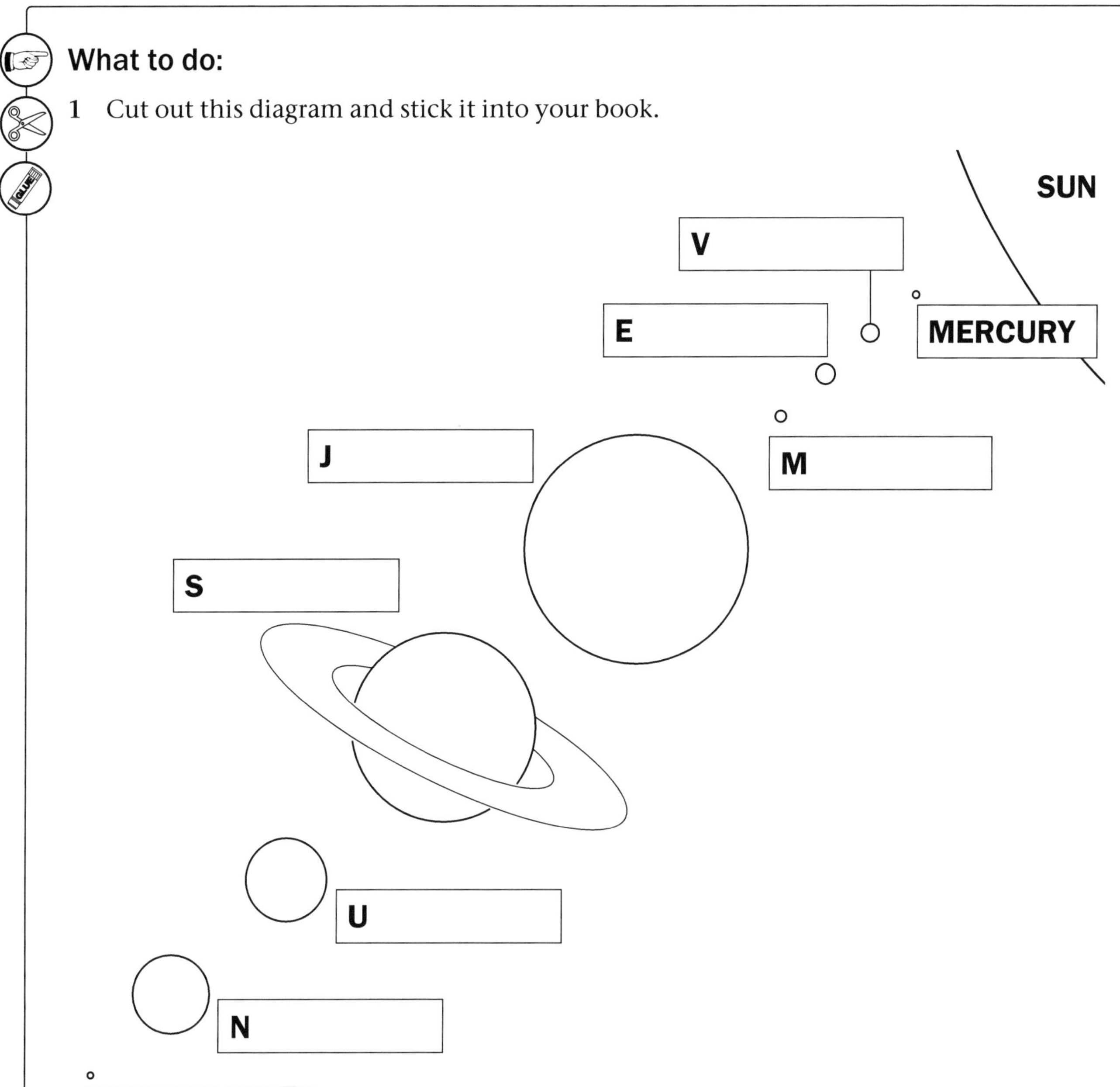

2 Write the names of the planets in the boxes.

11.1a
Practical notes

The Solar System

This simple activity helps pupils learn the relative positions of the planets in the Solar system.

11.1a
Technician's notes

The Solar System

Each pupil will need:

- a copy of worksheet 11.1a
- scissors
- glue.

Number of apparatus sets:

Number of pupils:

Number of groups:

Visual aids:

ICT resources:

Equipment/apparatus needed:

Safety notes

CLEAPSS/SSERC SAFETY REFERENCE:

11.1b Phases of the Moon

W/S

Name: **Date:** **Group:**

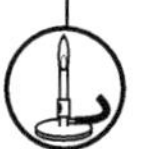

What you need:

Eight table tennis balls, black felt pen, masking tape, drinking straws, Blu-tack, scissors.

What to do:

1 Shade half of each table tennis ball with the felt pen. (Put some tape around the ball to help you).

Fix each ball to a straw with Blu-tack as shown in the diagram.

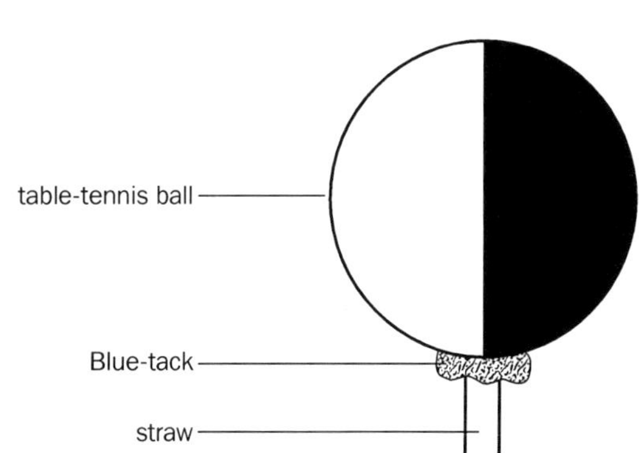

2 Stick each straw to the table with Blu-tack like this. Make sure all white sides face the same way.

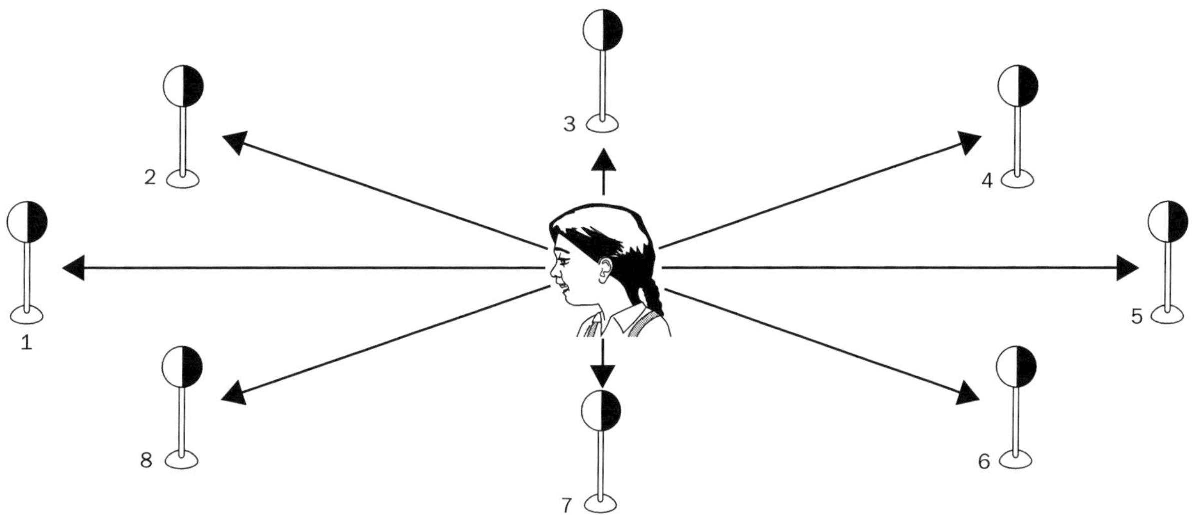

3 Look at each table tennis ball (in the direction shown by the arrows).

Draw the shape of the white side of the ball (that you can see) in the space below.

The first two have been done for you.

  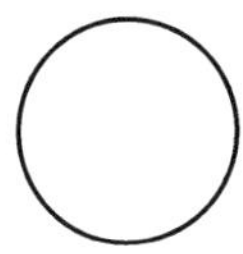 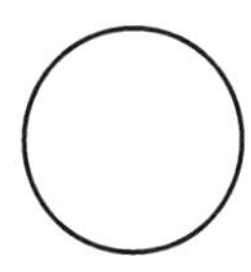 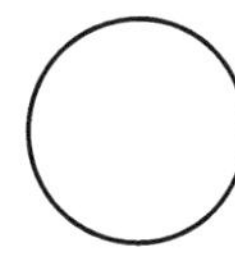 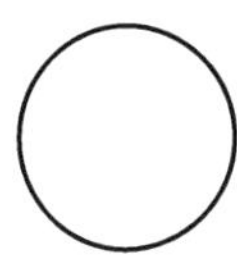 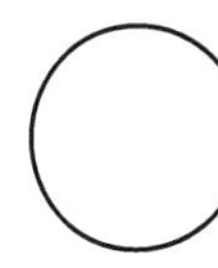 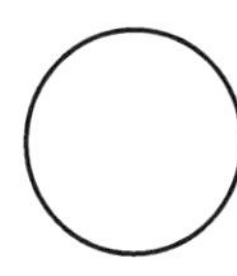

1 2 3 4 5 6 7 8

11.1b
Practical notes

Phases of the Moon

This activity gives pupils the opportunity of seeing how the shapes of the Moon change during a lunar month. An alternative would be for each group to use one table tennis ball, moving it to the eight positions in turn. One-sided lighting is essential for this activity.

11.1b
Technician's notes

Phases of the Moon

Each pupil will need:

- a copy of worksheet 11.1b
- eight table tennis balls
- black felt pen
- masking tape
- drinking straws
- Blu-tack or Plasticine
- scissors.

Safety notes

CLEAPSS/SSERC SAFETY REFERENCE:

Number of apparatus sets:

Number of pupils: .... Number of groups: ....

Visual aids:

....

....

ICT resources:

....

....

Equipment/apparatus needed:

....

....

11.2a How far is the Moon from Earth? W/S

Name: Date: Group:

What you need:

Metre stick, tape measure, 1p coin, Blu-tack or Plasticine, a clear night with a full Moon.

What to do:

1 Support the metre stick on a windowsill, fence, or wall so that your eye, the stick, and the Moon are in line.

2 Move the coin backwards and forwards along the stick until it just covers (eclipses) the Moon. Fix the coin in place with Blu-tack and check its position again. Adjust it if necessary.

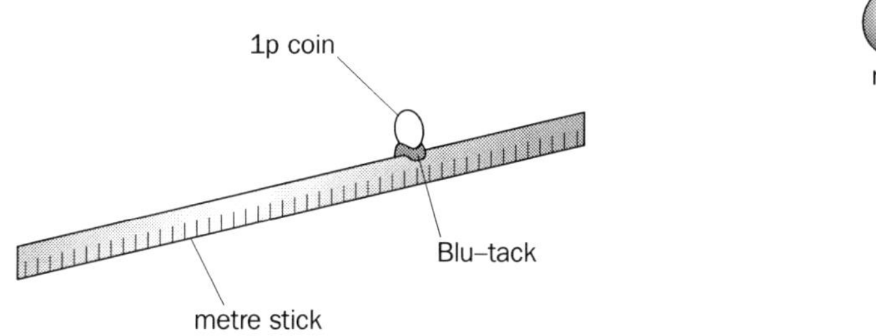

3 Measure the distance from the coin to the end of the stick nearest your eye and make a note of it.

4 The diameter of the Moon is about 3500 km.
The diameter of a 1p coin is 0.02 m
Use this formula to work out the distance of the Moon from Earth:

$$\text{Distance of Moon from Earth} = \frac{3\,500\,000 \text{ m} \times \text{distance of coin to eye in metres}}{0.02 \text{ m}}$$

Distance of Moon from Earth = ________ m

(You can change this to kilometres by dividing by 1000.)

11.2a
Practical notes

How far is the Moon from Earth?

This activity is best done as a home experiment. A broom handle makes a good substitute for a metre stick and most homes have some sort of tape measure. Blu-tack or Plasticine can be issued as necessary. Some pupils will have difficulty with the calculation and teachers may wish to leave this part of the activity until pupils are together in school.

11.2a
Technician's notes

How far is the Moon from Earth?

Each pupil will need:

- a copy of worksheet 11.2a
- metre stick
- tape measure
- 1p coin
- Blu-tack or Plasticine
- a clear night with a full Moon.

Number of apparatus sets:

Number of pupils: .... Number of groups: ....

Visual aids:

....

....

ICT resources:

....

....

Equipment/apparatus needed:

....

....

Safety notes

CLEAPSS/SSERC SAFETY REFERENCE:

11.2b Water rockets

W/S

Name: **Date:** **Group:**

What you need:

2 litre plastic bottle, rubber bung with hole, plastic tube, foot pump, thick card, sticky tape, tripod, scissors.

What to do:

1. Use the template to mark out 3 fin shapes on thick card.
2. Fix the three fins to the side of the bottle with sticky tape. Make sure they are evenly spaced around the bottle.
3. Half fill the plastic bottle with water.
4. Put in the bung and attach the foot pump to make a rocket.
5. Stand the rocket on a tripod.

6. STAND CLEAR and pump away!
7. Try making different shaped nose cones from cardboard and sticky tape.

What shape of nose cone makes your rocket fly best?

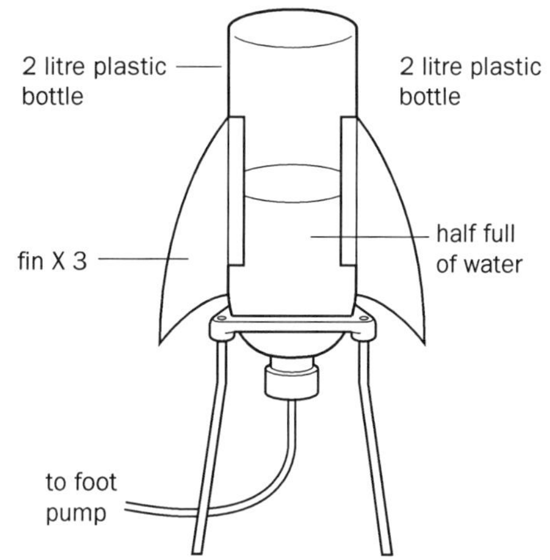

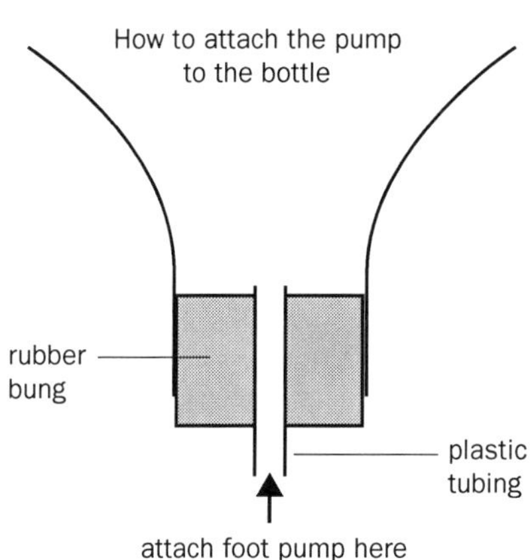

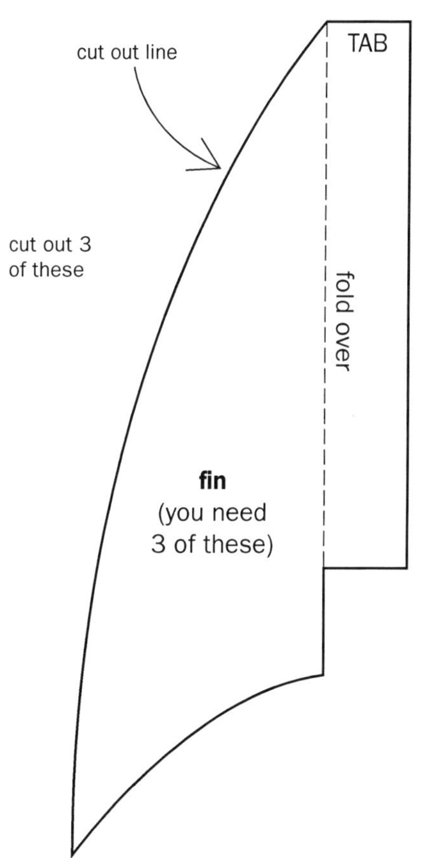

11.2b
Practical notes

Water rockets

This is a popular outdoor activity giving pupils the opportunity to investigate rocket design. Make sure to provide plenty of rubber/plastic tubing so that pupils are as far away as possible from their rockets, and don't get soaked or injured during launching. Tell pupils to thread the tubing up through the tripod before fixing the bung in the neck of the bottle, this way the rocket sits squarely on the tripod launch pad. Pupils will assume more water means more power; this is not so. The best water/air ratio is achieved by trial and error. Watch out for silly behaviour with water and make sure the rockets are pointing upwards.

11.2b
Technician's notes

Water rockets

Each group will need:

- a copy of worksheet 11.2b
- 2 litre plastic bottle (sparkling mineral water or similar)
- about 3–4 metres flexible rubber or plastic tubing
- bung to fit neck of bottle with hole fitted with rigid plastic tube to attach tubing
- foot pump, the bigger the better
- thick card
- sticky tape
- tripod
- scissors
- access to water, e.g. buckets with plastic cups/filter funnels for filling rockets.

Number of apparatus sets:

Number of pupils: ____________ Number of groups: ____________

Visual aids: ____________

____

____

ICT resources: ____________

____

____

Equipment/apparatus needed: ____________

____

____

Safety notes

- **Make sure that pupils stand well back during launching.**
- **Launch one rocket at a time.**
- **Make sure rockets are pointing upwards.**

CLEAPSS/SSERC SAFETY REFERENCE:

11.3 Star charts

Name: **Date:** **Group:**

What you need:

A clear night.

What to do:

1 Find the star chart for this time of year.

2 Hold the chart in front of you and look at the night sky. See if you can match the stars to your star chart. The bigger the 'blob' on the chart, the brighter the star.

3 Write the names of the stars you can recognise from the star chart.

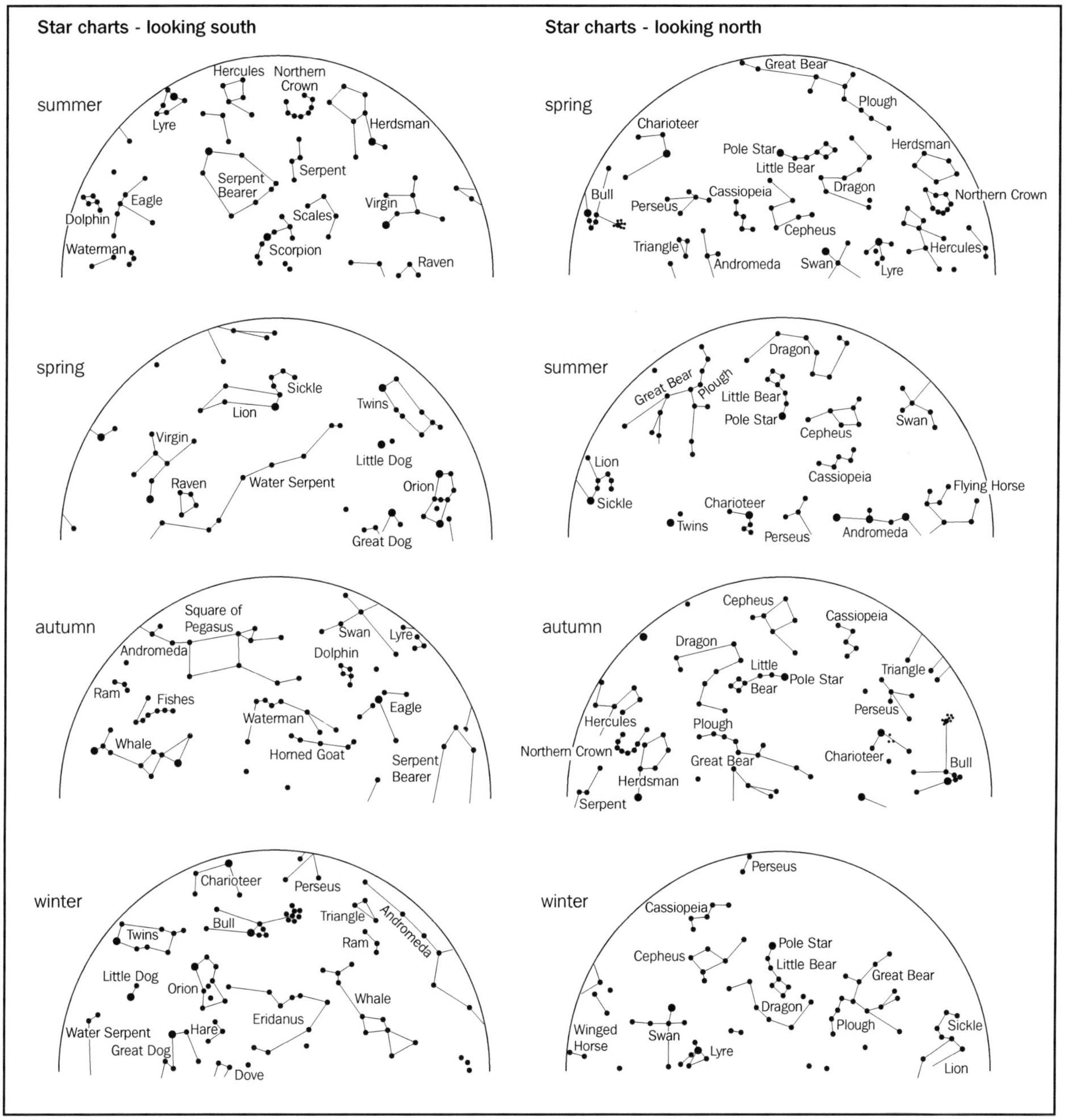

Star charts - looking south

summer: Hercules, Northern Crown, Lyre, Herdsman, Serpent, Serpent Bearer, Eagle, Virgin, Dolphin, Scales, Waterman, Scorpion, Raven

spring: Sickle, Lion, Twins, Virgin, Little Dog, Raven, Water Serpent, Orion, Great Dog

autumn: Square of Pegasus, Swan, Lyre, Andromeda, Dolphin, Ram, Fishes, Eagle, Waterman, Whale, Horned Goat, Serpent Bearer

winter: Charioteer, Perseus, Bull, Triangle, Andromeda, Twins, Ram, Little Dog, Orion, Whale, Eridanus, Water Serpent, Hare, Great Dog, Dove

Star charts - looking north

spring: Great Bear, Plough, Charioteer, Pole Star, Herdsman, Little Bear, Dragon, Bull, Cassiopeia, Northern Crown, Perseus, Cepheus, Triangle, Hercules, Andromeda, Swan, Lyre

summer: Dragon, Great Bear, Plough, Little Bear, Pole Star, Cepheus, Swan, Lion, Cassiopeia, Sickle, Charioteer, Flying Horse, Twins, Perseus, Andromeda

autumn: Cepheus, Cassiopeia, Dragon, Little Bear, Pole Star, Triangle, Perseus, Hercules, Plough, Northern Crown, Great Bear, Charioteer, Bull, Herdsman, Serpent

winter: Perseus, Cassiopeia, Cepheus, Pole Star, Little Bear, Great Bear, Dragon, Plough, Winged Horse, Swan, Sickle, Lyre, Lion

11.3
Practical notes

Star charts

This activity has to be done at night, so it could be used as an additional homework resource. Explain to pupils that they will be very lucky indeed to spot all of the stars on the chart. The more obvious star patterns such as the Plough and W-shape of Cassiopeia should be easily recognised.

11.3
Technician's notes

Star charts

Each pupil will need:

- a copy of worksheet 11.3.

Safety notes

CLEAPSS/SSERC SAFETY REFERENCE:

Number of apparatus sets:

Number of pupils:

Number of groups:

Visual aids:

ICT resources:

Equipment/apparatus needed:

11.1a The Solar System H/W

Name: **Date:** **Group:**

What you need to know …

Our Solar System is made up of the Sun, the planets, comets, and asteroids which travel round it. Earth is one of the planets. The Sun is at the centre of the Solar System. It provides all the planets with light and heat and holds the Solar System together. The Sun's gravity pulls on the planets and they travel around the Sun in paths called orbits. The table gives some information about the planets in our Solar System:

	Mercury	**Venus**	**Earth**	**Mars**	**Jupiter**	**Saturn**	**Uranus**	**Neptune**	**Pluto**
Average distance from Sun (in million km)	60	108	150	230	780	1400	2900	4500	5900
Time for one orbit (in Earth time)	88 days	225 days	365.25 days	687 days	12 years	29 years	84 years	165 years	248 years
Diameter (in km)	5000	12 000	12 750	7000	140 000	120 000	52 000	50 000	3000
Average temperature (in°C)	350 day 170 night	480	22	-23	-150	-180	-210	-220	-230

What to do:

Answer these questions about the Solar System:

1 **a** Which planet is
  - **i** closest to the Sun
  - **ii** furthest away from the Sun?

  b Which is the
  - **i** largest planet
  - **ii** smallest planet?

  c Which planet has the
  - **i** shortest year
  - **ii** longest year?

  d Which planet has the greatest temperature range?

2 What link can you see between the time for a planet's orbit and its distance from the Sun?

3 Some scientists believe that there is life on Mars. Give one reason why
  - **i** they could be right
  - **ii** they could be wrong.

4 Explain why there is no 'real daylight' on Neptune.

5 What evidence is there that Venus has an atmosphere made up mainly of carbon dioxide?

HANDY HINTS

Carbon dioxide is a greenhouse gas that causes global warming.

11.1b Day, months, years and seasons H/W

Name: **Date:** **Group:**

What you need to know …

The length of time for a planet to orbit the Sun once is called a year. Earth takes 365.25 days to orbit the Sun once. As well as travelling around the Sun, the Earth also spins on its axis. Earth takes one whole day or 24 hours to spin once. The Earth's axis is tilted, so at certain times in a year the top and bottom halves of the Earth get different amounts of sunlight. This causes the changing seasons. The diagram shows the Earth orbiting the Sun.

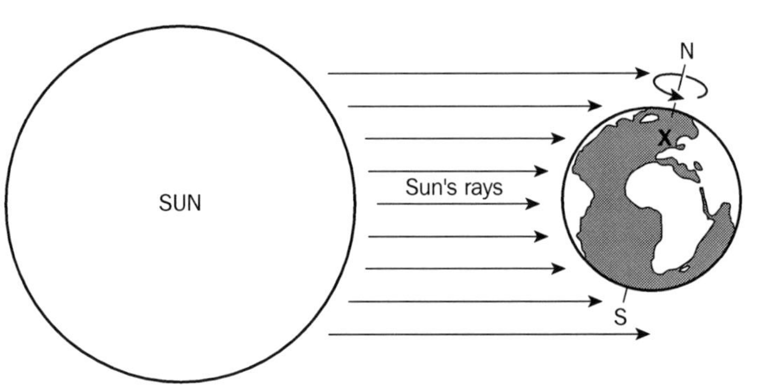

A/w A102

What to do:

1 How long does it take for the Earth to orbit the Sun?

2 Explain what a day is.

3 Copy the diagram of the Earth and shade the part where it is dark.

4 Explain why someone living at X will not always be in darkness.

5 **a** In the diagram, what season is it in
   i the northern hemisphere
   ii the southern hemisphere?

   b Explain your answer.

6 Explain why:

   a it never gets dark at the North Pole in summer

   b Australians often have Christmas dinner on the beach.

7 Use a diagram to explain why day and night are of equal lengths at the Equator.

HANDY HINTS

Try sketching the Earth as it would appear on the other side of the Sun. Remember to keep the Earth tilted in the same direction. Draw a few parallel lines to represent the Sun's rays.

11.1c Phases of the Moon H/W

Name: **Date:** **Group:**

What you need to know …

We see the Moon because it reflects the light of the Sun. But we don't always see the same shape of Moon. This is because we can only see the side that is lit by the Sun, and as the Moon orbits the Earth, the amount of the lit side of the Moon which we can see changes.

What to do:

Use this sheet to keep a record of the shape of the Moon over a five week period. Write the date and time underneath each of your sketches. If there are nights when you can't see the Moon, try and work out what shape it must have been.

HANDY HiNTS

Try to make your observations at the same time each day.
Binoculars will help but they are not essential.

	Week 1	Week 2	Week 3	Week 4	Week 5
Mondays					
date					
time					
Tuesdays					
date					
time					
Wednesdays					
date					
time					
Thursdats					
date					
time					
Fridays					
date					
time					
Saturdays					
date					
time					
Sundays					
date					
time					

11.3 Amazing space

Name: **Date:** **Group:**

What you need to know ...

We use the word Universe to mean everything which exists. Astronomers have estimated that the Universe contains 100 billion galaxies each with 100 billion stars. These are only guesses but we do know that the Universe is massive, so big that astronomers have had to invent a new unit to measure distances. That unit is the light year, the distance which light travels in one year (9.46 million million kilometres).

What to do:

1 Cut out these pictures and arrange them in order of size. Start with the picture of a pupil at their desk.

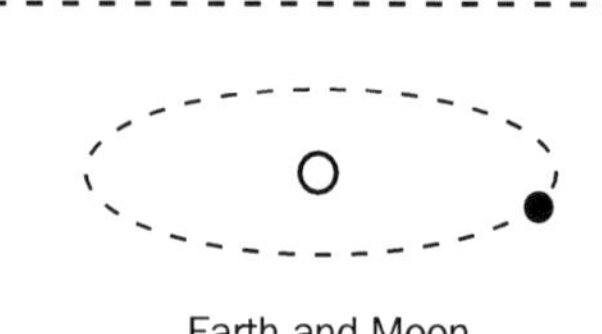

Earth and Moon

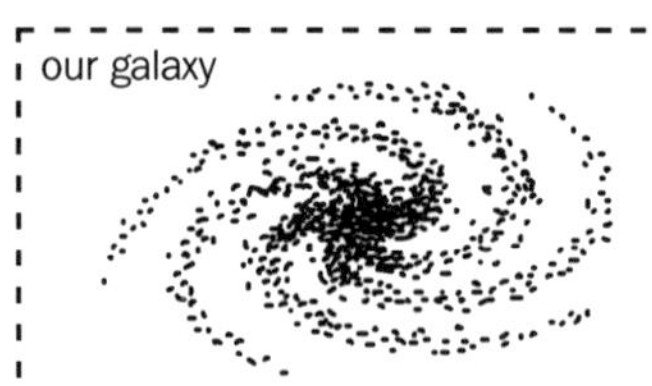

our galaxy

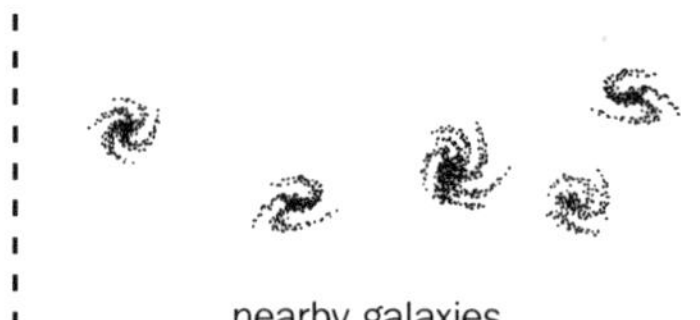

nearby galaxies

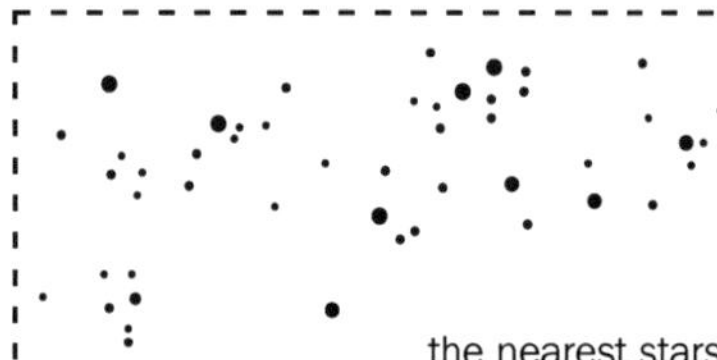

the nearest stars

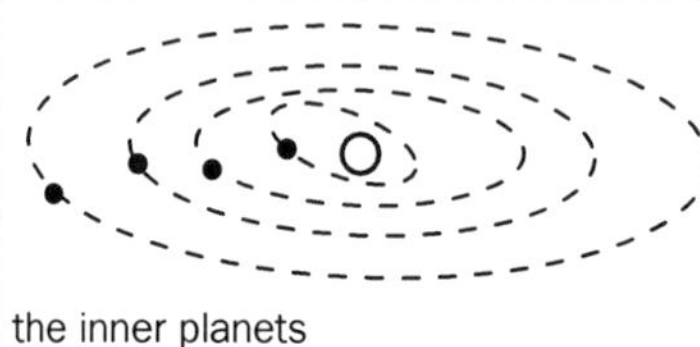

the inner planets

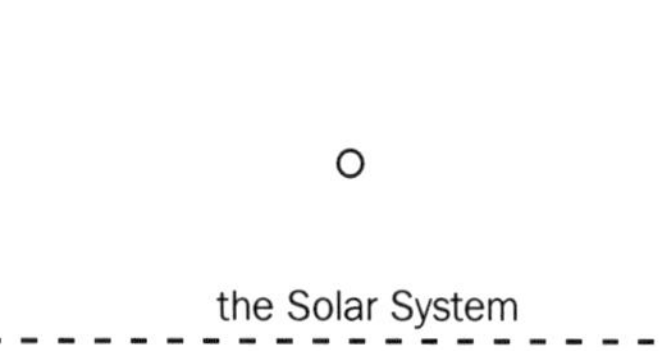

the Solar System

2 Stick the pictures into your book.

3 Match these measurements to the pictures. Write the measurement underneath the picture:

- 100 light years
- 150 million kilometres
- 1 metre
- 100 million light years
- 1000 kilometres
- 15 billion kilometres
- 1 kilometre
- 100 000 light years
- 13 000 kilometres.

4 Explain why distances in space are measured in light years and not kilometres.

HANDY HINTS

Arrange the pictures in order, then do the same with the distance measurements.

Chapter 11 ▶ Investigation 11F Orbiting planets

Planets are held in orbits around the Sun by a force. This force is the Sun's gravitational pull on the planets.

In this investigation: you are going to find out if there is a link between the speed of a planet and the force holding it in orbit.

Preparation: Predict

Finish the sentences in the box.

What I think will happen is…

I think this because…

Preparation: Plan

Write a short plan of your investigation.

Think about:

- the apparatus you are going to use
- how one variable depends upon another variable
- what you are going to measure and how you are going to measure it
- how many readings you are going to take
- how you are going to record your results
- how you are going to make your investigation fair
- how you are going to make your investigation safe.

Show your plan to your teacher before going on.

Carry out

Carry out your investigation and record your results.

Present your results in an appropriate way.

Report

Write a report on your investigation.

Here are some things you should include:

- a diagram of your apparatus
- what you did
- what happened
- explain your results
- if your prediction was correct or not
- how reliable your results were
- what you could have done if you had more time.

Chapter 11 ▶ Investigation 11E Orbiting planets

Planets are held in orbits around the Sun by a force. This force is the Sun's gravitational pull on the planets.

In this investigation: you are going to find out if there is a link between the speed of a planet and the force holding it in orbit.

Preparation: Predict

Finish the sentences in the box.

What I think will happen is...

I think this because...

Preparation: Plan

Write a short plan of your investigation.

Think about:

- the apparatus you are going to use
- what you are going to measure and how you are going to measure it
- how many readings you are going to take
- how you are going to record your results
- how you are going to make your investigation fair
- how you are going to make your investigation safe.

Show your plan to your teacher before going on.

Carry out

Carry out your investigation and record your results in a table.

Draw a bar graph of your results.

Report

Write a report on your investigation.

Here are some things you should include:

- what you did
- what happened
- explain your results
- if your prediction was correct or not
- what you could do to improve the investigation
- what you could have done if you had more time.

Chapter 11 ▶ Investigation 11D Orbiting planets

Planets are held in orbits around the Sun by a force. This force is the Sun's gravitational pull on the planets.

In this investigation: you are going to find out if there is a link between the speed of a planet and the force holding it in orbit.

Preparation: Predict

Finish the sentence in the box.

I think that there (is/is not) a link between the speed of a planet and the force holding it in orbit because...

You are going to use this equipment to find out if there is a link between the speed of a planet and the force holding it in orbit:

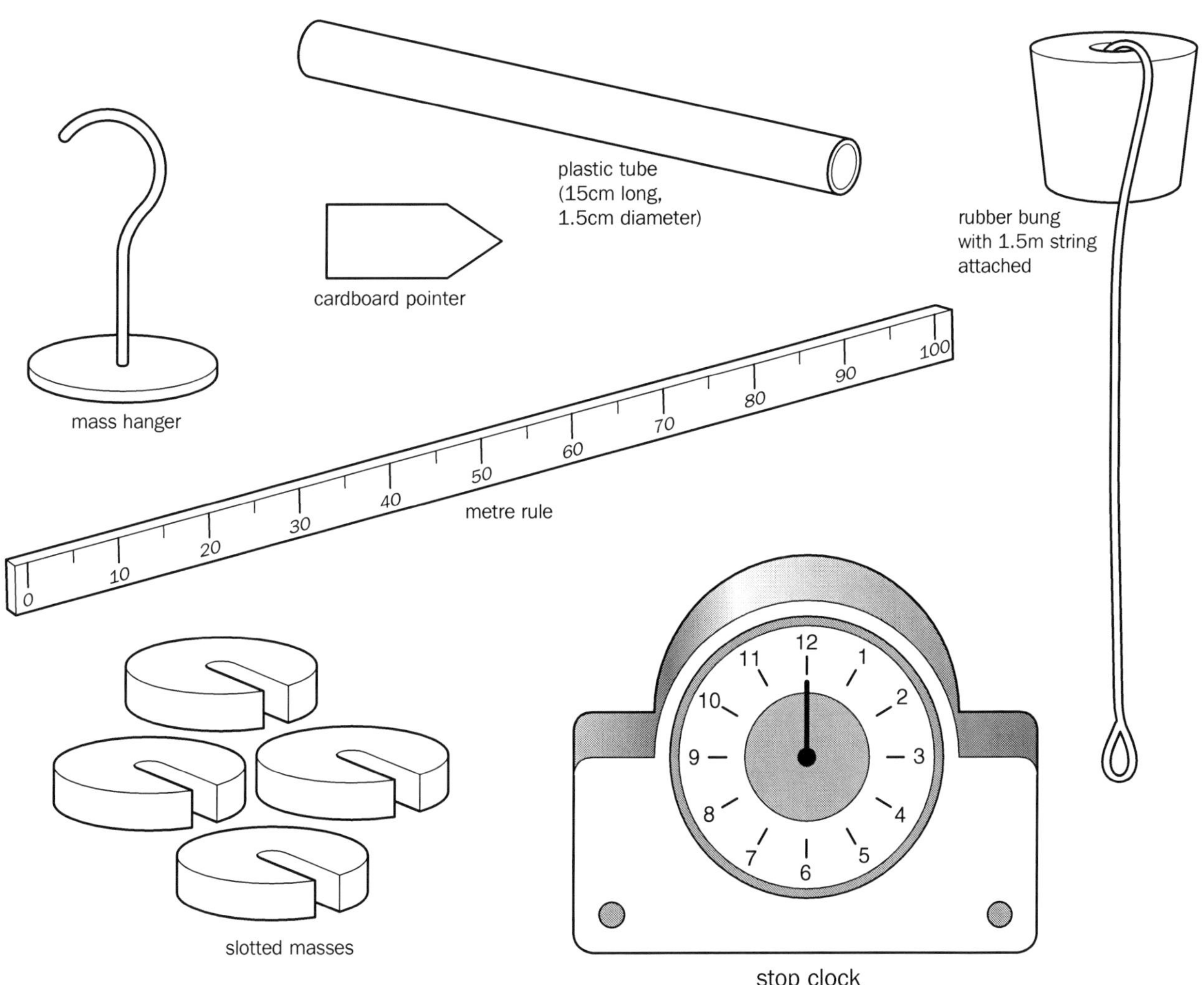

Chapter 11 ► Investigation 11D Orbiting planets

Preparation: Plan

Finish the sentences in the box.

I will measure…

Things I will keep the same are…

My investigation will be fair because…

My investigation will be safe because…

Carry out

Carefully whirl the bung around your head keeping the pointer in position. Try different masses on the mass hanger.

Force in grams	Time for ten orbits in seconds
100	
200	
300	
400	

Draw a line graph of your results on the same piece of graph paper. Label the axes like this:

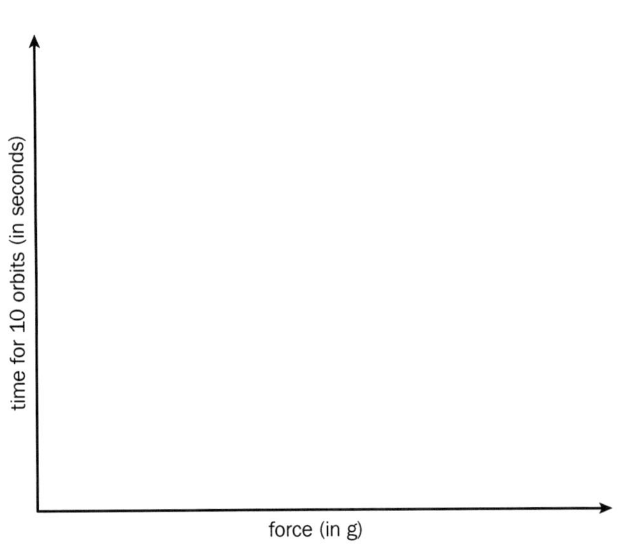

Report

Write a report on your investigation.

Here are some things you should include:

- what you did
- what happened
- explain your results
- if your prediction was correct or not
- what you could do to improve the investigation
- what you could have done if you had more time.

Chapter 11 ► Investigation 11C Orbiting planets

Planets are held in orbits around the Sun by a force. This force is the Sun's gravitational pull on the planets.

In this investigation: you are going to find out if there is a link between the speed of a planet and the force holding it in orbit.

Preparation: Predict

Finish the sentence in the box.

I think that there (is/is not) a link between the speed of a planet and the force holding it in orbit because...

You are going to use this equipment to find out if there is a link between the speed of a planet and the force holding it in orbit:

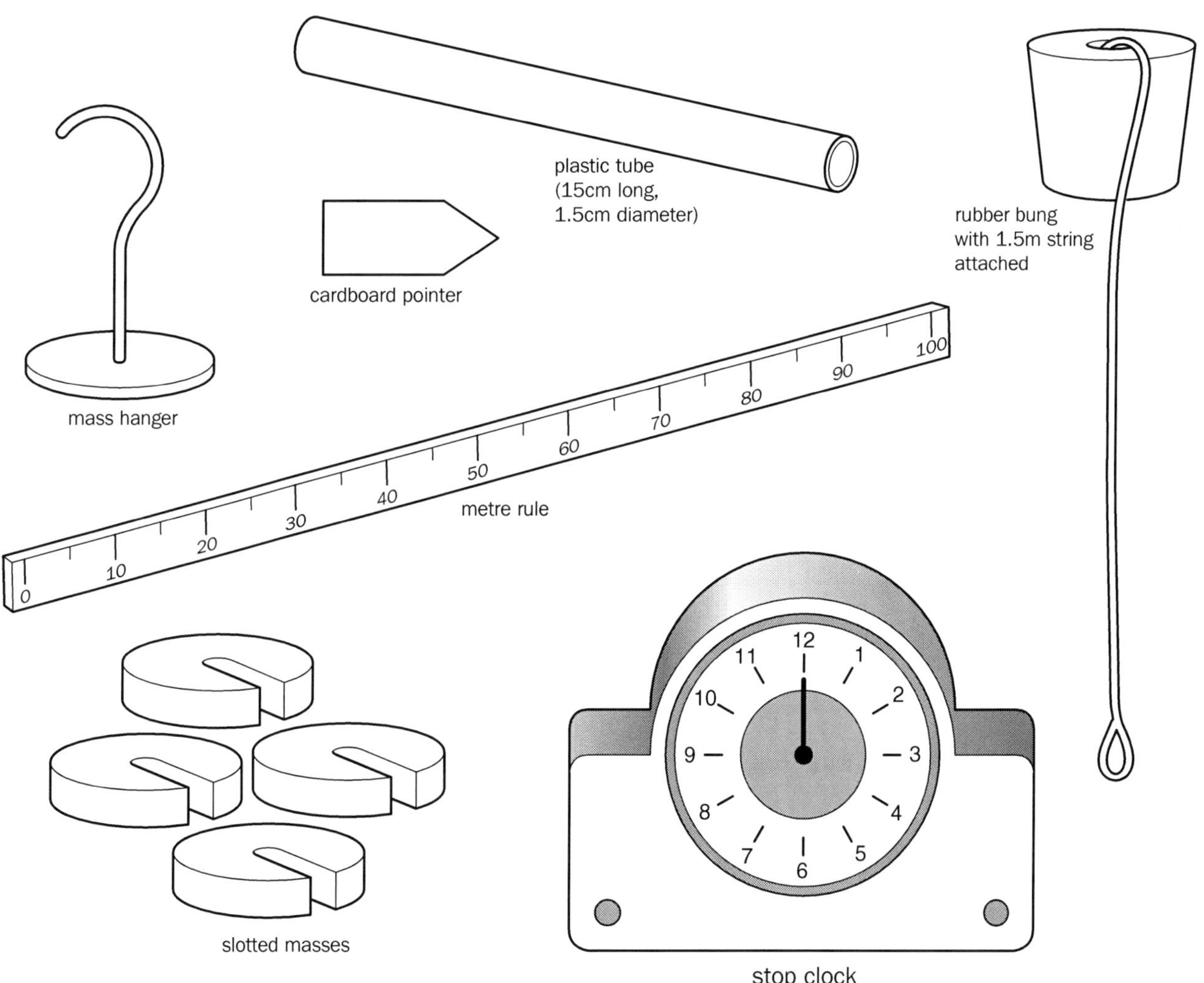

Chapter 11 ► Investigation 11C Orbiting planets

Preparation: Plan

Finish the sentences in the box.

I will measure…

Things I will keep the same are…

My investigation will be fair because…

My investigation will be safe because…

Carry out

- Thread the string through the tube and fix the pointer on to the string.
- Move the pointer so it just touches the bottom of the tube when there is 0.5 m of string between the tube and the bung.
- Hang the mass hanger with one 100 g mass from the loop.
- Find a large space so you can whirl the bung in a circle without hitting anything or anyone!
- Hold the tube and whirl the bung around your head in a horizontal axis. Get the speed just right to keep the pointer just below the tube.
- Time how long it takes for ten orbits.
- Do the same with 200 g, 300 g and 400 g hanging from the mass hanger.
- Put your results in a table like this:

Force in grams	Time for ten orbits in seconds
100	
200	
300	
400	

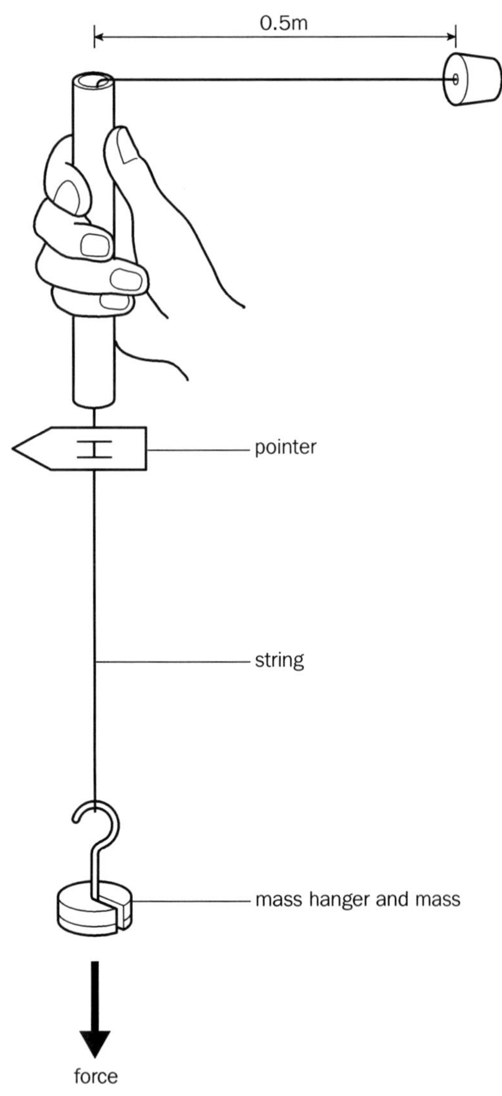

Chapter 11 ► Investigation 11C Orbiting planets

Draw a line graph of your results on this grid.

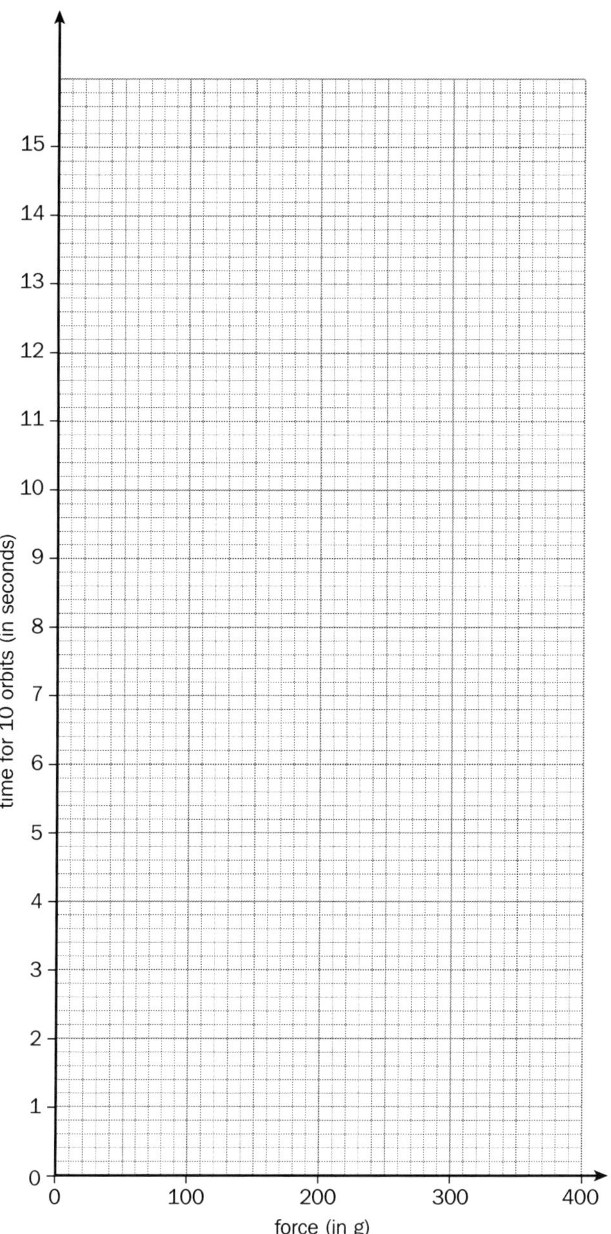

Report

Finish the sentences in the box.

What I did was…

What happened was…

From my results, I found out that there (is/is not) a link between the speed of a planet and the force holding it in orbit.

I know this because…

My prediction (was/wasn't) correct.

If I could do the investigation again I would…

Investigation 11
Practical notes

Orbiting planets

Investigation 11
Technician's notes

Orbiting planets

Each group will need:

Number of apparatus sets:

Number of pupils:

Number of groups:

Visual aids:

ICT resources:

Equipment/apparatus needed:

- rubber bung with hole and about 1.5 m string attached. The string should have a loop tied at the other end to hold a mass hanger
- mass hanger
- four 100 g slotted masses
- piece of rigid plastic tubing about 15 cm long and 1.5 cm diameter
- cardboard pointer (with two slits to thread string through)
- meter rule
- stop clock.

Safety notes

- **This activity must be carried out in a large area where there is no risk of pupils hitting anything or anyone.**
- **Warn pupils to keep plenty of space between themselves and others.**

CLEAPSS/SSERC SAFETY REFERENCE:

Chapter 11 ▶ Test

White

Earth in space

1 The diagram shows the Solar system

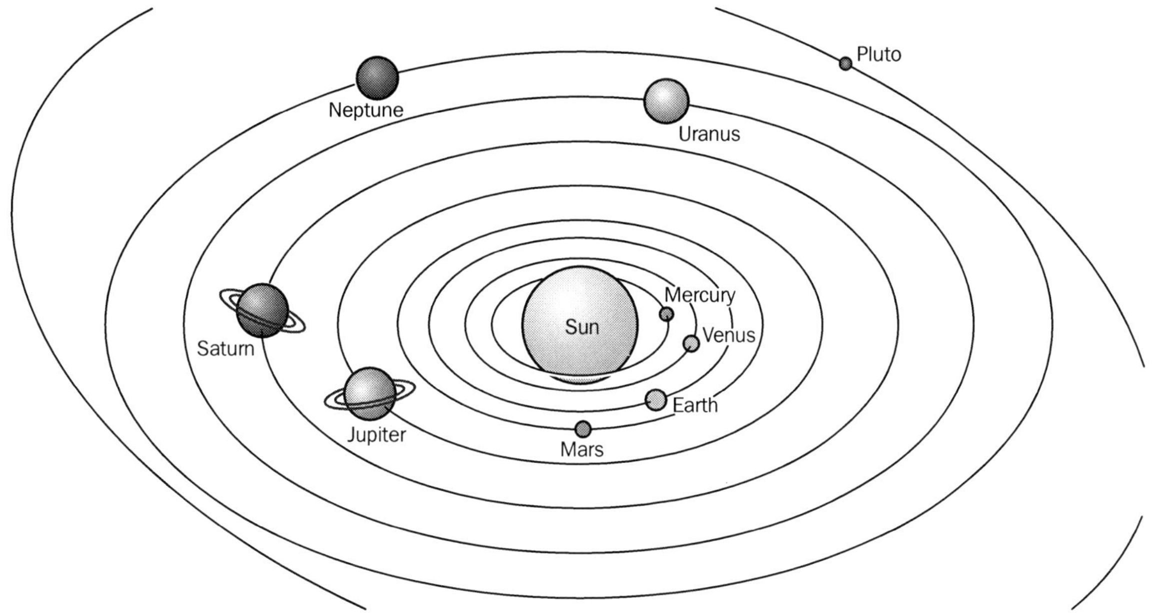

a What is the Solar system?

b Name two planets that are
i smaller
ii larger than Earth.

c i Name two planets are colder than Earth.
ii Explain why these planets are colder than Earth.

d Venus's atmosphere is mainly carbon dioxide. Give **two** reasons why Venus has a high surface temperature.

e i What force holds the planets in orbit around the Sun?
ii Venus and Earth have a similar mass. Which of these two planets has the stronger force on it?

f i What is a year?
ii Explain why Mercury has a shorter year than Pluto.

11 marks

2 The Sun is a star. It provides the planets with light and heat.

a Explain why astronomers can see
i stars
ii planets at night.

b An astronomer watches a bright light in the sky every night for several months. How can the astronomer tell whether the light is from a star or a planet?

c The diagram shows a time lapse photograph of some of the stars around the Pole Star.

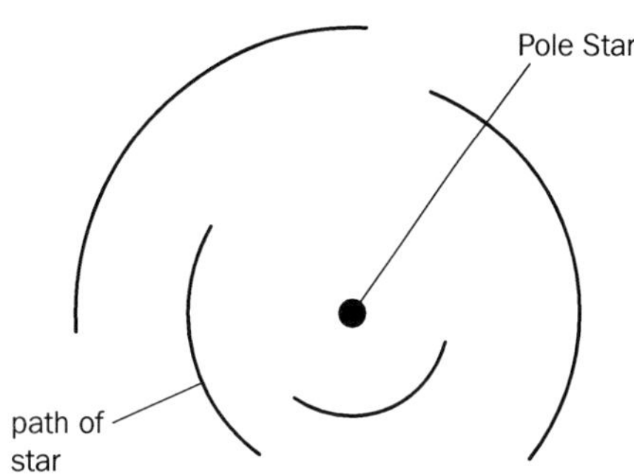

i Why do the paths of all the stars except the Pole Star appear as curved lines?
ii Suggest why the Pole Star does not appear to move.

5 marks

Chapter 11 ► Test

Earth in space

3 Much of our knowledge about Space has come from investigations made by people and machines. To get into Space, powerful rockets have been developed. Like rockets, the space shuttle can carry heavy loads into space but is less costly to use.

a Why do rockets need to be so powerful?

b Why do you think Space rockets carry liquid oxygen?

c Explain how the Space shuttle has cut the cost of space exploration.

d Name one machine, launched by rocket or Space shuttle, that gives us information about Space.

e Give one other way we can find out about Space without leaving Earth.

5 marks

4 The diagram shows a torch shining on a football. The torch gives out heat and light energy.

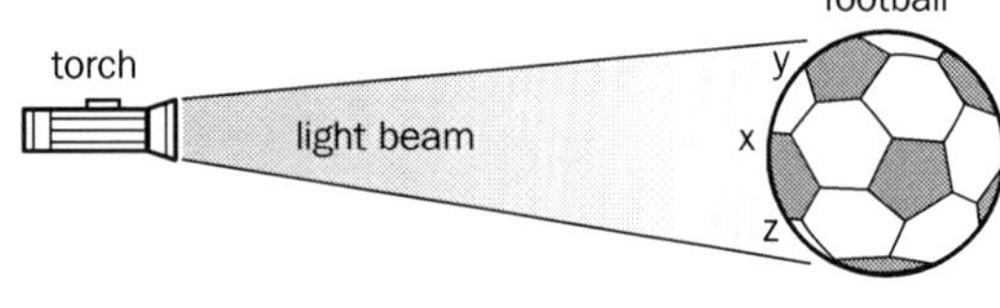

a Describe how the brightness changes over the surface of the football.

b Explain why it is likely to get hotter at point X than at Y and Z

c Draw a circle representing the football and shade the part of the football that is in shadow.

4 marks

5 The diagram shows the path of the Sun as it moves across the sky on a Spring day in Scotland.

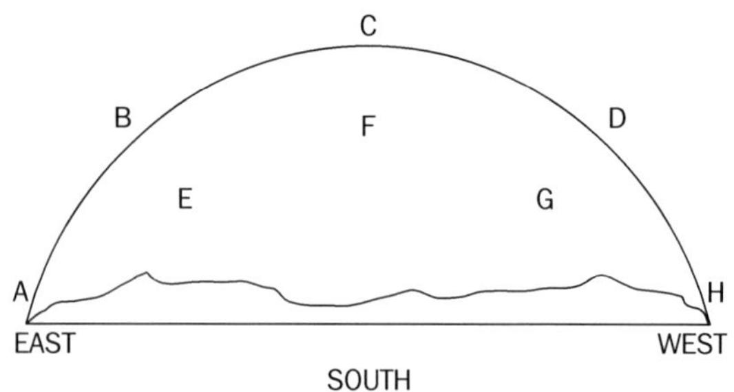

a Write the letter that shows

i where the Sun rises

ii where the Sun sets

iii where the Sun is on a summer morning

iv where the Sun is on a winter afternoon.

b Explain why the Sun appears to move across the sky.

5 marks

Chapter 11 ▶ Test

Blue

Earth in space

1 The Sun is a star. It provides the planets with light and heat.

a Explain why astronomers can see stars.

b An astronomer watches a bright light in the sky every night for several months. How can the astronomer tell whether the light is from a star or a planet?

c The diagram shows a time lapse photograph of some of the stars around the Pole Star.

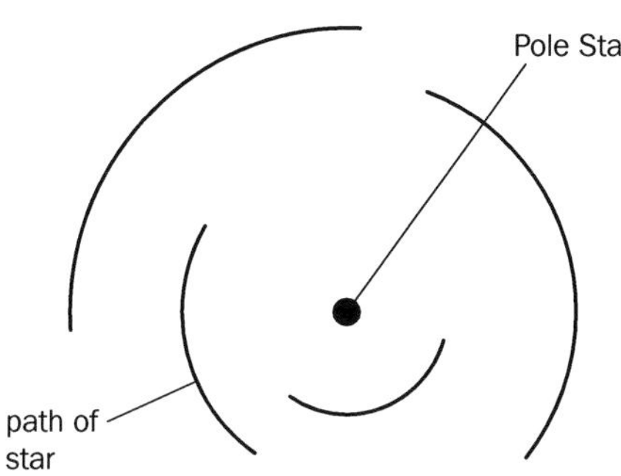

i Why do the paths of all the stars except the Pole Star appear as curved lines?

ii Suggest why the Pole Star does not appear to move.

4 marks

2 Much of our knowledge about space has come from investigations made by people and machines. To get into Space, powerful rockets have been developed. Like rockets, the Space shuttle can carry heavy loads into Space but is less costly to use.

a Why do rockets need to be so powerful?

b Why do you think Space rockets carry liquid oxygen?

c Explain how the Space shuttle has cut the cost of Space exploration.

d Name one machine, launched by rocket or Space shuttle, that gives us information about Space.

e Give one other way we can find out about Space without leaving Earth.

5 marks

3 The diagram shows a torch shining on a football. The torch gives out heat and light energy.

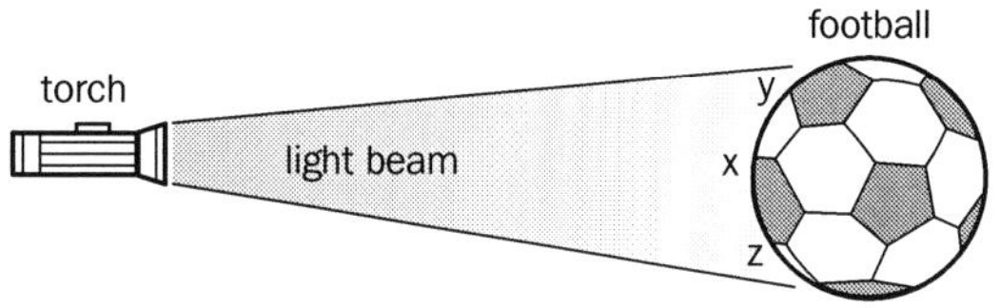

a Describe how the brightness changes over the surface of the football.

b Explain why it is likely to get hotter at point X than at Y and Z

c Draw a circle representing the football and shade the part of the football that is in shadow.

d Explain how the torch and the football can be used to demonstrate night and day on Earth.

6 marks

4 The diagram shows the path of the Sun as it moves across the sky on a Spring day in Scotland.

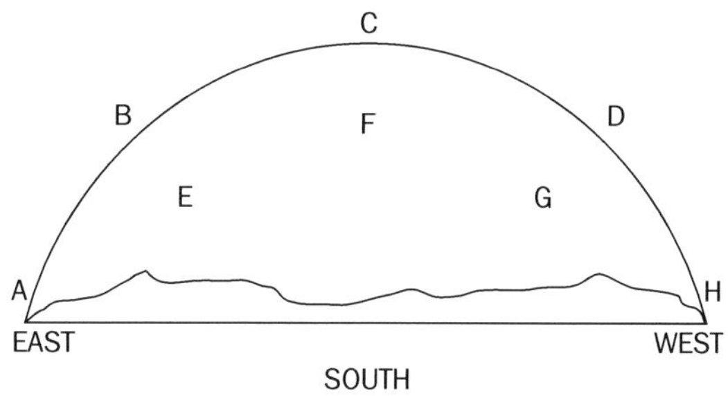

a Write the letter that shows

i where the Sun rises

ii where the Sun sets

iii where the Sun is on a summer morning

iv where the Sun is on a winter afternoon.

Earth in space

b Explain why the Sun appears to move across the sky.

c Describe the difference you would see in the path of the Sun across the sky if you lived on the Equator?

6 marks

5 Put the following in order of size starting with the **smallest**.

galaxy
Earth
Solar system
the Moon
Universe

4 marks

6 The American astronomer Edwin Hubble made some important observations that give us an idea about the origin of the Universe. Hubble noticed that as a distant object moves away from us its light is distorted showing up as a reddening of the light. The more reddening of the light, the further away the object is. Galaxies furthest away from us have the greatest reddening of light.

a What did Hubble call the reddening of light produced when distant objects move away from us?

b Explain how these observations support the 'big bang' theory of the origin of the Universe.

5 marks

Chapter 11 ► Mark scheme Earth in space

White

Question	Answer	Marks	Level
1 a	Sun and planets (comets and asteroids)	1	C
b i	any two from Mercury, Venus, Mars, Pluto	1	
ii	any two from Jupiter, Saturn, Uranus, Neptune	1	C
c i	any two from Mars, Jupiter, Saturn, Uranus, Neptune, Pluto	1	
ii	further from Sun	1	C
d	close to Sun (1) global warming (1)	2	C
e i	gravity	1	
ii	Venus	1	D
f i	time taken for Earth to orbit Sun once	1	
ii	closer to Sun/ smaller radius	1	C
		11	
2 a i	produce own light	1	
ii	reflect light from Sun	1	D
b	planets move	1	D
c i	they are moving	1	
ii	it is directly overhead	1	D
		5	
3 a	to escape/overcome gravity	1	D
b	no atmosphere/ oxygen in space	1	D
c	can be reused	1	D
d	satellite, etc.	1	D
e	telescope, etc.	1	D
		5	

Question	Answer	Marks	Level
4 a	bright in middle, less so towards top and bottom	1	E
b	closer to heat/torch	1	E
c	back of ball in shadow (1) vertical line through centre of circle (1)	2	E
		4	
5 a i	A	1	
ii	H	1	
iii	B	1	
iv	G	1	E
b	rotation of Earth	1	E
		5	
		TOTAL 30 marks	

Suggested grade/level boundaries

C 10/30

D 18/30

E 27/30

Chapter 11 ▶ Mark scheme

Earth in Space

Question	Answer	Marks	Level
1 a	produce own light	1	D
b	planets move	1	D
c i	they are moving	1	
ii	it is directly overhead	1	D
		4	
2 a	escape/overcome gravity	1	D
b	no atmosphere/ oxygen in space	1	D
c	can be reused	1	D
d	satellite, etc.	1	D
e	telescope, etc.	1	D
		5	
3 a	bright in middle, less so towards top and bottom	1	E
b	closer to heat/torch	1	E
c	back of ball in shadow (1) vertical line through centre of circle (1)	2	E
d	torch represents Sun (1) shining on one side of Earth (1)	2	E
		6	
4 a i	A	1	
ii	H	1	
iii	B	1	
iv	G	1	E
b	rotation of Earth	1	E
c	no change in path of Sun	1	E
		6	

Question	Answer	Marks	Level
5	Moon, Earth, Solar System, galaxy, Universe (all correct = 4; deduct 1 mark per error)	4	E
		4	
6 a	red shift	1	F
b	super atom exploded (1) bits in all directions (1)	2	F
	increasing red shift (1) means galaxies are still moving away from each other (1)	2	F
		5	
		TOTAL 30 marks	

Suggested grade/level boundaries

D 10/30

E 22/30

F 27/30

12 Forces

This chapter investigates what forces are, what they can do, and what happens if there are no forces. The work is related to pupils' experience wherever possible. Scientific terms such as force, newton, friction, mass, weight, gravity, and pressure are used and some pupils may find them difficult. However, it is hoped that pupils will become familiar with them as they are used in practical situations.

Assessment opportunities

Formative assessment opportunities are provided by worksheets, homework sheets, and an investigation.

The **worksheets** cover material at levels C, D, E, and F for attainment targets for knowledge and understanding. Teachers may wish to use these worksheets not only as part of practical activities but also to provide evidence of pupil achievement.

Worksheet	Level
12.1	C
12.2	E
12.3a	D
12.3b	E
12.4a	F
12.4b	F

The **homework sheets** cover material at levels C, D, E, and F for attainment targets for knowledge and understanding. These homework sheets can be used individually as a follow-up to work done in class or assembled into a homework booklet allied closely to schemes of work.

Homework sheet	Level
12.1	C
12.2a	C/D/E
12.2b	E
12.4	F

The **investigation** covers all three skill areas at levels C, D, E, and F. It is written in a way that allows for pupils to be assessed in all three skill areas at one level. Alternatively, customised assessments can be constructed enabling pupils to be assessed at different levels in all three skills. The latter approach is more time consuming, but it does provide the opportunity for pupils to show evidence of achievement at different levels in different skills in the same investigation. Teachers will need to use their professional judgement when deciding which level is appropriate to individual pupils. It is envisaged that pupils will show progression through the levels as they work through their science course.

Summative tests are provided at two levels, white and blue. The white test contains questions covering attainment target levels C and D. The blue test contains questions covering attainment target levels E and F. Each test has a total of 30 marks and will take about 30 minutes for pupils to complete, although this can be varied depending on pupil ability. Mark schemes are provided together with suggested grade/level boundaries.

It is envisaged that these tests will be given to pupils on completion of the material covered in Chapter 12.

ICT opportunities

The use of data loggers/remote sensors can extend the range, speed, and sensitivity of measurements in many of the worksheets for this chapter. Once downloaded onto a PC, data-handling programs can be used to analyse information gathered, data can be manipulated, and appropriate graphs etc. presented. The Internet provides pupils with access to a huge range of scientific information. A list of suitable websites is included in this Teacher's Guide.

Students' book chapter 12 contents and guide levels

12.1	The force of friction	*Starting off*	C
	Investigating friction	*Going further*	C
	Friction in fluids	*For the enthusiast*	C/D
12.2	Balanced forces	*Starting off*	E
	See-saws and levers	*Going further*	E
	Forces and movement	*For the enthusiast*	E
12.3	Gravity and measuring weights	*Starting off*	D/E
	Weight in space	*Going further*	D/E
	Mass and weight	*For the enthusiast*	D/F
12.4	The effects of pressure	*Starting off*	F
	Measuring pressure	*Going further*	F
	Pressure in action	*For the enthusiast*	F

12.1 Friction

W/S

Name: **Date:** **Group:**

What you need:

Wooden block (one smooth surface, one surface with fine sandpaper and one surface with course sandpaper), two thick rubber bands, force meter, beaker of wet sand.

What to do:

1 Copy this table:

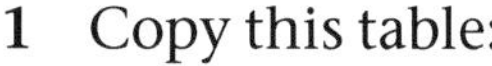

Surface	Reading on force meter in Newtons
Smooth surface	
Fine sandpaper	
Coarse sandpaper	
Thick rubber bands	

2 Put the wooden block smooth side down on the bench.

3 Stand the beaker of wet sand on the block.

4 Pull the block along the bench with the force meter.

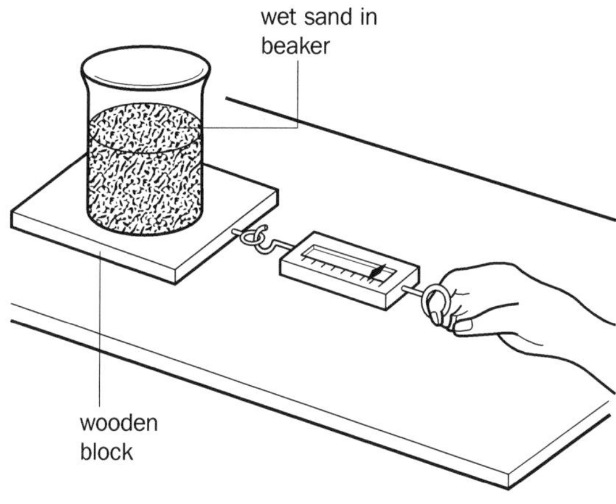

5 Look at the reading on the force meter and write it in the table.

6 Do the same with the fine sand paper surface and the course sandpaper surface.

Which surface has most friction? Why is this?

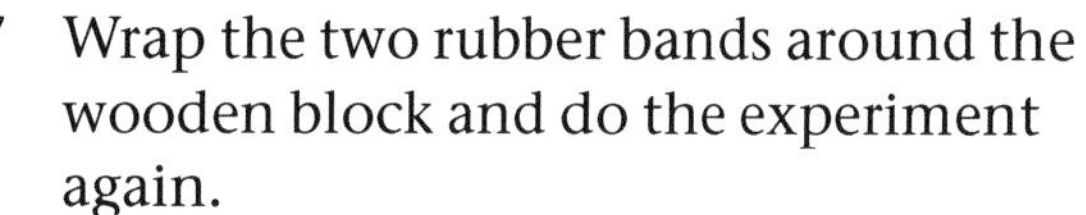

7 Wrap the two rubber bands around the wooden block and do the experiment again.

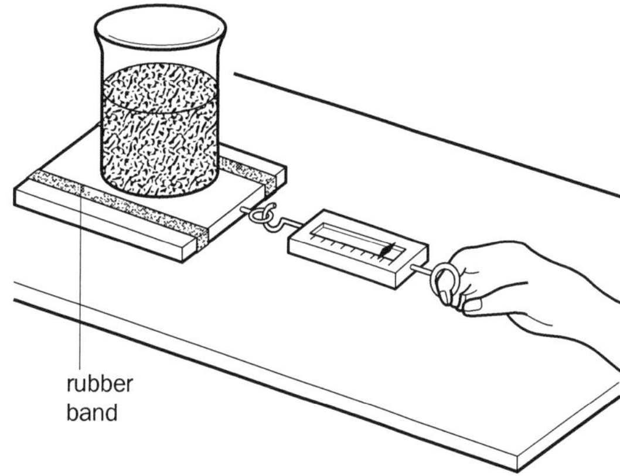

Explain how your experiment models bicycle tyres.

12.1 Practical notes

Friction

This is a straightforward activity giving pupils the opportunity to compare friction between different surfaces. Masses could be used instead of beakers of wet sand depending on availability. Plastic pots, e.g. margarine tubs could be used instead of beakers if there is a risk of breakage. The rubber bands link nicely with pupils' experiences of bicycle tyres. Safety aspects of bald car tyres could be introduced, including their relative performance on dry and wet roads.

12.1 Technician's notes

Number of apparatus sets: ....

Number of pupils:

Number of groups:

Visual aids: ....

ICT resources: ....

Equipment/apparatus needed: ....

Friction

Each group will need:

- a copy of worksheet 12.1
- block of wood approximately 10 cm × 5 cm × 5 cm; one side smooth, one side with fine sandpaper glued to surface and another side with course sandpaper glued firmly to surface. Attach an eye to leading edge for attachment of force meter
- 100 cm^3 beaker or plastic pot
- access to damp sand or masses
- two thick elastic bands to fit around wooden block
- force meter/Newton meter
- paper towels (if water used).

Safety notes

CLEAPSS/SSERC SAFETY REFERENCE:

12.2 Balancing forces

W/S

Name: **Date:** **Group:**

What you need:
Metre rule, clamp stand with clamp, string, scissors, two mass hangers, 10 × 100 g masses.

What to do:

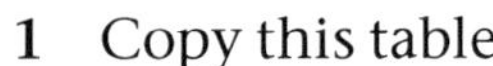

1 Copy this table:

Left hand side of pivot			**Right hand side of pivot**		
Force on hanger in N	Distance of hanger from pivot in cms	Force × distance	Force on hanger in N	Distance of hanger from pivot in cms	Force × distance
3	20	60	2		

2 Hang the metre rule from the clamp stand with string. This is the pivot.

3 Make two loops of string. Slide one loop over each end of the metre rule.

4 Hang a mass hanger and three 100 g masses 20 cm from the pivot on one side.

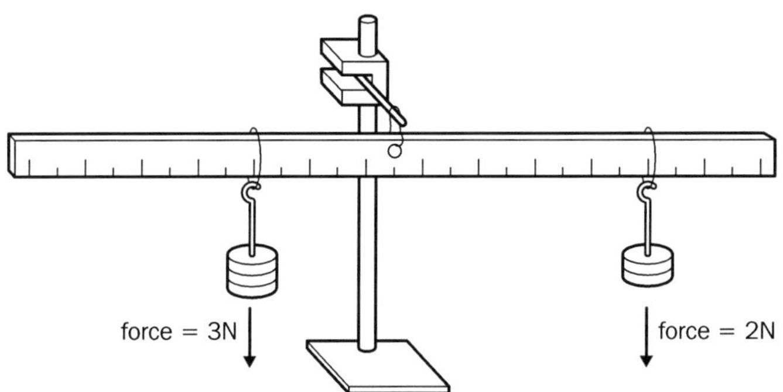

5 Hang the second mass hanger and two 100 g masses on the other side of the pivot. Move the hanger backwards and forwards until the metre rule balances.

6 Measure how far the second hanger is from the pivot and write it in the table.

7 Find the size of the unknown force in each of these diagrams. Put your results in the table.

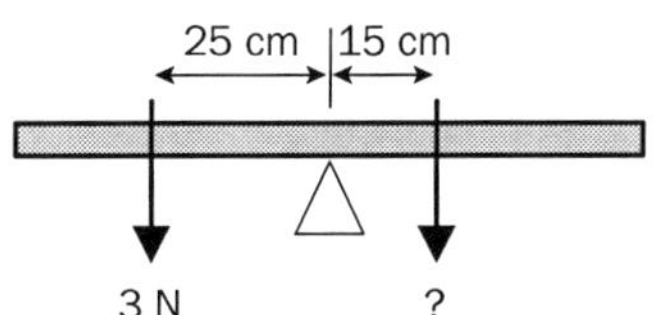

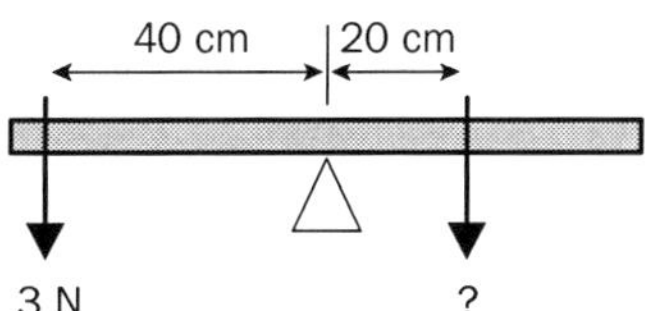

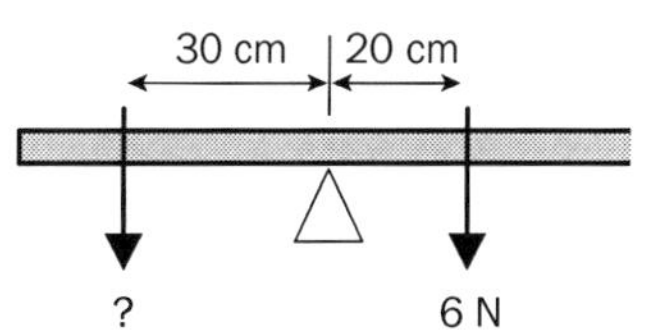

What is the pattern in your results?

8 Hang a 4 N weight on one side of the pivot and a 2 N weight on the other side.

Find three different positions for the weights to balance the metre rule.

Write your results in the table.

12.2 Practical notes

Balancing forces

Careful drilling of metre rules at the 50 cm mark will give a satisfactory balance. Pupils should be encouraged to learn and remember the 'balancing rule':

force × distance on left side of pivot = force × distance on right side of pivot.

12.2 Technician's notes

Balancing forces

Each group will need:

- a copy of worksheet 12.2
- metre rule with hole drilled at 50 cm mark
- string or thread to make loops
- scissors
- clamp stand and clamp or similar to hang balance beam
- two mass hangers
- 10 × 100 g masses.

Number of apparatus sets:

Number of pupils:

Number of groups:

Visual aids:

ICT resources:

Equipment/apparatus needed:

Safety notes

CLEAPSS/SSERC SAFETY REFERENCE:

12.3a Gravity and acceleration W/S

Name: **Date:** **Group:**

What you need:

Two metre rules, data logger, two timing gates, clamps and stands, table tennis ball, golf ball, pot of sand.

What to do:

1 Copy this table:

	Time in seconds for balls to fall 2m					
	1	2	3	4	5	Average
Table tennis ball						
Golf ball						

2 Set up the timing gates as shown in the diagram and connect them to a data logger.

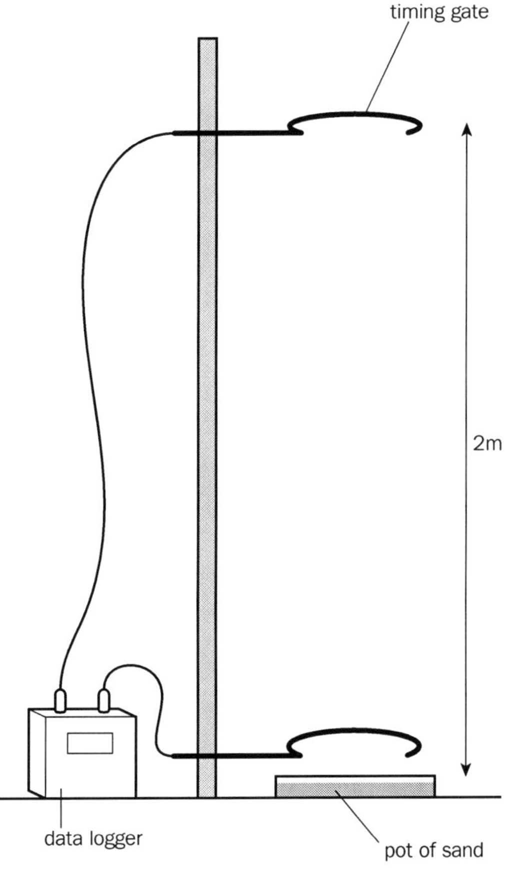

Make sure the distance between the gates is 2 metres and the bottom gate is exactly beneath the top gate. (Use a piece of string with a small weight attached to line the gates up.)

3 Hold the table tennis ball just above the top gate, make sure the data logger is switched on, then release the ball.

Record the time from the data logger in the table.

4 Repeat this four more times, each time recording your results in the table.

5 Now do the same with the golf ball recording all five results in the table.

6 Work out the average times for the two balls to fall 2 m and write them in the table.

What do you notice about the times?

Can you suggest a reason for any differences?

7 If you have time, try changing the distance between the two gates and getting some more results.

What do you notice this time?

12.3a
Practical notes

Gravity and acceleration

Using data loggers can be a tricky and time consuming business. Many pupils will need help in setting up the equipment. The most difficult part is making sure that the two gates are vertically in line. The equipment can generate some good results and pupils should be able to appreciate that gravity makes all objects accelerate towards the ground at the same rate. The plastic pot of sand is optional; it prevents the unnecessary proliferation of bouncing balls.

12.3a
Technician's notes

Gravity and acceleration

Each group will need:

- a copy of worksheet 12.3a
- data logger
- two timing gates
- two clamps and stands
- access to two metre rules
- table tennis ball
- golf ball
- plastic margarine tub filled with dry sand or similar to catch balls
- access to string with weight attached to act as plumb line.

Number of apparatus sets:

Number of pupils:

Number of groups:

Visual aids: ....

ICT resources: ....

Equipment/apparatus needed: ....

Safety notes

CLEAPSS/SSERC SAFETY REFERENCE:

12.3b Mass and weight W/S

Name: **Date:** **Group:**

What you need:

Clamp stand and clamp, G clamp, spring, mass hanger, five 100 g masses.

What to do:

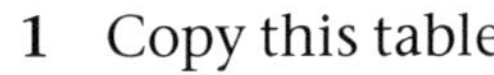

1 Copy this table:

Mass in g	Stretched length of spring in mm	Unstretched (starting) length of spring in mm	Increase in length of spring in mm

2 Fix the clamp stand firmly to the bench with a G clamp. Hang the spring from the clamp.

3 Measure the length of the **unstretched** spring and write it in the table (five times).

4 Hang the mass hanger on the end of the spring and put a 100 g mass on to the hanger.

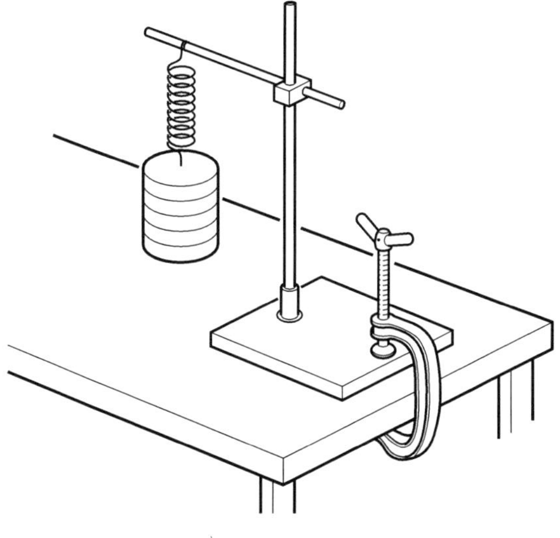

5 Measure the spring and write down the length in the table.

6 Add another 100 g mass, measure the spring again and write the length in the table.

7 Keep adding 100 g masses. Each time measure the length of the spring and write the length in the table.

8 Stop when the mass hanger is so heavy that it touches the table.

9 Work out the increase in length of the spring for each of the masses. Write these in the table.

10 Draw a line graph of your results. Label your axes like this.

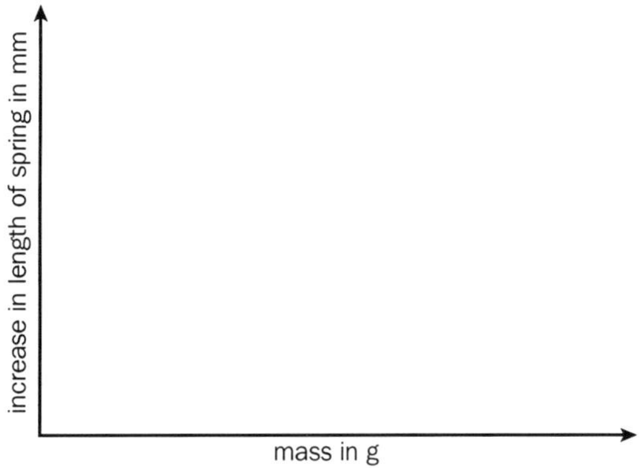

12.3b
Practical notes

Mass and weight

This is a simple activity which should enable pupils to distinguish between mass and weight. Depending on the strength of spring used, good results should be obtained to enable pupils to draw a line graph illustrating the relationship between mass and spring extension due to gravity. Warn pupils of the dangers of falling masses.

12.3b
Technician's notes

Mass and weight

Each group will need:

- a copy of worksheet 12.3b
- clamp stand and clamp
- G clamp
- spring
- mass hanger
- five 100 g masses.

Number of apparatus sets:

Number of pupils:

Number of groups:

Visual aids:

ICT resources:

Equipment/apparatus needed:

Safety notes

CLEAPSS/SSERC SAFETY REFERENCE:

12.4a Pressure

W/S

Name: **Date:** **Group:**

What you need:

Plasticine, four 1 cm^3 cubes, 1 kg mass, strip of card.

What to do:

1 Use the lump of Plasticine to make a thick disc about 2 cm thick.

2 Put the four cubes onto the disc but don't press them in.

3 Carefully rest the 1 kg mass on top of the cubes so its weight acts evenly over the four cubes.

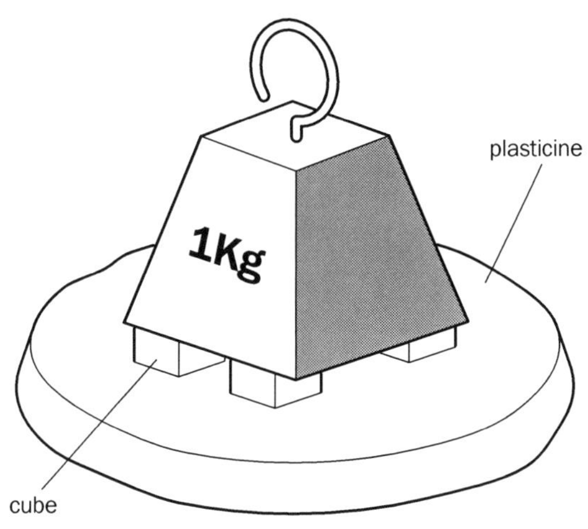

4 Take off the mass and remove one of the cubes. Carefully replace the mass on the remaining three cubes.

5 Do the same with two cubes and, finally one cube.

6 Lift off the mass and look carefully at the dents in the Plasticine.

Use a strip of card and a ruler to measure the depth of the four dents.

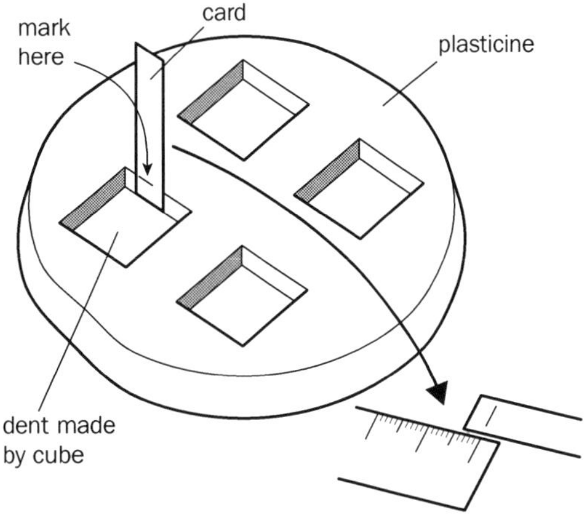

Is there any difference in the depth of the four dents?

Explain why this is.

7 Use this formula: $\text{pressure} = \frac{\text{force (N)}}{\text{area}}$

to work out the pressure on the Plasticine when the 1 kg mass rests on:

a one
b two
c three
d four blocks.

(Remember: 1 kg weighs 10 N)

12.4a
Practical notes

Pressure

This simple activity establishes the quantitative relationship between pressure, force and area. Make sure pupils use correct units in their calculation, i.e. N/cm^2.

12.4a
Technician's notes

Pressure

Each group will need:

- a copy of worksheet 12.4a
- Plasticine (enough to make a disc approximately 10 cm diameter and 2 cm thick)
- four 1 cm^3 cubes
- 1 kg mass with flat base
- thin strip of card to measure depth of dents in Plasticine
- ruler.

Number of apparatus sets:

Number of pupils:

Number of groups:

Visual aids:

ICT resources:

Equipment/apparatus needed:

Safety notes

CLEAPSS/SSERC SAFETY REFERENCE:

12.4b Hydraulic systems W/S

Name: **Date:** **Group:**

What you need:

One small syringe (10 cm^3), one large syringe (50 cm^3), plastic tubing, bowl of water, paper towels.

What to do:

1 Put the plastic tubing into the bowl of water. Check to see that the tubing is full of water and that there are no air bubbles.

2 Push the plunger of the small (10 cm^3) syringe down so there is no air inside it. Put the syringe in the water and pull out the plunger so the syringe holds 10 cm^3 of water. Do the same with the large (50 cm^3) syringe so that it holds 20 cm^3 of water.

3 Connect the two syringes with plastic tubing as shown in the diagram. Check that there are no air bubbles.

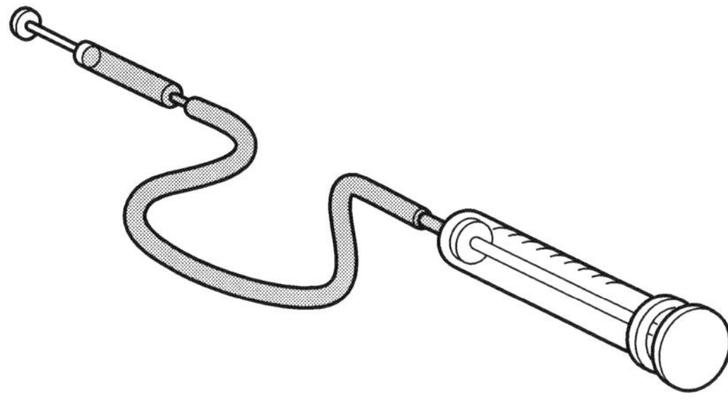

4 Put the equipment on a paper towel and remove the surface water.

5 Carefully empty the small syringe by pushing the plunger slowly down.

What happens to the plunger in the large syringe?

6 Measure the distances moved by the plungers in both of the syringes.

What have you discovered?

7 Carefully dismantle the equipment and dry it.

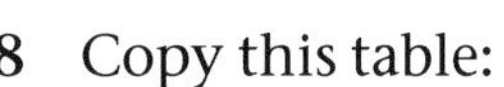

8 Copy this table:

Small syringe	Large syringe
Diameter of plunger =	Diameter of plunger =
Radius of plunger =	Radius of plunger =
Area =	Area =

9 Measure the diameter of the two syringe plungers.

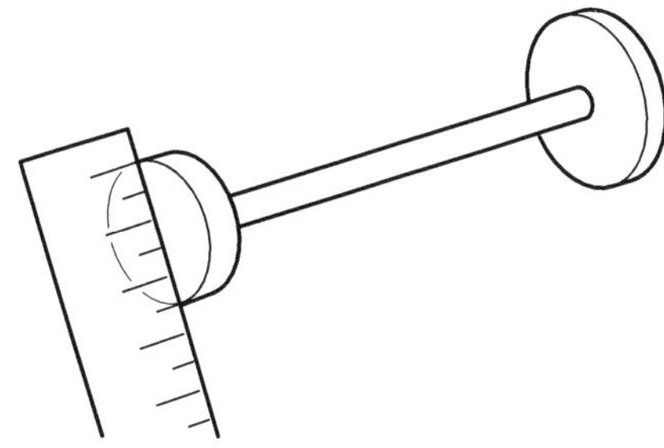

Divide the diameter by 2 to get the radius of each plunger. Write these in the table.

10 Use this formula to work out the area of each plunger:

Area = $\pi \times r \times r$ (πr^2)

Write the area of each syringe in the table.

11 If the small syringe is pushed with a force of 5 N, what pressure is produced in the small syringe?

(Remember: pressure = $\dfrac{\text{force (N)}}{\text{area}}$)

Work out the pressure on the plunger of the large syringe.

12.4b
Practical notes

Hydraulic systems

This is a simple yet effective way of demonstrating hydraulic systems. The activity works with air as well as water, so providing the opportunity to compare the hydraulic and pneumatic systems. Some pupils will find the mathematics a bit difficult but there are good opportunities for differentiation here. Watch out for silliness with syringes and water.

12.4b
Technician's notes

Hydraulic systems

Each group will need:

- a copy of worksheet 12.4b
- one small syringe ($10\,cm^3$)
- one large syringe ($50\,cm^3$)
- piece of clear plastic tubing to fit syringes (about 20 cm long)
- access to bowl of water
- paper towels.

Number of apparatus sets:

Number of pupils:

Number of groups:

Visual aids:

ICT resources:

Equipment/apparatus needed:

Safety notes

CLEAPSS/SSERC SAFETY REFERENCE:

12.1 Forces and friction

H/W

Name: **Date:** **Group:**

What you need to know ...

Friction is the force produced when two surfaces rub on each other. Friction can be useful when grip is required, but it can cause problems in machinery. This is why machines have to be lubricated.

What to do:

1 Read the information given on the drawing of a bicycle.

a Work out how forces are transmitted between the different parts of the bicycle. Write your answers in a table like this. The first one has been done for you.

Where the force is applied	What it does there	Where the force is transmitted to	What it does there	How the force is transmitted
The ends of the handlebars	Turns the handlebars	The front forks	Turns the front wheel	Metal tubes (handlebars and forks)
The brake levers				
The pedals				

b The cyclist in the diagram is using friction and trying to reduce friction at the same time. Give two ways in which the cyclist

i uses friction

ii tries to reduce friction.

c What happens to the force of friction as the cyclist goes faster, and why does this happen?

2 The diagrams show some examples of where the force of friction acts:

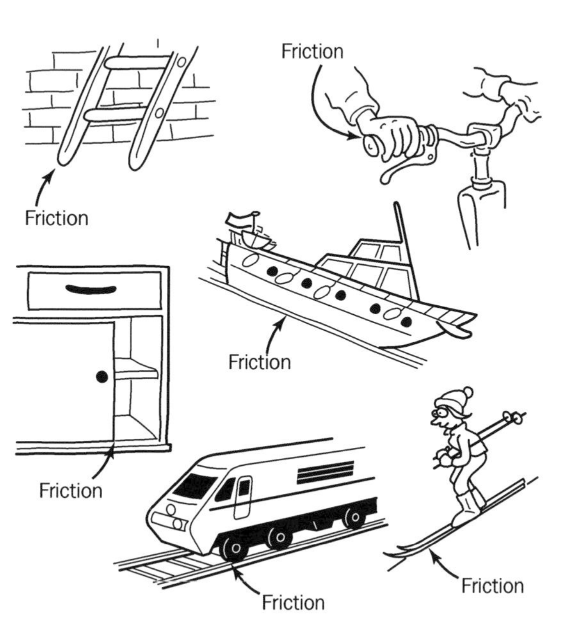

For each example:

a Would it be useful for the force of friction to be large or small?

b Explain why you made your choice.

12.2a Falling and floating H/W

Name: Date: Group:

What you need to know ...

When it is free to fall, an object always falls downwards. This is due to the gravitational attraction of the Earth. Objects seem to lose weight when they are put in water. This is due to the upthrust of the water.

What to do:

1 The diagrams show a skydiver at various stages on her journey from a plane to the ground.

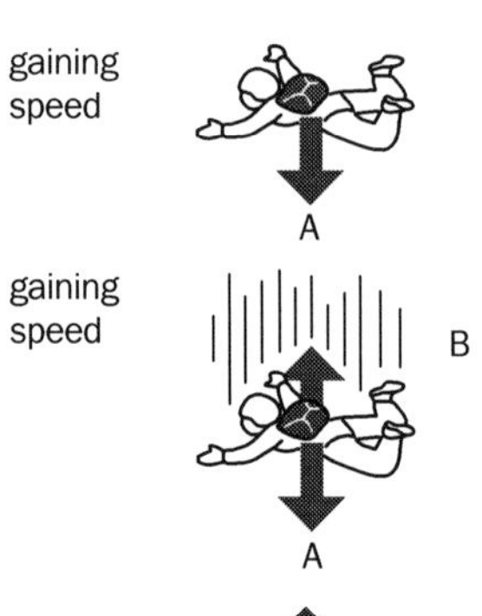

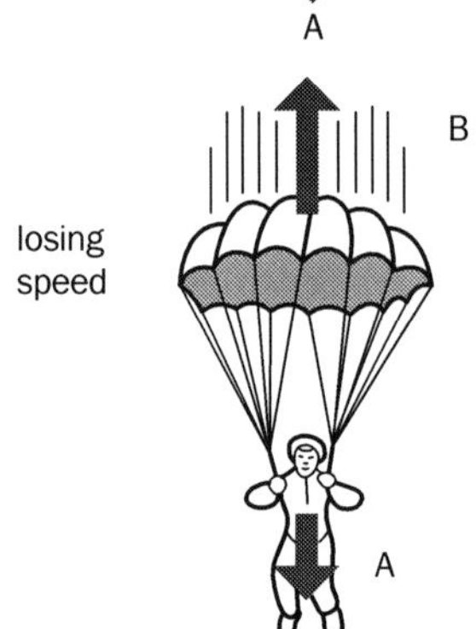

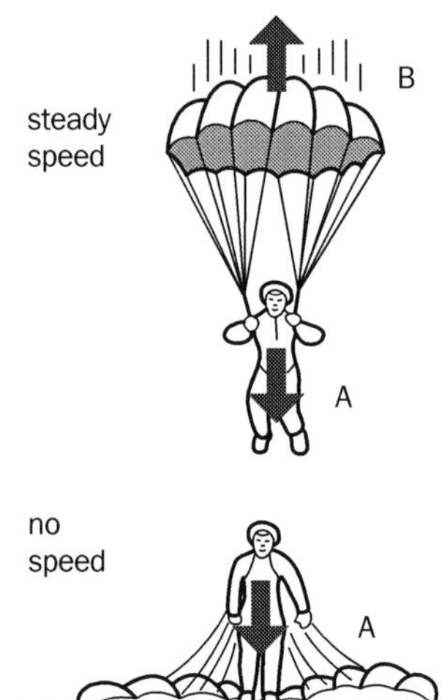

a What is force A which causes the skydiver to accelerate when she jumps from the plane?

b As she gains speed, what is force B that gets stronger and stronger as she falls?

c Why does the skydiver suddenly lose speed when she opens her parachute?

d What stops the skydiver from sinking into the ground when she lands?

2 A block of metal hanging from a force meter weighs 40 N in the air. When the object is lowered into water it only weighs 30 N.

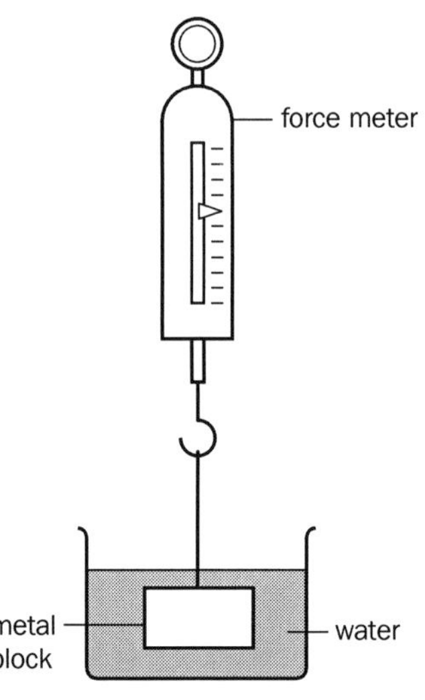

a What is the mass of the metal block?

b What is the upthrust on the block?

c What weight of water is displaced by the block?

d What mass of water is displaced by the block?

3 Explain the following:

a A life jacket helps you float in water.

b A steel block sinks but a ship made of steel floats.

c A ship sinks deeper in the water as more cargo is put into it.

d Whales are unable to move when they get beached on land.

e Walking on pebbles with bare feet does not hurt as much when you get into the sea.

HANDY HINTS

Look at the diagrams carefully and use the information at the top of the page. Remember, mass is the amount of matter, weight is a force, and 1 kg = 10 N.

12.2b Turning forces

H/W

Name: Date: Group:

What you need to know ...

A turning force is measured in Newton metres (Nm). It can be calculated using the formula:
Turning force = force × distance from turning point.
If two turning forces on opposite sides of a pivot are equal then they are balanced.

What to do:

1 The diagram shows two different spanners being used to unscrew a nut.

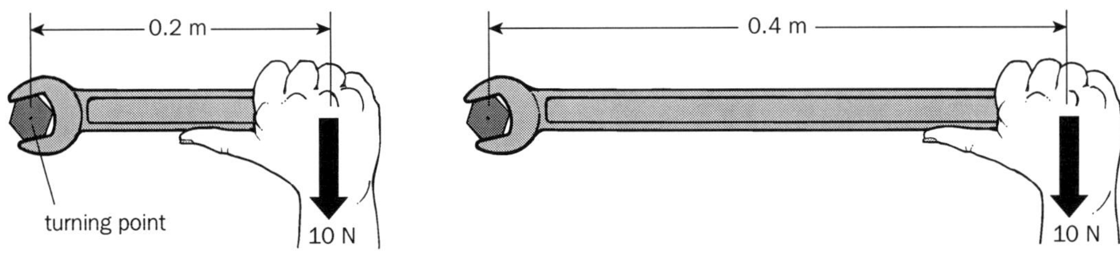

a Which spanner would make the job of unscrewing the nut easier and why?

b Calculate the turning force on the nut using each spanner.

2 Explain how using a spoon to lever off a tin lid is a similar situation to using a spanner to unscrew a nut. Use a diagram to help you in your answer.

3 The diagram shows some see-saws with a person sitting at each end.

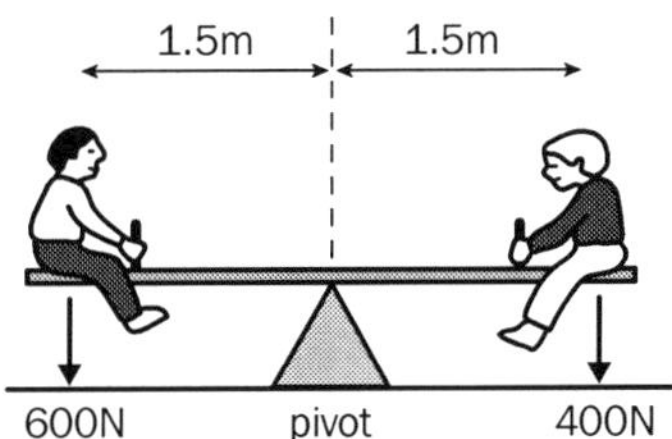

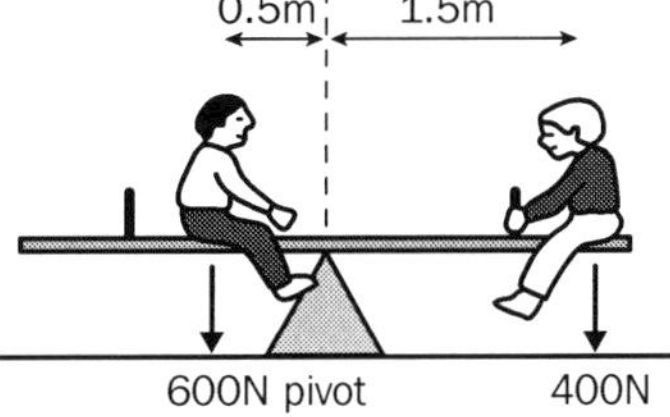

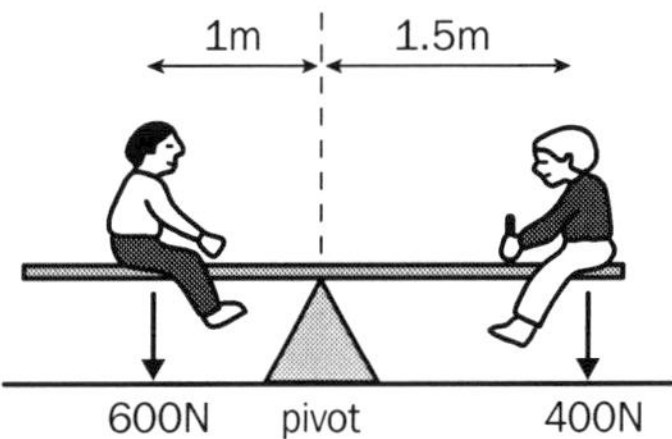

For each see-saw:

a Calculate the turning force on the
  i left of the pivot
  ii right of the pivot.

b Say whether the see-saw is balanced or unbalanced.

HANDY HINTS

Keep using the formula and remember which way the turning forces are being applied.

12.4 Pressure in action

H/W

Name: Date: Group:

What you need to know ...

Pressure is measured in Newtons per metre squared (N/m^2). It can be calculated using the formula:

$$\text{pressure} = \frac{\text{force}}{\text{area}}$$

It is sometimes more useful to measure small areas in cm^2 and pressure in N/cm^2.

What to do:

1 A rectangular block measures 25 cm × 4 cm × 2 cm and has a weight of 20 N. The block stands on one side on a level bench top.

a Draw a diagram to show the position of the block when the pressure is at its

i highest

ii lowest.

b Calculate the pressure of the block in both cases.

2 Explain the following:

- Stiletto heels are more likely to damage floors than trainers.
- Inuits wear snow shows.
- Tractors have large tyres.
- Football boots have studs.
- An Indian fakir can lie on a bed of nails.
- Mountain bike tyres are wide with a deep tread.
- It hurts to carry a heavy plastic carrier bag from the supermarket.
- It is more comfortable to sit on a chair than a fence.
- Camels have got big, wide feet.
- Some lorries have eight wheels on their trailer.

Chapter 12 ► Investigation 12F Forces

Skydivers use parachutes to slow their rate of fall before landing on the ground. A parachute increases air resistance, a sort of friction between the parachute and the air.

In this investigation: you are going to find out if the size of the parachute and the weight of the object attached to it affect the rate of fall.

Preparation: Predict

Finish the sentences in the box.

What I think will happen is…

I think this because…

Preparation: Plan

Write a plan of your investigation.

Think about:

- the apparatus you are going to use
- how one variable depends upon another variable
- what you are going to measure and how you are going to measure it
- how many readings you are going to take
- how you are going to record your results
- how you are going to make your investigation fair
- how you are going to make your investigation safe.

Show your plan to your teacher before going on.

Carry out

Carry out your investigation and record your results.

Present your results in an appropriate way.

Report

Write a report on your investigation.

Here are some things you should include:

- what you did
- what happened
- explain your results
- if your prediction was correct or not
- how reliable your results were
- what you could have done if you had more time.

Chapter 12 ▶ Investigation 12E Forces

Skydivers use parachutes to slow their rate of fall before landing on the ground. A parachute increases air resistance, a sort of friction between the parachute and the air.

In this investigation: you are going to find out if the size of the parachute and the weight of the object attached to it affect the rate of fall.

Preparation: Predict

Finish the sentences in the box.

What I think will happen is…

I think this because…

Preparation: Plan

Write a plan of your investigation.

Think about:

- the apparatus you are going to use
- what you are going to measure and how you are going to measure it
- how many readings you are going to take
- how you are going to record your results
- how you are going to make your investigation fair
- how you are going to make your investigation safe.

Show your plan to your teacher before going on.

Carry out

Carry out your investigation and record your results in a table.

Draw a bar graph of your results.

Report

Write a report on your investigation.

Here are some things you should include:

- what you did
- what happened
- explain your results
- if your prediction was correct or not
- what you could do to improve the investigation
- what you could have done if you had more time.

Chapter 12 ► Investigation 12D Forces

Skydivers use parachutes to slow their rate of fall before landing on the ground. A parachute increases air resistance, a sort of friction between the parachute and the air.

In this investigation: you are going to find out if the size of the parachute and the weight of the object attached to it affect the rate of fall.

Preparation: Predict

Finish the sentence in the box.

I think that the size of the parachute and the weight of the object attached to it (will/will not) affect the rate of fall because…

You are going to use this equipment to find out if the size of the parachute and the weight of the object attached to it affect the rate of fall.

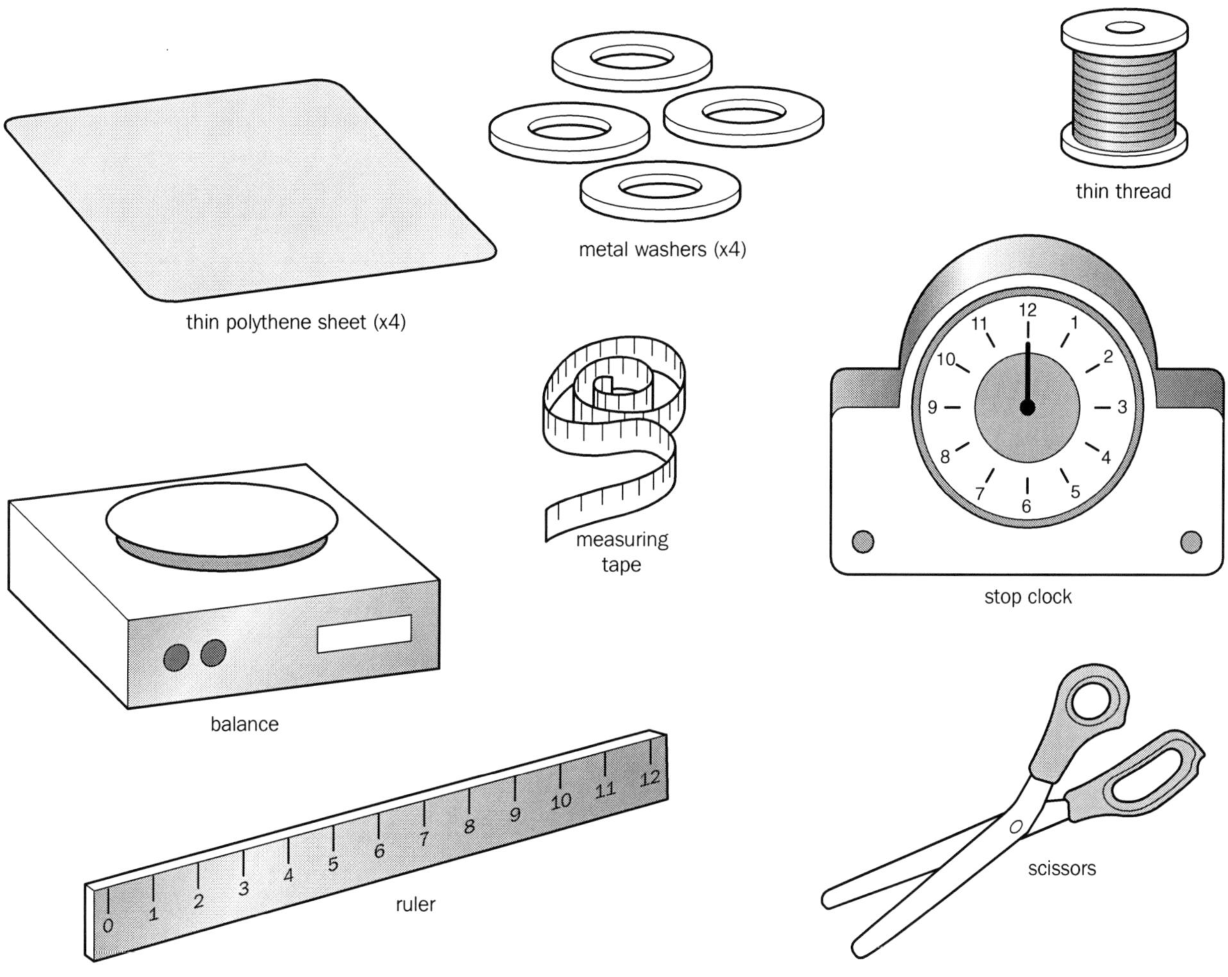

Chapter 12 ► Investigation 12D Forces

Preparation: Plan

Finish the sentences in the box.

I will measure…
Things I will keep the same are…
My investigation will be fair because…
My investigation will be safe because…

Carry out

Do your experiment first with a large parachute then with a smaller parachute.
Try each parachute with the same light weight then the same heavier weight.

Put your results in a table like this:

Parachute size______________				
Number of washers	**Mass of washers**	**Height of fall in m**	**Time taken in seconds**	**Average time in seconds**

Draw a bar graph of your results on a piece of graph paper.

Use a key with different colours for each result. Label the axes like this:

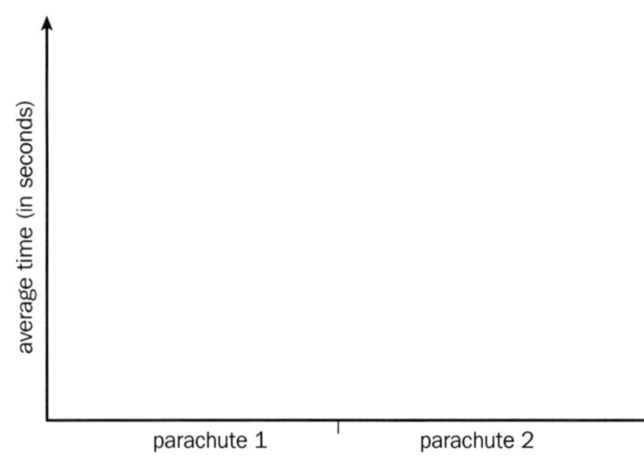

Report

Write a report on your investigation.

Here are some things you should include:

- what you did
- what happened
- explain your results
- if your prediction was correct or not
- what you could do to improve the investigation
- what you could have done if you had more time.

Chapter 12 ► Investigation 12C Forces

Skydivers use parachutes to slow their rate of fall before landing on the ground. A parachute increases air resistance, a sort of friction between the parachute and the air.

In this investigation: you are going to find out if the size of the parachute and the weight of the object attached to it affect the rate of fall.

Preparation: Predict

Finish the sentence in the box.

I think that the size of the parachute and the weight of the object attached to it (will/will not) affect the rate of fall because...

You are going to use this equipment to find out if the size of the parachute and the weight of the object attached to it affect the rate of fall:

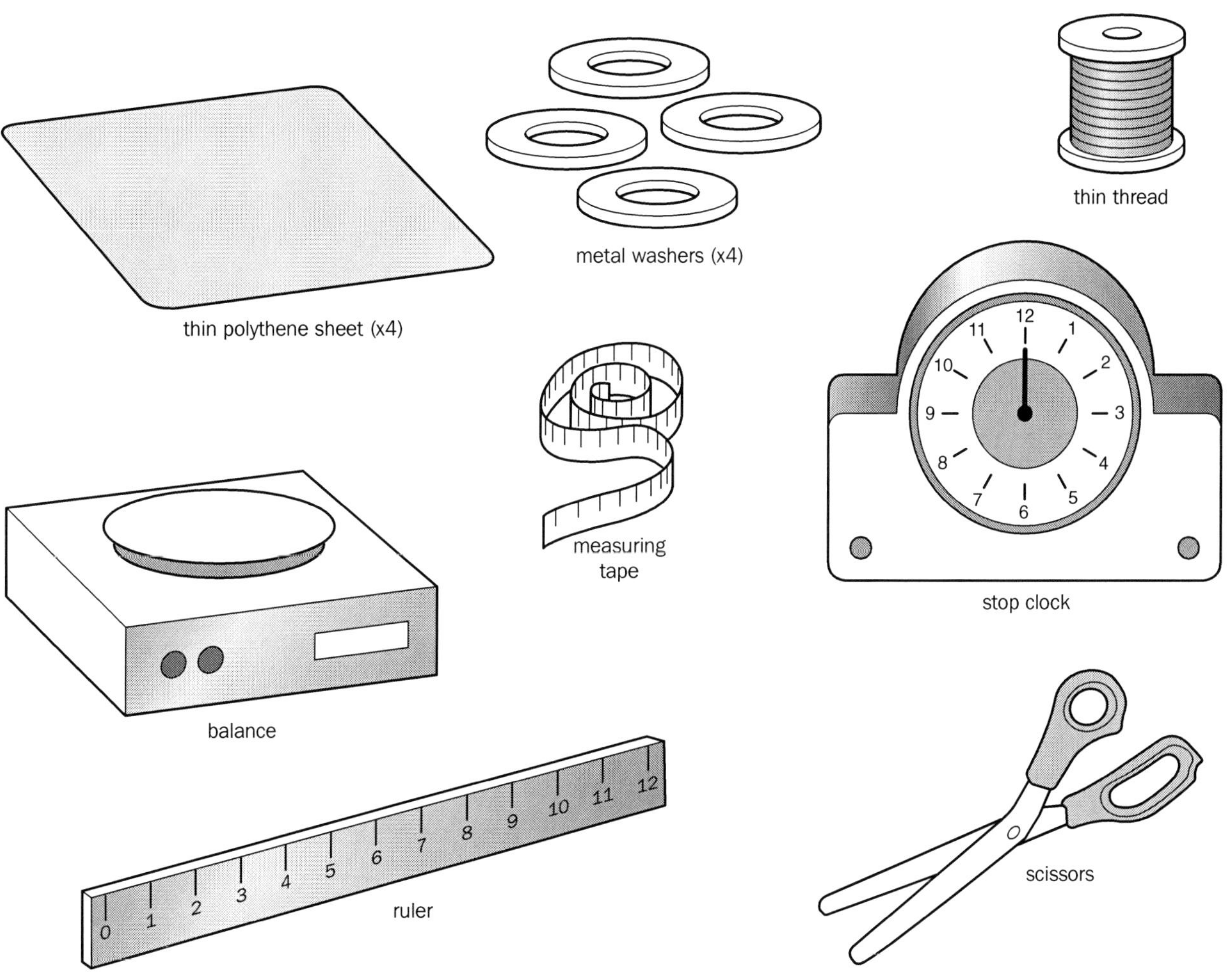

Chapter 12 ► Investigation 12C Forces

Preparation: Plan

Finish the sentences in the box.

I will measure…

Things I will keep the same are…

My investigation will be fair because…

My investigation will be safe because…

Carry out

- Cut a piece of polythene sheet about 40 cm × 40 cm.
- Weigh one of the washers.
- Make a parachute as shown in the diagram using one metal washer.
- Do a test drop to see that your parachute falls steadily. If necessary adjust the length of the strings and/or add another washer.
- Measure a fixed height, then time how long it takes for your parachute to fall to the ground.
- Do this **three** times.
- Add one more washer to your parachute and time three more drops from the same height.
- Make a smaller parachute (about 20 cm × 20 cm) and repeat the steps above.
- Put your results in a table like this:

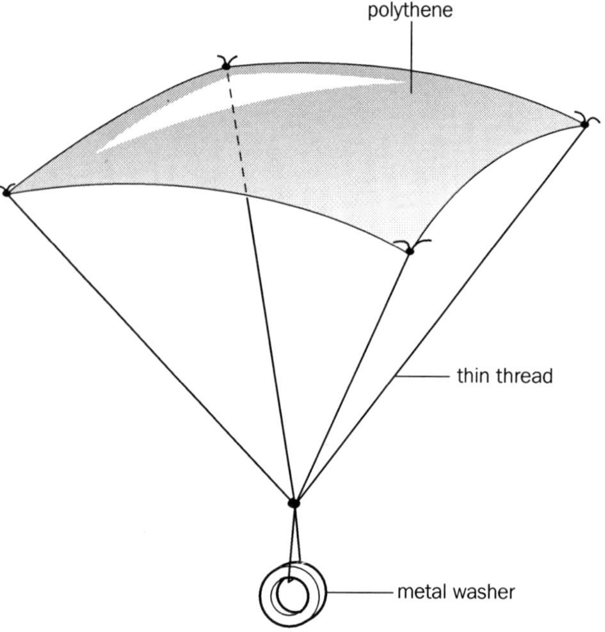

Parachute size__________				
Number of washers	**Mass of washers**	**Height of fall in m**	**Time taken in seconds**	**Average time in seconds**

Chapter 12  Investigation 12C Forces

Draw a graph of your results on this grid.

Use different colours for each bar.

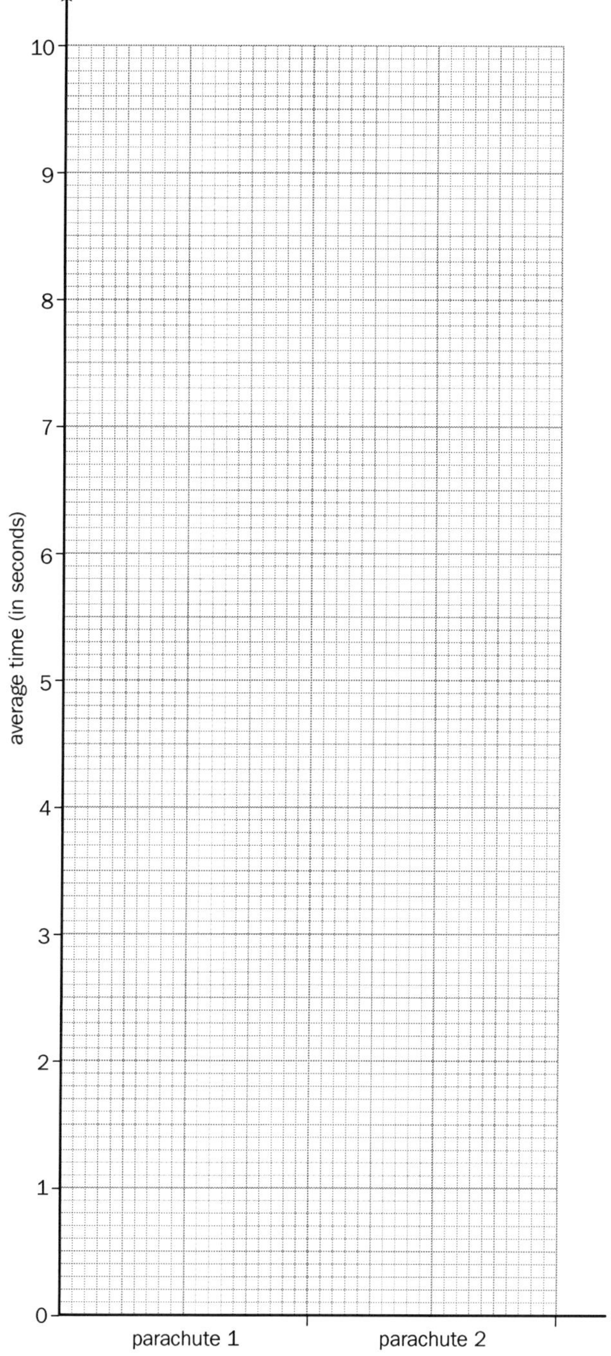

Report

Finish the sentences in the box.

What I did was...

What happened was...

From my results I found out that the size of the parachute and the weight of the object attached to it (did/did not) affect the rate of fall. I know this because...

My prediction (was/wasn't) correct. If I could do the investigation again I would...

Investigation 12
Practical notes

Forces

Investigation 12
Technician's notes

Forces

Each group will need:

- four pieces of thin polythene sheet (supermarket bags are ideal)
- thin thread, e.g. reel of cotton
- some metal washers (about four should do depending on size/mass)
- stop clock
- access to balance
- ruler
- scissors
- measuring tape.

Number of apparatus sets:

Number of pupils:

Number of groups:

Visual aids:

ICT resources:

Equipment/apparatus needed:

Safety notes

CLEAPSS/SSERC SAFETY REFERENCE:

Chapter 12 ▶ Test Forces

White

1 Friction is a force that tries to stop things moving. In the following, is friction **useful** or a **problem**?

a walking on a dry footpath

b sliding down a slide

c pedalling a bicycle

d skating on ice

e using chalk on a chalk board.

5 marks

2 The diagram shows two pieces of metal as seen under a microscope.

a Explain why friction is high as one piece of metal slides over the other.

b How can this friction be reduced?

c Explain why this reduces the friction between the two pieces of metal.

5 marks

3 The diagrams show a cyclist riding a bike at two different times. The arrows F represent the force due to the cyclist pedalling as hard as he can. (The wider the arrow the bigger the force.)

a What do arrows X1 and X2 represent?

b What will happen to the size of X2 if a wind blows in the cyclist's face?

c i In which direction will the force of gravity act on the cyclist?

ii Explain why the force of gravity does not affect the motion of the bicycle moving over flat ground.

d i Which diagram shows the cyclist speeding up?

ii Explain your answer.

e What happens to the effort put in by the cyclist in overcoming force X2?

8 marks

4 The diagram shows a parachutist falling to the ground. The arrows represent the forces acting on the parachutist.

a What is the name of
i force A **ii** force B?

b The parachutist lets go of a heavy rucksack she is holding.
i How does this affect force A?
ii How does this affect force B?
iii How does it affect the parachutist's downward speed?
iv Explain your answer.

7 marks

5 A bag of cement weighs more than the same size bag of potatoes.

a What is meant by weight?

b Explain why a bag of cement weighs more than a bag of potatoes.

c Explain why a bag of potatoes weighs less on the Moon.

d Explain why a bag of potatoes weighs less when it put into water.

5 marks

Chapter 12 ▶ Test Forces

1 Explain the following.

a You need less force to cut with a sharp knife than with a blunt knife.

b Tracked vehicles are better than wheels for driving over muddy ground.

c You can push one end of a drawing pin but not the other.

3 marks

2 The diagram shows two children, Joel and Jolene, on a see-saw. The see-saw is balanced.

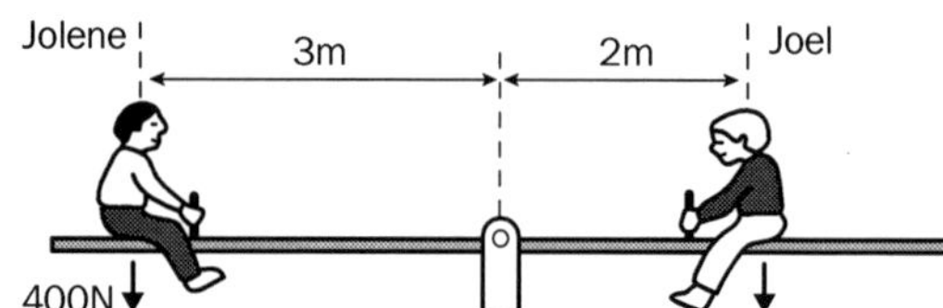

a Balancing a see-saw depends on two things, what are they?

b Work out Joel's weight. Show all your working and use the correct units.

c Jolene moves 1 m towards the pivot. Where does Joel have to move to make the see-saw balance again?

8 marks

3 The gravitational field on the Moon is one sixth that on Earth. On Earth each kilogram of matter weighs 10 N.

a Work out the weights of these masses on Earth:

i 5 kg bag of potatoes

ii 25 kg bag of cement.

b The diagram shows the path of a spacecraft from Earth to the Moon. The spacecraft carries one astronaut. The astronaut has a mass of 60 kg.

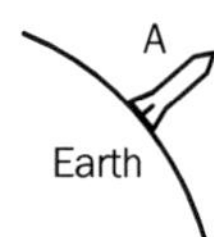

What will the astronaut weigh at

i A

ii B

iii C?

5 marks

4 The diagram shows a block of metal. The block weighs 18 N.

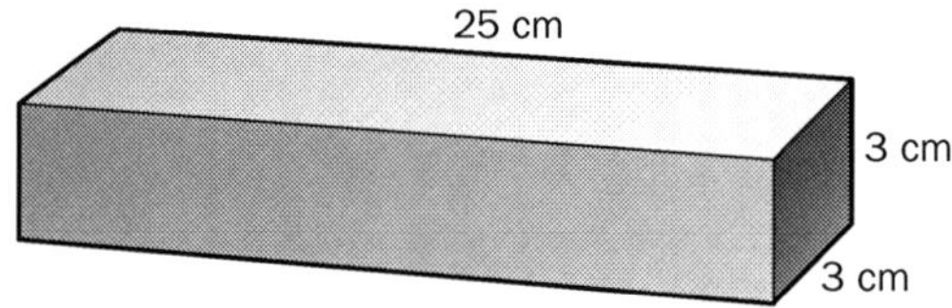

a Work out

i the area of the face it is lying on

ii the area of the end face.

b Which face must it stand on to give maximum pressure?

c Work out the maximum pressure.

7 marks

5 The diagram shows a hydraulic car jack. A handle is used to apply a force to the small piston. This enables a large load to be lifted by the large piston.

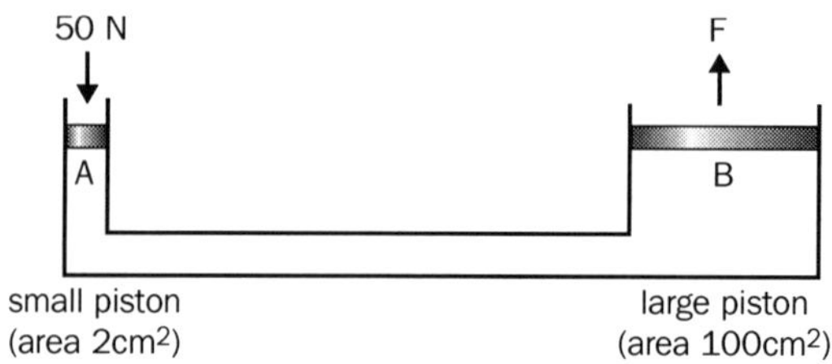

a If the small piston is pushed down with a force of 50 N, what pressure is produced in the hydraulic fluid at A?

b What pressure is exerted by the hydraulic fluid at B?

c Work out the upward force on the large piston.

7 marks

Chapter 12 ▶ Mark scheme White

Forces

Question	Answer	Marks	Level
1 a	useful	1	C
b	problem	1	C
c	problem	1	C
d	either (friction produces heat which melts ice, etc.)	1	C
e	useful	1	C
		5	
2 a	high points catch (1) stop surfaces sliding (1)	2	C
b	lubrication/oil, etc.	1	C
c	lubricant forces surfaces apart (1) high points no longer catch (1)	2	C
		5	
3 a	air resistance/friction	1	C
b	increases	1	C
c i	downwards/towards centre of Earth	1	
ii	no effect on horizontal movement	1	C
d i	1	1	
ii	F is greater than X1	1	C
e	lost (1) as heat (1)	2	C
		8	
4 a i	air resistance	1	
ii	gravity	1	D
b i	no effect	1	
ii	gets less	1	
iii	slows	1	
iv	answer to show appreciation of link between mass, weight, and gravity, e.g. mass gets less (1) so weight gets less (1)	2	D
		7	
5 a	downward force of gravity	1	D
b	bag of cement has greater mass/more matter	1	D
c	force of gravity less on Moon	1	D
d	effect of gravity is reduced (1) by upthrust (1)	2	D
		5	
		TOTAL 30 marks	

Suggested grade/level boundaries

C = 16/30

D = 24/30

Chapter 12 ► Mark scheme

Blue

Forces

Question	Answer	Marks	Level
1 a	small surface area so pressure high	1	F
b	large surface area so pressure low	1	F
c	sharp end has small surface area so pressure high	1	F
		3	
2 a	size of force/mass/weight (1) distance (from pivot) (1)	2	E
b	$\frac{400 \times 3}{2}$ (1) = 600 (1) Nm (1)	3	E
c	$\frac{400 \times 2}{600}$ (1) = 1.33 (1) Nm (1)	3	E
		8	
3 a i	50 N (must have units)	1	
ii	250 N (must have units)	1	E
b i	600 N (must have units)	1	
ii	0/weightless	1	
iii	100 N (must have units)	1	E
		5	
4 a i	75 (1)	1	
ii	9 (1)	1	
	correct units in i) or ii) or both (1)	1	F
b	end face/9 cm^2	1	F
c	$\frac{18}{9}$ (1) = 2 (1) cm^2 (1)	3	F
		7	

Question	Answer	Marks	Level
5 a	$\frac{50}{2}$ (1) = 25 (1) N/cm^2 (1)	3	F
b	25 N/cm^2	1	F
c	100 × 25 (1) = 2500 (1) N (1)	3	F
		7	
		TOTAL 30 marks	

Suggested grade/level boundaries

E = 12/30

F = 21/30

13 Keeping the body working

This chapter looks at how the human body obtains and deals with the food and water it needs to function properly. The composition of different foods and the importance of their constituents is considered along with details of how food is broken down into simple molecules. Breathing and respiration are covered in a way that enables pupils to differentiate between them.

Assessment opportunities

Formative assessment opportunities are provided by worksheets, homework sheets, and an investigation.

The **worksheets** cover material at levels D and F for attainment targets for knowledge and understanding. Teachers may wish to use these worksheets not only as part of practical activities but also to provide evidence of pupil achievement.

Worksheet	Level
13.2	D
13.3a	D
13.3b	F
13.3c	F
13.4a	D
13.4b	F

The **homework sheets** cover material at level D for attainment targets for knowledge and understanding. These homework sheets can be used individually as a follow-up to work done in class or assembled into a homework booklet allied closely to schemes of work.

Homework sheet	Level
13.1	D
13.2	D
13.3	D
13.4	D

The **investigation** covers all three skill areas at levels C, D, E, and F. It is written in a way that allows for pupils to be assessed in all three skill areas at one level. Alternatively, customised assessments can be constructed enabling pupils to be assessed at different levels in all three skills. The latter approach is more time consuming, but it does provide the opportunity for pupils to show evidence of achievement at different levels in different skills in the same investigation. Teachers will need to use their professional judgement when deciding which level is appropriate to individual pupils. It is envisaged that pupils will show progression through the levels as they work through their science course.

Summative tests are provided at two levels, white and blue. The white test, although only containing questions covering attainment target level D, can be used to assess at level C and D. The blue test contains questions covering attainment target levels D and F but can be used for assessing at levels D, E, and F. Each test has a total of 30 marks and will take about 30 minutes for pupils to complete, although this can be varied depending on pupil ability. Mark schemes are provided together with suggested grade/level boundaries. It is envisaged that these tests will be given to pupils on completion of the material covered in Chapter 13.

ICT opportunities

The use of data loggers/remote sensors can extend the range, speed, and sensitivity of measurements in many of the worksheets for this chapter. Once downloaded onto a PC, data-handling programs can be used to analyse information gathered, data can be manipulated, and appropriate graphs etc. presented. The Internet provides pupils with access to a huge range of scientific information. A list of suitable websites is included in this Teacher's Guide.

Students' book chapter 13 contents and guide levels

13.1	Why do you have to eat?	*Starting off*	D
	Supplying every cell	*Going further*	D
13.2	Teeth	*Starting off*	D
	Oral hygiene	*Going further*	D
13.3	Digestion: breaking down food	*Starting off*	D
	Breaking up molecules	*Going further*	F
	A closer look at enzymes	*For the enthusiast*	F
13.4	Breathing	*Starting off 1*	D
	A closer look at the lungs	*Starting off 2*	D
	Breathing: energy production	*Going further*	F
	Enzymes: chemical controllers	*For the enthusiast*	F

13.2 Sugar and tooth decay

W/S

Name: Date: Group:

What you need:

Cotton buds, neutral buffer solution, droppers, universal indicator solution, universal indicator chart, sugar, glass dish.

What to do:

1 Wash your hands.

2 Use a clean cotton bud to collect some plaque from your teeth. (Try behind your front teeth.)

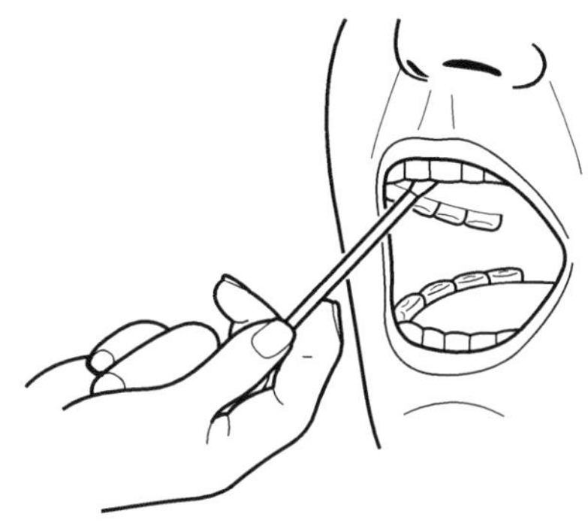

3 Put a few drops of neutral buffer solution on to the bud.

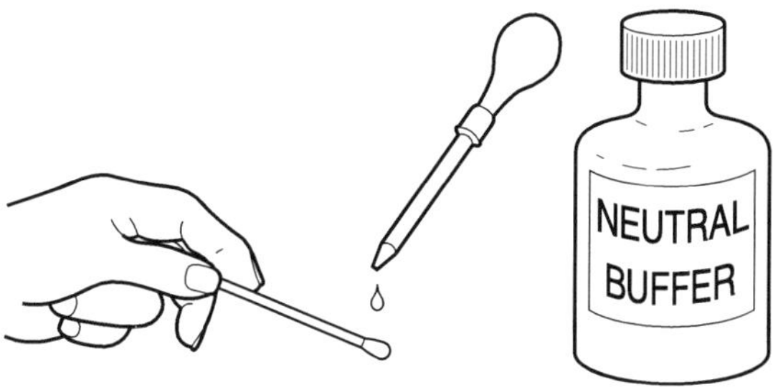

4 Put a few drops of universal indicator onto the bud.

Wait for 1 minute.

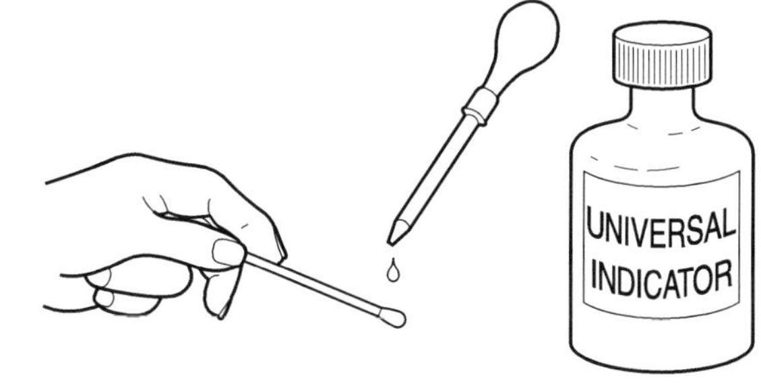

5 Dip the bud into some sugar. Wait a few minutes and watch closely for any colour change.

- What is the pH of the bud after you collected some plaque?
- What is the pH of the bud after it had been dipped in sugar?
- How do sugary foods affect the pH of plaque in your mouth?

SAFETY WARNING

Make sure your hands are clean.

Use a clean, sterile cotton bud.

13.2
Practical notes

Sugar and tooth decay

This activity shows pupils how sugar lowers the pH of plaque around their teeth. Bacteria in the plaque feed on the sugar, producing acid as a waste product. It is this acid that dissolves tooth enamel. Pupils may need advising individually where best to obtain plaque from their teeth. Modern dental hygiene standards are high, but even the cleanest teeth will yield some plaque, especially in-between the lower front teeth. Pupils should be reminded about the importance of cleanliness whilst doing this activity.

13.2
Technician's notes

Number of apparatus sets:

Number of pupils:

Number of groups:

Visual aids:

ICT resources:

Equipment/apparatus needed:

Sugar and tooth decay

Each group will need:

- a copy of worksheet 13.2
- cotton buds
- neutral buffer solution
- droppers
- universal indicator solution
- universal indicator chart
- sugar
- glass dish, alternatively use a clean, plastic Petri dish
- access to warm water, soap, and tissues for hand washing
- large beaker for collecting used cotton buds.

Safety notes

- **Pupils must wash their hands before starting the activity.**
- **Only clean, sterile cotton buds should be used.**
- **Arrange for the safe disposal of used buds.**

CLEAPSS/SSERC SAFETY REFERENCE:

13.3a Making a model intestine W/S

Name: Date: Group:

What you need:

Piece of visking tubing, large test tube, two test tubes, droppers, starch and glucose solution, cotton thread, iodine solution, Benedict's solution, Bunsen burner, tripod, gauze, heatproof mat, safety goggles.

What to do:

1 Put on safety goggles.

2 Tie a knot in one end of the visking tubing. Fill the tubing with starch and glucose solution.

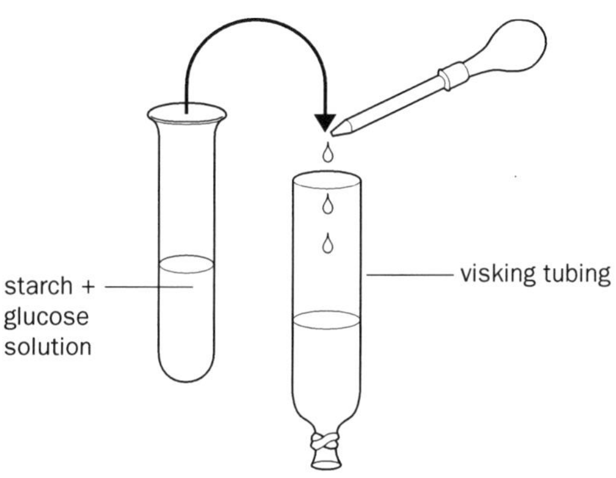

3 Tie the other end of the tubing tightly with cotton thread. Put the tubing into the test tube and add some water.

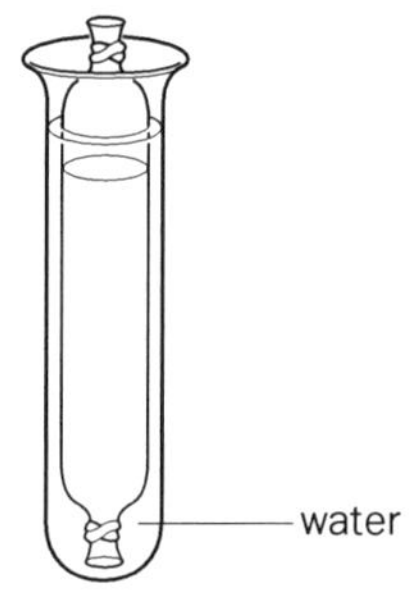

Leave the apparatus like this for 10 minutes.

4 After 10 minutes take some of the water from around the tubing and put it into two test tubes.

5 Add a few drops of iodine solution to one test tube. A blue/black colour shows starch is present.

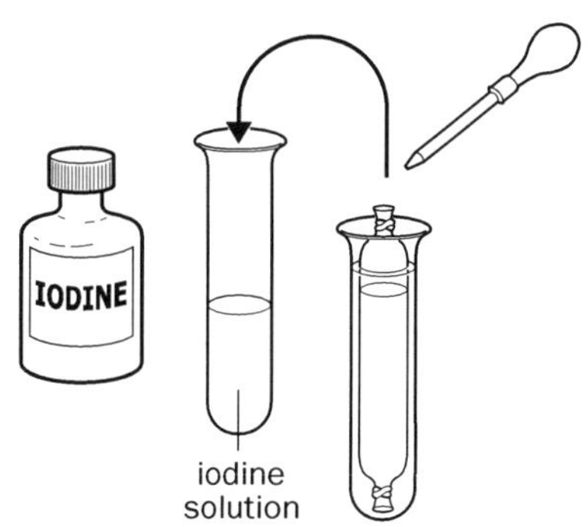

5 Add some Benedict's solution to the other test tube (enough to make the solution blue). Stand the test tube into a beaker of water and heat it over a Bunsen burner. An orange colour shows glucose is present.

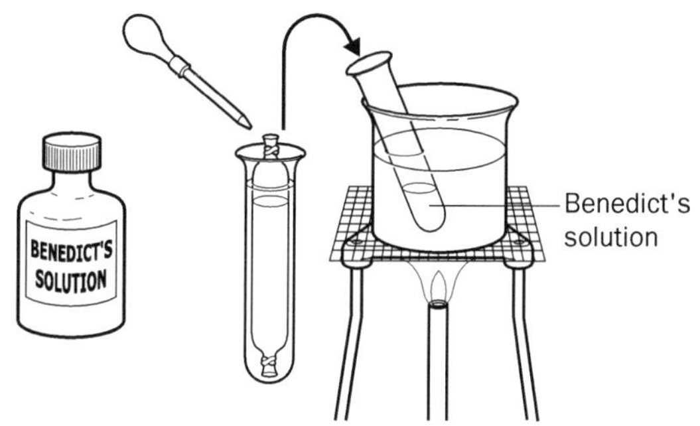

- What is the colour of the solution after testing with iodine solution?
- What is the colour of the solution after testing with Benedict's solution?
- What do these results mean?

SAFETY WARNING

Iodine and Benedict's solution are harmful to skin and eyes.

Wear safety goggles.

13.3a
Practical notes

Making a model intestine

The aim of this activity is to show pupils how only small (digested) molecules can pass through the walls of the small intestine. Visking tubing acts as a 'microscopic sieve' allowing only the tiny glucose molecules through its pores. Starch molecules are too large to pass through. Depending upon ability it may be necessary for some pupils to practice the tests for starch and glucose before doing this activity. This activity could be extended by putting a mixture of starch and amylase (or saliva) into a visking 'sausage' instead of starch and glucose. This will provide pupils with the opportunity of modelling the processes of digestion and absorption at the same time.

PRACTICAL HINTS

Visking tubing is easier to handle when it is wet. Tell pupils to wash the outside of their 'visking sausages' before putting them into the test tubes. This will avoid any contamination.

13.3a
Technician's notes

Making a model intestine

Each group will need:

- a copy of worksheet 13.3a
- piece of visking tubing about 15 cm long
- large test tube/boiling tube
- two test tubes
- droppers
- mixture of starch and glucose solution; make a paste of 1 g of soluble starch with water. Add this to 100 cm^3 of boiling water stirring all the time. Add 5 g of glucose, stir until it is dissolved and leave to cool
- cotton thread (enough to tie the neck of the visking tubing)
- iodine solution
- Benedict's solution
- Bunsen burner
- tripod
- gauze
- heatproof mat
- safety goggles.

Number of apparatus sets:

Number of pupils:

Number of groups:

Visual aids:

ICT resources:

Equipment/apparatus needed:

Safety notes

- **See HAZCARDS for iodine and Benedict's solution.**
- **Remind pupils that the apparatus will get hot.**
- **Pupils must wear safety goggles.**

CLEAPSS/SSERC SAFETY REFERENCE:

13.3b How temperature affects enzymes W/S

Name: Date: Group:

What you need:

Enzyme (amylase), starch solution, six test tubes, three beakers, labels, syringe, three droppers, Bunsen burner, tripod, gauze, heatproof mat, thermometer, three spotting tiles, iodine solution, ice, stop clock, safety goggles.

What to do:

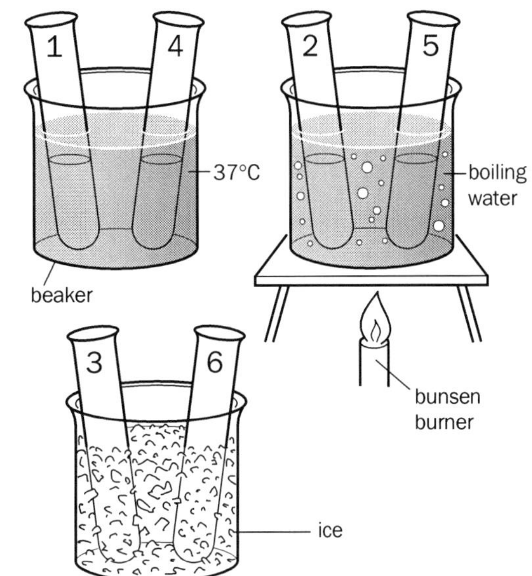

1 Put on safety goggles.

2 Label the test tubes 1–6. Put 5 cm^3 of starch solution into test tubes 1, 2, and 3. Rinse the syringe and put 1 cm^3 of enzyme into test tubes 4, 5, and 6.

3 Heat some water in a beaker to 37 °C.

Heat some water in another beaker until it is boiling, then adjust the Bunsen burner to keep the water simmering.

Fill a third beaker with ice.

4 Put test tubes 1 and 4 into the water at 37 °C.

Put test tubes 2 and 5 into the boiling water.

Put test tubes 3 and 6 into the beaker of ice.

Leave for 5 minutes.

5 While you are waiting, put one spot of iodine solution onto each space on the spotting tiles. Label the tiles '37 °C', 'Boiling' and 'Ice'.

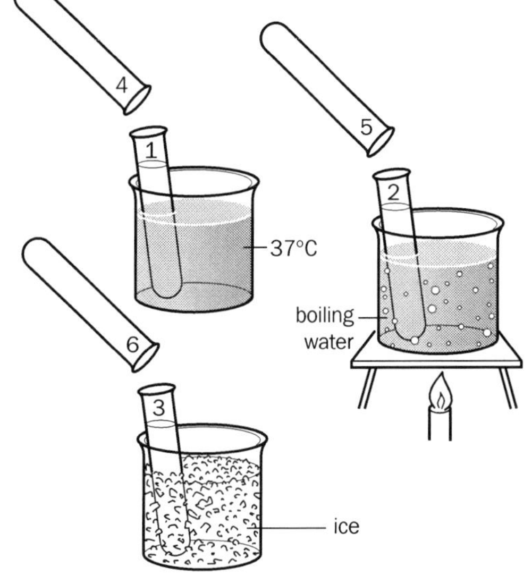

6 After 5 minutes pour the enzyme in tube 4 into the starch solution in tube 1. Keep tube 1 in the beaker.
Do the same with tubes 5 and 2 and 6 and 3.

Start timing.

7 Every minute put one drop of the mixture in test tube 1 on to a spot of iodine on tile '37°C'. Do the same with test tube 2 using the 'Boiling' tile and test tube 3 using the 'Ice' tile.

- How long does it take for the iodine to stop turning blue/black on each of the spotting tiles?
- Which of these three temperatures does this enzyme work best at?

SAFETY WARNING

Enzymes and iodine are harmful to skin and eyes.

Wear safety goggles.

13.3b
Practical notes

How temperature affects enzymes

Amylase is used as an easier and more socially acceptable option than gathering saliva. However, some teachers may prefer to use saliva. If so, the saliva should be diluted with water in the ratio of 1 part saliva to 10 parts water before starting the activity. Depending upon pupil ability and apparatus availability teachers may wish to divide the activity up. Some groups of pupils could investigate enzyme activity at one particular temperature while others do the rest. Syringes are suggested as a simple and effective means of measuring and dispensing liquids but teachers may need to watch out for their misuse. Pupils should be reminded to wear safety goggles.

13.3b
Technician's notes

How temperature affects enzymes

Each group will need:

Number of apparatus sets:

Number of pupils:

Number of groups:

Visual aids:

ICT resources:

Equipment/apparatus needed:

- a copy of worksheet 13.3b
- 15 cm^3 of 5% amylase solution, presented in a large beaker labelled 'Enzyme (amylase)'
- 3 cm^3 of 1% starch solution, presented in a large beaker labelled 'Starch solution'
- six test tubes
- three 250 cm^3 beakers
- labels
- syringe(s) or alternative to measure appropriate volumes
- three droppers
- Bunsen burner
- tripod
- gauze
- heatproof mat
- thermometer
- three spotting tiles
- iodine solution
- supply of ice cubes
- stop clock
- safety goggles.

Safety notes

- **See HAZCARDS for amylase and iodine.**
- **Pupils must wear safety goggles.**
- **Watch for silly behaviour with syringes.**

CLEAPSS/SSERC SAFETY REFERENCE:

13.3c How pH affects enzymes

W/S

Name: Date: Group:

What you need:

Enzyme (amylase), starch solution, acid (pH4), alkali (pH11), pure water (pH7), three test tubes, labels, syringe, three droppers, three spotting tiles, iodine solution, stop clock, safety goggles.

What to do:

1 Put on safety goggles.

2 Label the test tubes 1–3. Put 5 cm^3 of starch solution into each test tube. Rinse the syringe.

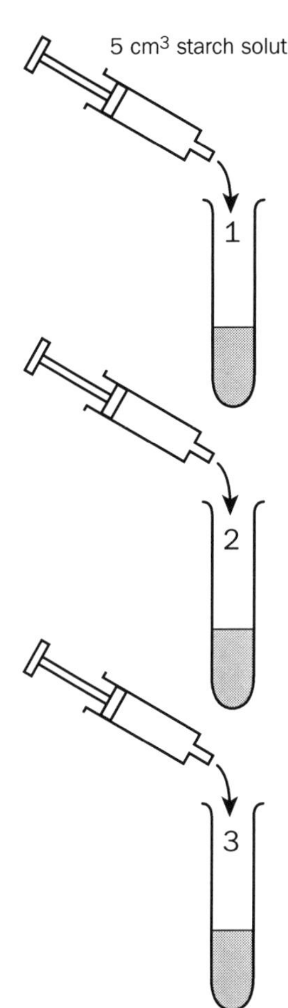

3 Put 1 cm^3 of acid into test tube 1, 1 cm^3 of water into test tube 2 and 1 cm^3 of alkali into test tube 3.

Make sure you rinse the syringe each time.

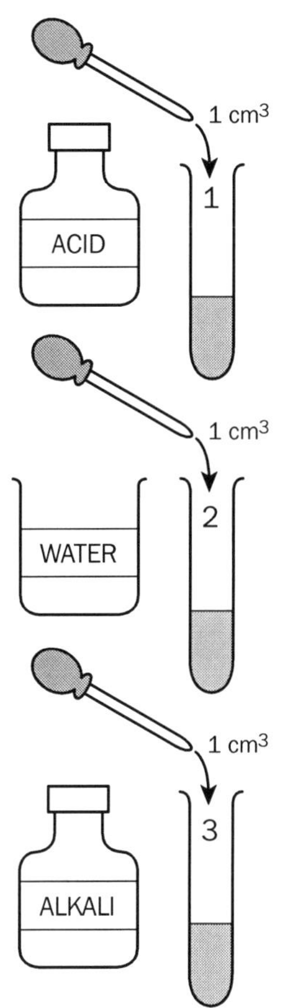

4 Put one spot of iodine solution into each space on the spotting tiles. Label the tiles 'pH4', 'pH7' and 'pH11'.

5 Put 1cm^3 of enzyme into each test tube.

Start timing.

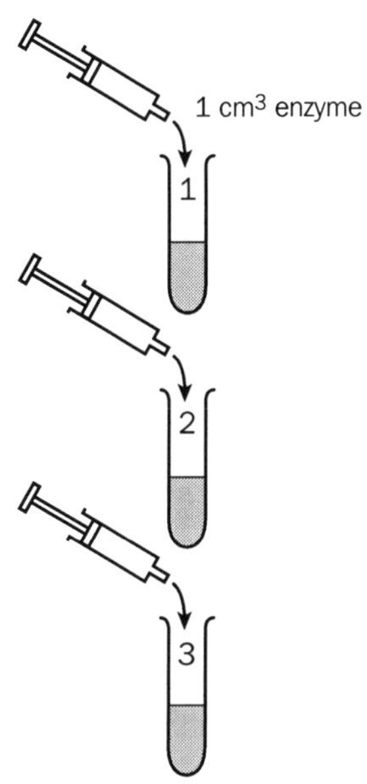

6 Every minute put a spot of the mixture in test tube 1 on to a spot of iodine on tile 'pH4'. Do the same with test tube 2 using the 'pH7' tile and test tube 3 using the 'pH11' tile.

- How long does it take for the iodine to stop turning blue/black in each of the spotting tiles?
- Which of these three pH levels does this enzyme work best at?

SAFETY WARNiNG

Acids, alkalis, enzymes and iodine are harmful.

Wear safety goggles.

13.3c
Practical notes

How pH affects enzymes

Amylase is used as an easier and more socially acceptable option than gathering saliva. However, some teachers may prefer to use saliva. If so, the saliva should be diluted with water in the ratio of 1 part saliva to 10 parts water before starting the activity. Depending upon pupil ability and apparatus availability teachers may wish to divide the activity up. Some groups of pupils could investigate enzyme activity at one particular pH while others do the rest. Syringes are suggested as a simple and effective means of measuring and dispensing liquids, but teachers may need to watch out for their misuse especially with acid and alkali being dispensed. Pupils should be reminded to wear safety goggles at all times.

13.3c
Technician's notes

How pH affects enzymes

Each group will need:

Number of apparatus sets:

Number of pupils:

Number of groups:

Visual aids:

ICT resources:

Equipment/apparatus needed:

- a copy of worksheet 13.3c
- 15 cm^3 of 5% amylase solution, presented in a large beaker labelled 'Enzyme (amylase)'
- 3 cm^3 of 1% starch solution, presented in a large beaker labelled 'Starch solution'
- 1 cm^3 hydrochloric acid (or alternative) at pH4, presented in a large beaker labelled 'Acid pH4'
- 1 cm^3 sodium hydroxide (or alternative) at pH11, presented in a large beaker presented in a large beaker labelled 'Alkali pH11'
- 1 cm^3 pure/distilled water (bottled water is usually good enough) presented in a large beaker labelled 'Pure water pH7'
- three test tubes
- labels
- syringe(s) or alternative to measure appropriate volumes
- three droppers
- three spotting tiles
- iodine solution
- stop clock
- safety goggles.

Safety notes

- **See HAZCARDS for hydrochloric acid (or alternative), sodium hydroxide (or alternative), amylase and iodine.**
- **Pupils must wear safety goggles.**
- **Watch for silly behaviour with syringes, acids and alkalis.**

CLEAPSS/SSERC SAFETY REFERENCE:

13.4a Inhaled and exhaled air

W/S

Name: Date: Group:

Part 1

What you need:

Candle, glass jar, bowl of water, tubing, heatproof mat, stop clock, matches, safety goggles.

What to do:

1 Put on safety goggles.

2 Stand the candle on a heatproof mat. Light the candle and put the jar over it. The jar contains ordinary air, i.e. the air that we breathe into our lungs.

Time how long the candle burns.

3 Lie the jar on its side under water in a bowl to get rid of all of the air.

4 Stand the jar upside down in the water and put some tubing inside the opening. Breathe out slowly through the tubing and trap the exhaled air in the jar.

5 Light the candle again. This time put the jar of exhaled air over the candle.

Time how long the candle burns.

Explain the difference in your results.

Part 2

What you need:

Two large test tubes, two bungs with tubing, lime water, safety goggles.

What to do:

1 Put on safety goggles.

2 Put the same amount of lime water into each test tube, about half full will do. Attach the bungs like this:

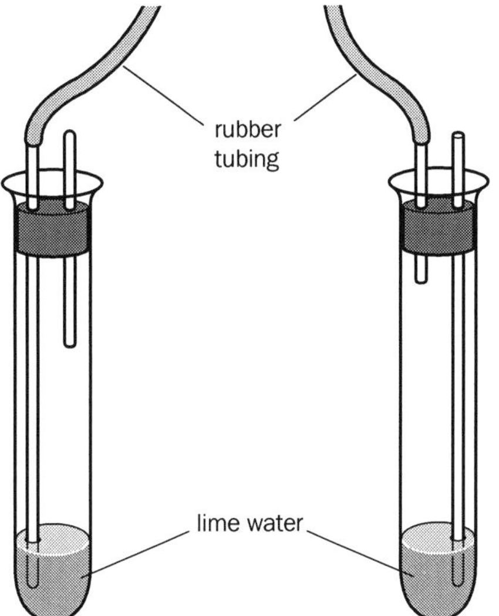

3 Put the ends of both tubes into your mouth and breathe **gently** in and out through the tubes.

Keep doing this until the lime water in one test tube goes milky.

- Which lime water goes milky first?
- What does this mean?

SAFETY WARNING

If you get lime water in your mouth, rinse your mouth with drinking water.

13.4a
Practical notes

Inhaled and exhaled air

These are two straightforward activities designed to show pupils the main differences between inhaled and exhaled air. Further work could include breathing onto a cold mirror or test tube and using a temperature sensor to compare the temperature of inhaled and exhaled air. Foil 'cups' could be made to hold the candles in Part 1. These will prevent the messy wax deposits on heatproof mats. In Part 2 pupils will need reminding to breath gently through the tube. A practice run with tap water might be advisable in some cases if unnecessary spillages etc. are to be avoided.

13.4a
Technician's notes

Inhaled and exhaled air

Each group will need:

Number of apparatus sets:

Number of pupils:

Number of groups:

Visual aids:

ICT resources:

Equipment/apparatus needed:

Part 1
- a copy of worksheet 13.4a
- candle
- glass jar
- bowl of water
- rubber or plastic tubing (must be clean/sterile; rinse with antiseptic mouthwash before use by each pupil)
- heatproof mat
- stop clock
- matches
- safety goggles.

Part 2
- a copy of worksheet 13.4a
- two boiling tubes
- two bungs with tubing as shown
- lime water
- safety goggles.

Safety notes
- **See HAZCARDS for lime water.**
- **Pupils must wear safety goggles.**
- **Watch for silly behaviour while breathing through tubing.**

CLEAPSS/SSERC SAFETY REFERENCE:

13.4b Respiration in maggots

W/S

Name: Date: Group:

What you need:

Large test tube, bung with one hole, soda lime, spatula or spoon, capillary tubing, card ring, muslin or 'J' cloth, maggots, ruler, stop clock.

What to do:

1 Put some soda lime into the large test tube. Soda lime absorbs carbon dioxide.

2 Make a cradle for the maggots by pushing the card ring on to the muslin and down into the test tube.

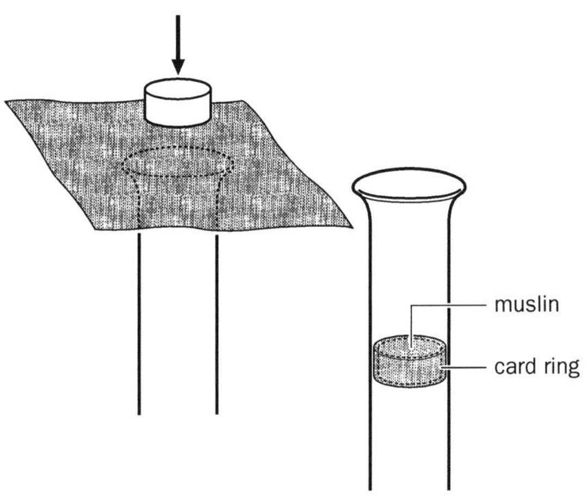

3 Put some maggots onto the muslin cradle and put the bung and capillary tube in place.

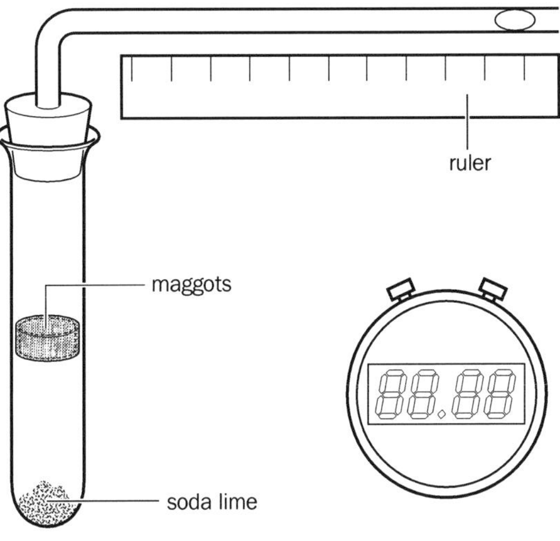

4 Put a drop of water on to the end of the capillary tube.

5 Time how long it takes for the water 'bubble' to move 5 cm along the tube.

13.4b
Practical notes

Respiration in maggots

This activity shows pupils that respiration involves the uptake of oxygen and production of carbon dioxide. Once pupils have got over their initial reaction to working with maggots they usually settle down and enjoy this activity. The more squeamish may wish to use tweezers to handle the maggots. Rubber gloves could also be provided for those who refuse to touch the larvae. Watch out for silliness, there is a huge potential for it here.

13.4b
Technician's notes

Respiration in maggots

Each group will need:

- a copy of worksheet 13.4b
- boiling tube
- bung with one hole fitted with capillary tubing as shown
- soda lime
- spatula or spoon
- card ring made to fit loosely inside boiling tube
- piece of muslin or 'J' cloth approximately 5 cm square
- 10 maggots, count them out and count them in
- ruler
- stop clock.

Number of apparatus sets:

Number of pupils:

Number of groups:

Visual aids:

ICT resources:

Equipment/apparatus needed:

Safety notes

- **See HAZCARDS for soda lime.**
- **Watch out for silly behaviour with maggots.**

CLEAPSS/SSERC SAFETY REFERENCE:

13.1 What's in a food? H/W

Name: Date: Group:

What you need to know …

Most foods are mixtures of carbohydrates, proteins, fats, vitamins, minerals and water. Different foods have different amounts of these in them. Some foods may not have all of these things in them. The diagram shows the amounts of carbohydrates, proteins, fat, vitamins, minerals and water in some foods.

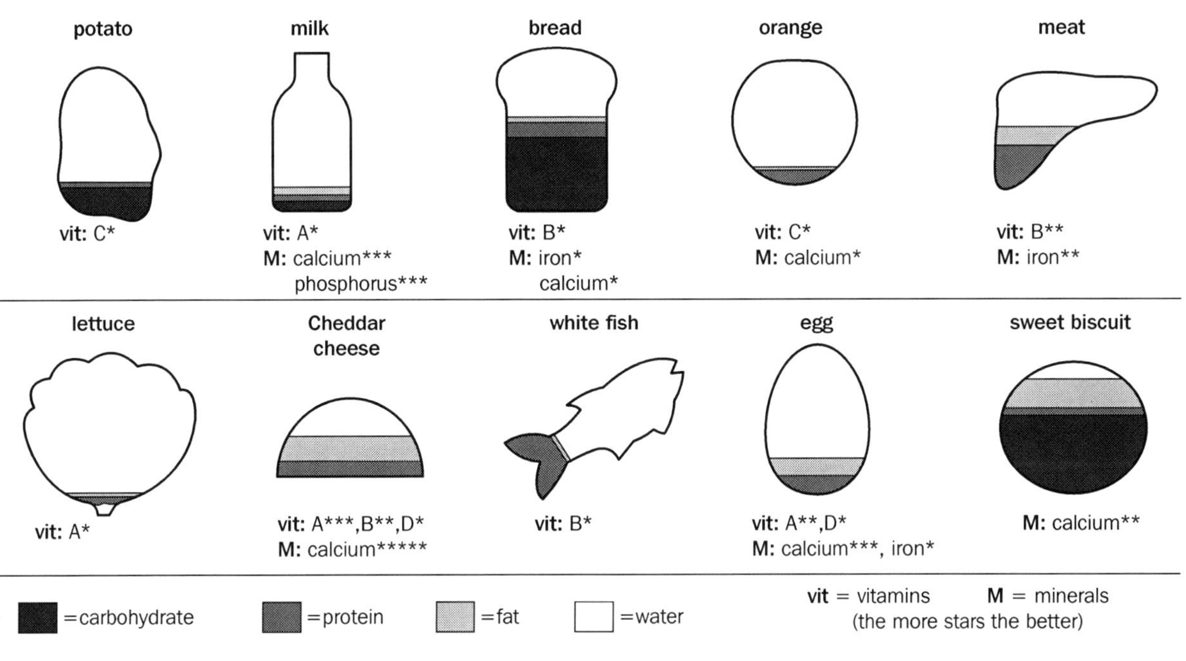

What to do:

Answer these questions:

1 Which food gives the best supplies of

a protein

b vitamins

c minerals?

2 Which foods are mostly made up of

a carbohydrate

b water?

3 If a doctor told you not to eat fats, which of the foods should you cut out?

4 a Explain why you can eat lots of green vegetables when you are slimming.

b Which of the foods should you only eat in small quantities when on a diet?

5 Cheddar cheese is made from milk. What differences are there between the two?

6 A chocolate bar contains 75 per cent carbohydrate, 5 per cent protein and 19 per cent fat

a Draw a diagram to represent this information.

b What do you think the remaining 1 per cent of the chocolate bar is made up of?

HANDY HINTS

Use the same colours/shading and style of diagram as the one shown above.

13.2 Teeth

H/W

Name: **Date:** **Group:**

What you need to know ...

Food has to be broken down into small pieces before your body can use it. This is the job of your teeth. Incisors are sharp, biting teeth, canines are also biting teeth, molars and premolars grind the food into tiny bits. Eating sweet, sticky food can cause tooth decay. The diagram shows the teeth of an adult human.

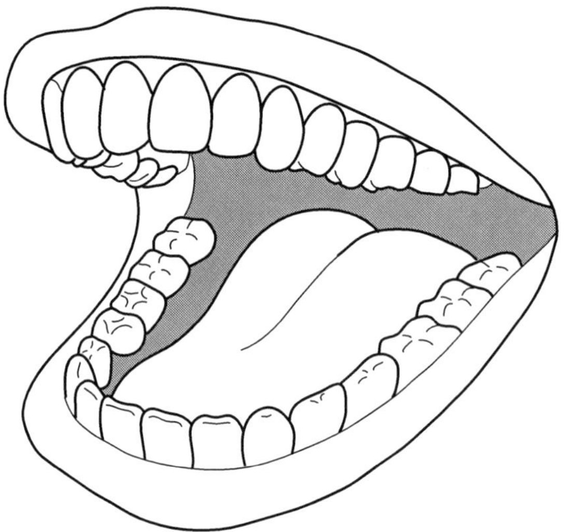

What to do:

1. Cut out and stick or copy the diagram into your book.
   - **a** Label the incisors, canines, premolars and molars.
   - **b** How many teeth are their in the mouth of an adult human?
   - **c** Where is the enamel?
   - **d** What is inside a tooth?
2. The following statements explain how tooth decay happens, but they are in the wrong order. Write the statements in their correct order.
   - When you eat sweet, sticky, food lots of bacteria grow in the plaque on your teeth.
   - The bacteria change sugar to acid.
   - The acid eats through the enamel making a hole.
   - Bacteria get into the hole and start eating the living tissue inside the tooth.
   - When bacteria start eating nerve tissue you get toothache.
   - If you go to a dentist quickly enough they can stop the rot by drilling out the dead tissue and filling the hole in the tooth.

HANDY HINTS

Look at the diagram and other information carefully.
Cut out the statements or write them on bits of paper before trying to arrange them in the right order.

13.3 Keep fit with fibre

Name: **Date:** **Group:**

What you need to know ...

To keep your digestive system in good working order you need to eat plenty of fibre (roughage). Food is kept moving through the digestive system by a set of muscles. Fibre gives these muscles something to push on. How much fibre did you eat yesterday? Scientists say we should be eating at least 30 g of fibre each day.

What to do:

1 Copy this table into your book:

What I ate yesterday	How much fibre is in the food

2 Make a list of all the food you ate yesterday.

3 Use the food labels to find out how much fibre each food contains.

4 Work out how much fibre you ate altogether.

5 Did you eat enough fibre?

6 Which of the foods shown below contain the fibre you need?

HANDY HINTS

If you can't get enough information from food labels at home have a look next time you go to the supermarket.

13.4 Breathing can be dangerous

H/W

Name: Date: Group:

What you need to know ...

When we breathe, air is taken into the lungs through a branching network of tubes. These tubes get smaller and smaller, eventually ending in tiny, thin walled air sacs called alveoli. Alveoli are surrounded by blood capillaries. Lining the air passages are cells which produce sticky mucus and cells with moving hairs called cilia. These cells help to take dirt from the air you breathe in. Tobacco smoke contains chemicals which kill the cilia so the natural protection against dirt and bacteria is lost. The tar in cigarette smoke can cause lung cancer.

Part 1
What to do:

The diagram shows the structure of the lungs

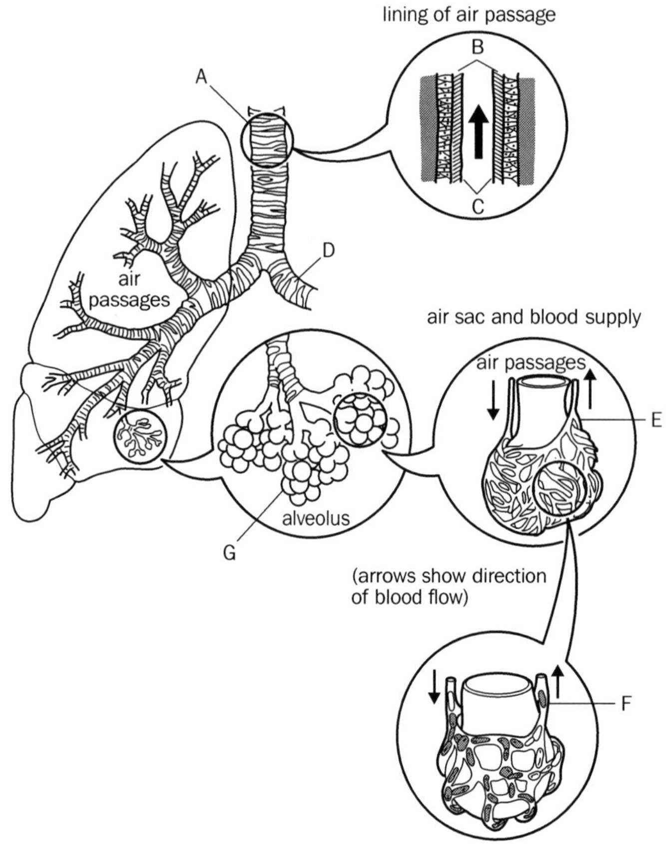

Match the letters on the diagram to the correct labels from this list:

- alveolus
- blood capillaries
- cilia
- windpipe.
- blood cells
- bronchus
- mucus

Write the letter followed by the correct label.

Part 2
What to do:

The table shows the percentage of males and females having lung cancer in some European countries.

Country	Males	Females
Scotland	80%	34%
Holland	73%	13%
Germany	71%	10%
France	67%	7%
England and Wales	62%	23%
Switzerland	61%	13%
Spain	48%	3%

a Which country has the highest percentage of lung cancer in men and women?

b Which country has the lowest percentage of lung cancer in women?

c Roughly, how much bigger is the percentage of women with lung cancer in Scotland than in Spain?

d What does this information tell you about lung cancer in men and women?

e What does this information tell you about the smoking habits of men and women in Scotland?

Chapter 13 ▶ Investigation 13F Temperature and breathing rate

Respiration is the release of energy from food. Oxygen is used up in the process and carbon dioxide gas is produced. Animals exchange oxygen and carbon dioxide with the air when they breathe.

In this investigation: you are going to find out if there is a link between temperature and breathing rate in animals

Preparation: Predict

Finish the sentences in the box.

What I think will happen is…

I think this because…

Preparation: Plan

Write a short plan of your investigation.

Think about:

- the apparatus you are going to use
- how one variable depends upon another variable
- what you are going to measure and how you are going to measure it
- how many readings you are going to take
- how you are going to record your results
- how you are going to make your investigation fair
- how you are going to make your investigation safe.

Show your plan to your teacher before going on.

Carry out

Carry out your investigation and record your results.

Present your results in an appropriate way.

Report

Write a report on your investigation.

Here are some things you should include:

- what you did
- what happened
- explain your results
- if your prediction was correct or not
- how reliable your results were
- what you could have done if you had more time.

Chapter 13 ▶ Investigation 13E
Temperature and breathing rate

Respiration is the release of energy from food. Oxygen is used up in the process and carbon dioxide gas is produced. Animals exchange oxygen and carbon dioxide with the air when they breathe.

In this investigation: you are going to find out if there is a link between temperature and breathing rate in animals

Preparation: Predict

Finish the sentences in the box.

What I think will happen is...

I think this because...

Preparation: Plan

Write a short plan of your investigation.

Think about:

- the apparatus you are going to use
- what you are going to measure and how you are going to measure it
- how many readings you are going to take
- how you are going to record your results
- how you are going to make your investigation fair
- how you are going to make your investigation safe.

Show your plan to your teacher before going on.

Carry out

Carry out your investigation and record your results in a table.

Draw a line graph of your results.

Report

Write a report on your investigation.

Here are some things you should include:

- what you did
- what happened
- explain your results
- if your prediction was correct or not
- what you could do to improve the investigation
- what you could have done if you had more time.

Chapter 13 ▶ Investigation 13D Temperature and breathing rate

Respiration is the release of energy from food. Oxygen is used up in the process and carbon dioxide gas is produced. Animals exchange oxygen and carbon dioxide with the air when they breathe.

In this investigation: you are going to find out if there is a link between temperature and breathing rate in animals

Preparation: Predict

Finish the sentence in the box.

I think there (is/is not) a link between temperature and breathing rate in animals because...

You are going to use this equipment to find out if there is a link between temperature and breathing rate in animals:

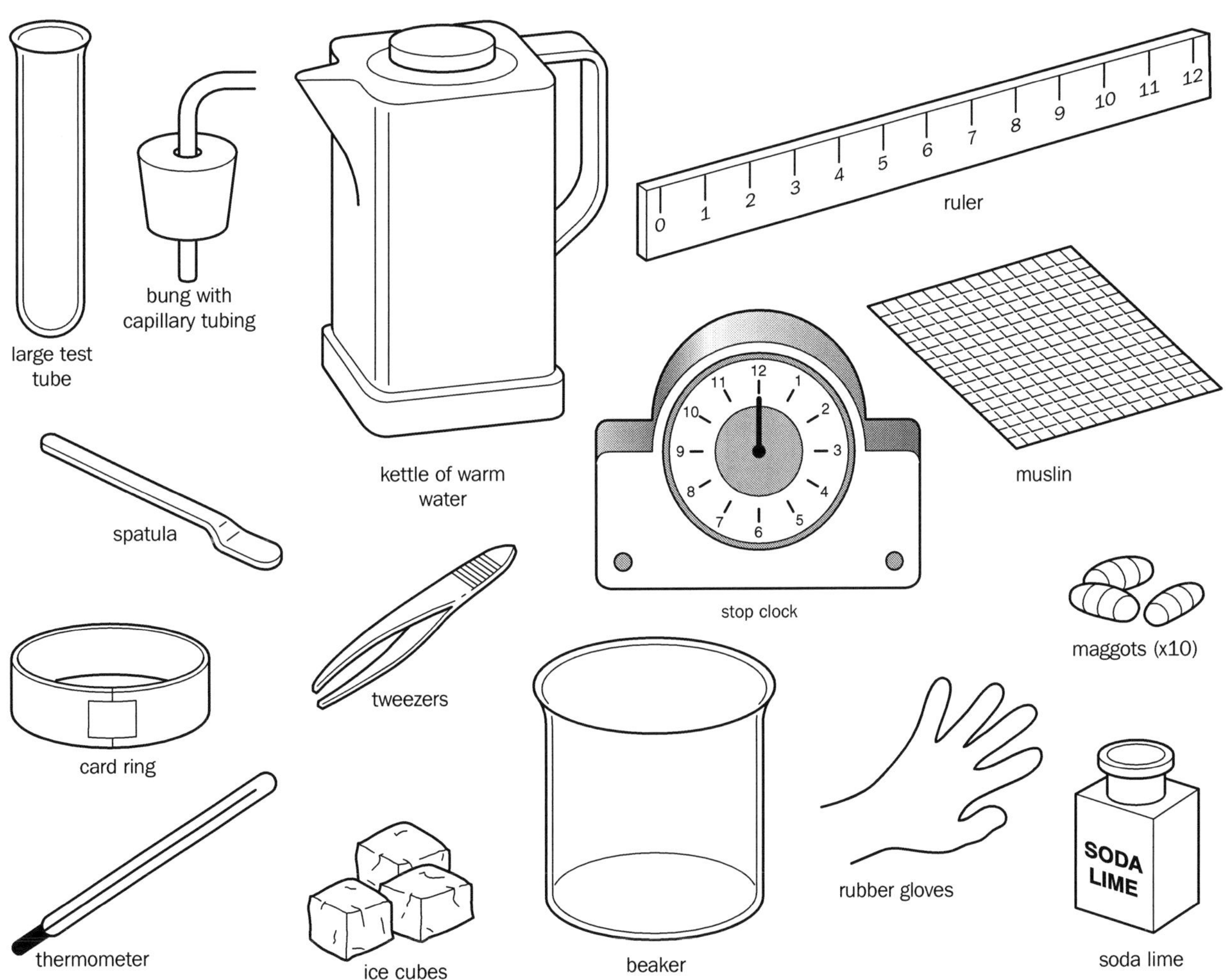

Chapter 13 ► Investigation 13D Temperature and breathing rate

Preparation: Plan

Finish the sentences in the box.

I will measure...

Things I will keep the same are...

My investigation will be fair because...

My investigation will be safe because...

Carry out

See how quickly or slowly a water bubble moves along the capillary tube when you stand the test tube of maggots in water at different temperatures.

Warning: Do not go above 35 °C.

Put your results in a table like this:

Temperature in °C	Time for bubble to move 5 cm along tube
5	
15	
25	
35	

Draw a line graph of your results on a piece of graph paper. Label the axes like this:

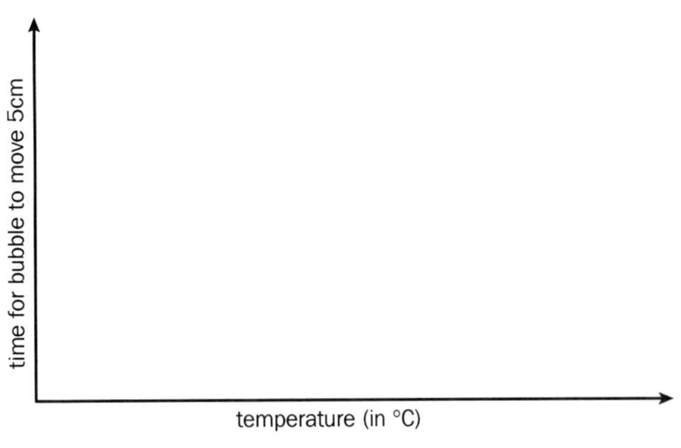

Report

Write a report on your investigation.

Here are some things you should include:

- what you did
- what happened
- explain your results
- if your prediction was correct or not
- what you could do to improve the investigation
- what you could have done if you had more time.

Chapter 13 ▶ Investigation 13C Temperature and breathing rate

Respiration is the release of energy from food. Oxygen is used up in the process and carbon dioxide gas is produced. Animals exchange oxygen and carbon dioxide with the air when they breathe.

In this investigation: you are going to find out if there is a link between temperature and breathing rate in animals

Preparation: Predict

Finish the sentence in the box.

I think there (is/is not) a link between temperature and breathing rate in animals because…

You are going to use this equipment to find out if there is a link between temperature and breathing rate in animals:

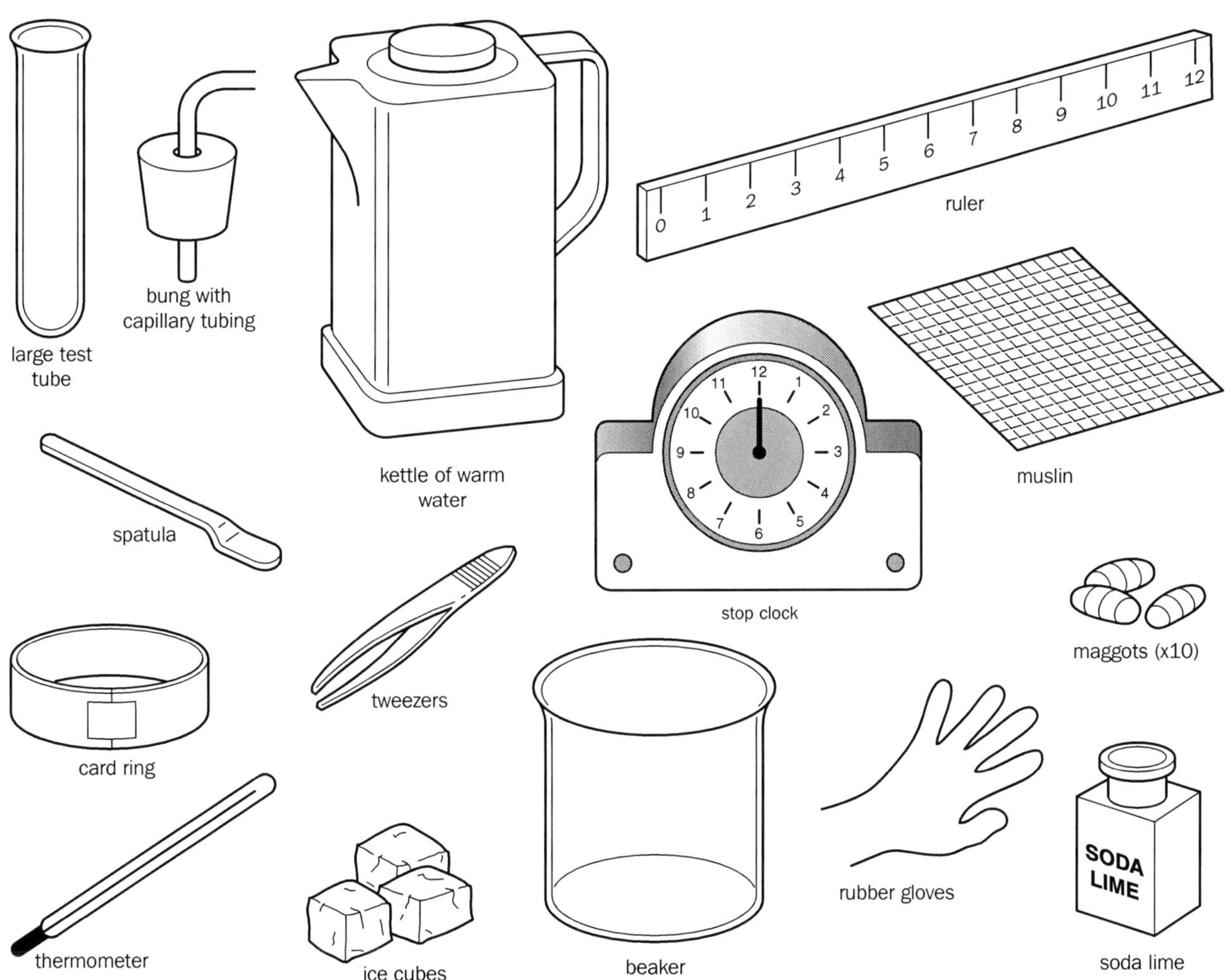

Chapter 13 ► Investigation 13C Temperature and breathing rate

Preparation: Plan

Finish the sentences in the box.

I will measure…
Things I will keep the same are…
My investigation will be fair because…
My investigation will be safe because…

Carry out

- Put some soda lime into the test tube. Soda lime absorbs carbon dioxide.
- Make a cradle for the maggots by pushing the card ring on to the muslin and down into the test tube.

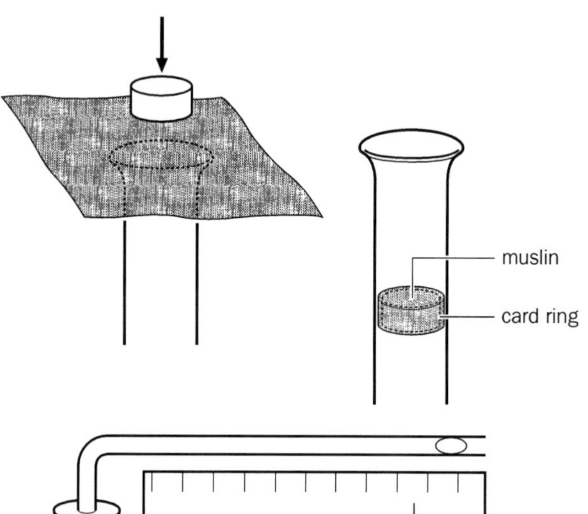

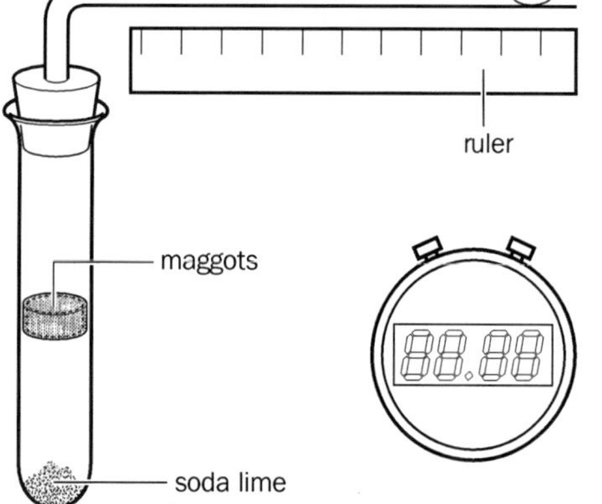

- Put some maggots on to the muslin cradle and put the bung and capillary tube in place.
- Put a drop of water on to the end of the capillary tube.
- Stand the test tube in a beaker of water at 5 °C (use some ice) and leave it for a few minutes for the maggots to get used to the temperature.
- Time how long it takes for the water 'bubble' to move 5 cm along the tube.
- Do the experiment again using water at 15 °C, 25 °C, and 35 °C.

Put your results in a table like this:

Temperature in °C	Time for bubble to move 5 cm along tube
5	
15	
25	
35	

Chapter 13 ► Investigation 13C
Temperature and breathing rate

Draw a line graph of your results on this grid.

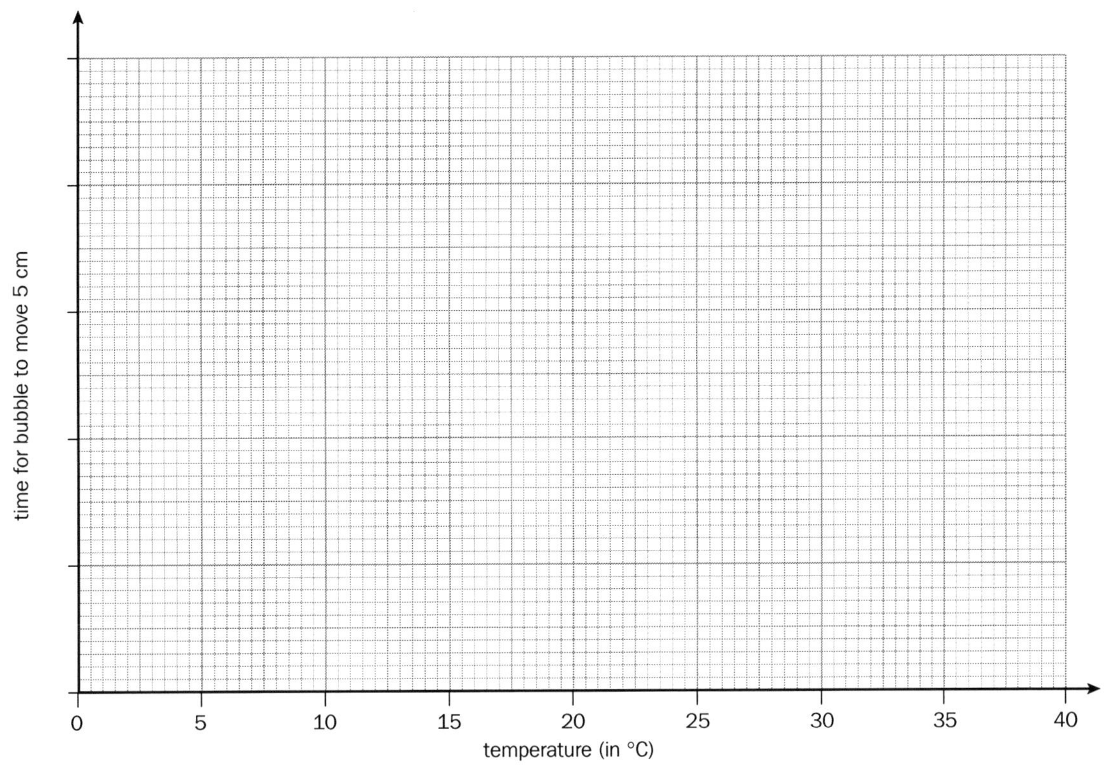

Report

Finish the sentences in the box.

What I did was...

What happened was...

From my results, I found out that there (is/is not) a link between temperature and breathing rate.
I know this because...

My prediction (was/wasn't) correct.
If I could do the investigation again I would...

Investigation 13
Practical notes

Temperature and breathing rate

Investigation 13
Technician's notes

Temperature and breathing rate

Each group will need:

Number of apparatus sets:

Number of pupils:

Number of groups:

Visual aids:

ICT resources:

Equipment/apparatus needed:

- large test tube
- bung with capillary tubing (capillary tubing attached at right angles, i.e. horizontal)
- soda lime
- spatula
- card ring
- piece of muslin or 'J' cloth (approximately 5 cm square)
- 10 maggots; count them out and count them in
- ruler
- stop clock
- beaker
- thermometer
- access to kettle of warm water
- access to ice cubes
- rubber gloves
- tweezers.

Safety notes

- **See HAZCARDS for soda lime.**
- **Watch out for silly behaviour with maggots**

CLEAPSS/SSERC SAFETY REFERENCE:

Chapter 13 ► Test

White

Keeping the body working

1 Food gives us all the things we need to keep the body working properly. The things we eat contain different types of food. These are carbohydrates, protein, fat, minerals, and vitamins.

a Which food types
- **i** give us energy
- **ii** are needed to build new cells and tissue?

b Name something that you can buy in the shops that contains
- **i** carbohydrate
- **ii** fat.

c Explain why a balanced diet must also contain plenty of roughage (fibre).

5 marks

2 The diagram shows the human digestive system.

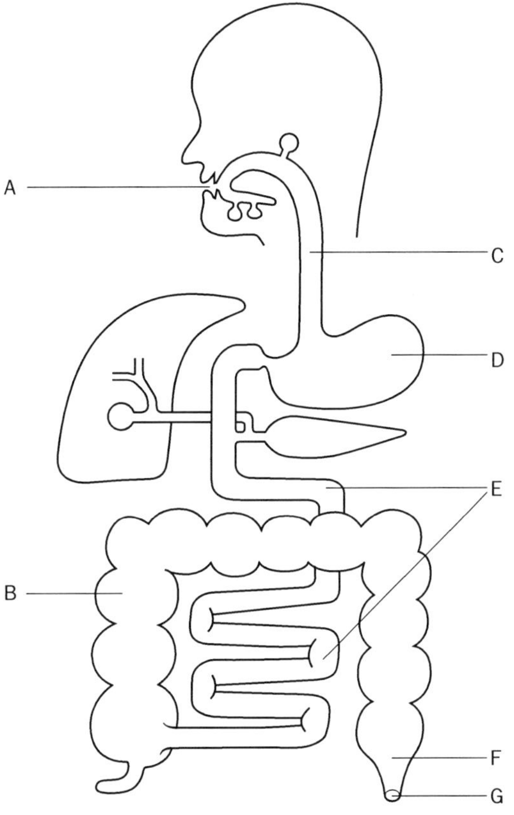

a Match these words to the letters on the diagram. Write the letter then the word.

anus
gullet
large intestine
mouth
rectum
small intestine
stomach

b Which letter labels a part which
- **i** contains acid
- **ii** chews food
- **iii** stores faeces
- **iv** absorbs digested food
- **v** produces enzymes (chemicals that break down food)?

12 marks

3 Which of these words match the gaps in the passage which follows?

bacteria
dentine
enamel
plaque
pulp
cavity

Each tooth is covered with a layer of white______(a)______. This is a very hard, dead substance which protects the tooth. Inside this is a layer of ______(b)______. This is living material, a bit like bone. At the centre of the tooth is the ______(c)______, containing soft living tissue.

Although the covering of the tooth is hard, it can be attacked by acid. Teeth usually have a sticky substance called ______(d)______ on them. This contains ______(e)______ which change sugary food into acid.

5 marks

Keeping the body working

4 The diagram shows human lungs inside the ribcage. Breathing air in and out of the lungs involves the rib muscles and the diaphragm.

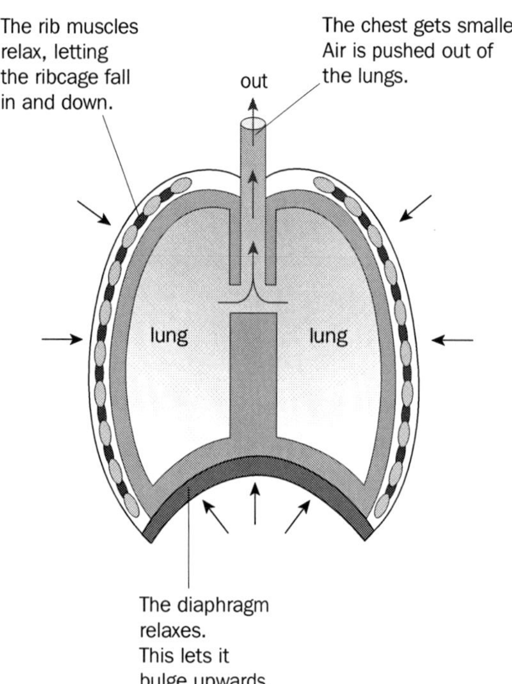
The rib muscles relax, letting the ribcage fall in and down.
out
The chest gets smaller. Air is pushed out of the lungs.
lung
lung
The diaphragm relaxes. This lets it bulge upwards.

a Write these two headings: 'Breathing in' and 'Breathing out'. Write these statements under the correct heading and in the right order:

- air pushed out of lungs
- diaphragm relaxes
- rib muscles contract
- air sucked into lungs
- chest gets bigger
- rib muscles relax
- diaphragm contracts
- chest gets smaller

b Name the two gases that are exchanged in the lungs when we breathe.

c The lining of the lungs is very thin. Explain how this helps the lungs do their job.

8 marks

Chapter 13 ► Test

Blue

Keeping the body working

1 The diagram shows human lungs inside the ribcage. Breathing air in and out of the lungs involves the rib muscles and the diaphragm.

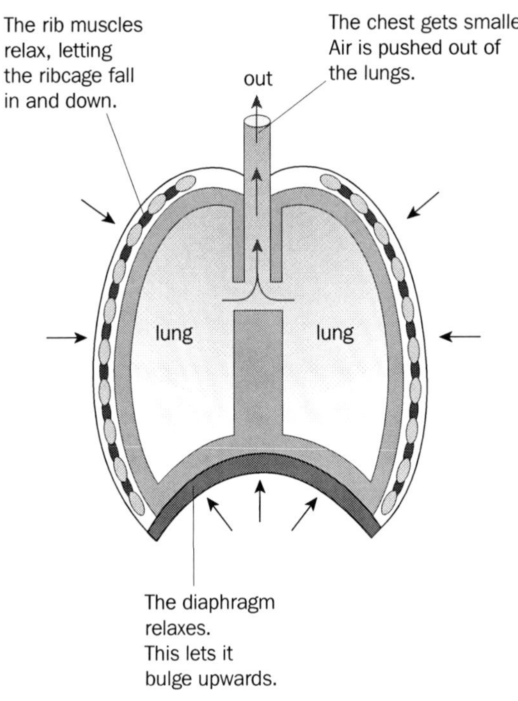

The rib muscles relax, letting the ribcage fall in and down.
out
The chest gets smaller. Air is pushed out of the lungs.
lung
lung
The diaphragm relaxes. This lets it bulge upwards.

a Write these two headings: 'Breathing in' and 'Breathing out'. Write these statements under the correct heading and in the right order:

- air pushed out of lungs
- diaphragm relaxes
- rib muscles contract
- air sucked into lungs
- chest gets bigger
- rib muscles relax
- diaphragm contracts
- chest gets smaller

b Name the two gases that are exchanged in the lungs when we breathe.

c The lining of the lungs is very thin. Explain how this helps the lungs do their job.

7 marks

2 Two students set up this apparatus to investigate respiration.

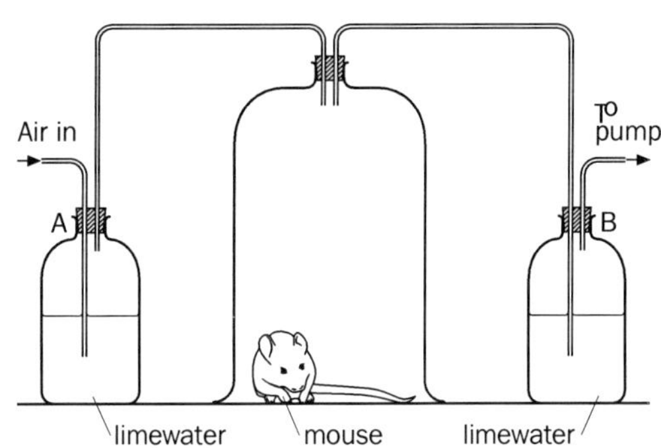

Air in
To pump
A
B
limewater
mouse
limewater

a What does lime water test for?

b The lime water in bottle B went cloudy before the lime water in bottle A. What does this show?

c Condensation appeared on the inside of the mouse container. Suggest where this might have come from.

d What do the letters X and Y stand for in this word equation for respiration?

Glucose + oxygen ⟶ __X__ + __Y__

e Explain the difference between respiration and breathing.

f Explain why it is important that the students stop the experiment and release the mouse as soon as they have got their results.

8 marks

3 The diagram shows the human digestive system.

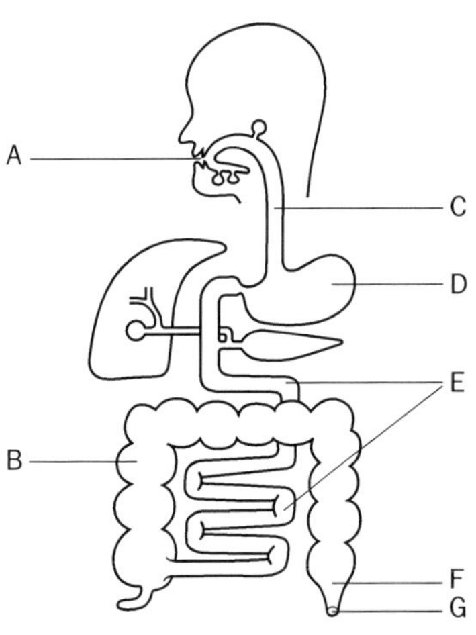

A
B
C
D
E
F
G

Chapter 13 ► Test

Blue

Keeping the body working

a Match these words to the letters on the diagram.

anus	rectum
gullet	small intestine
large intestine	stomach
mouth	

During digestion in the stomach, protein molecules are broken down into amino acids by the enzyme protease.

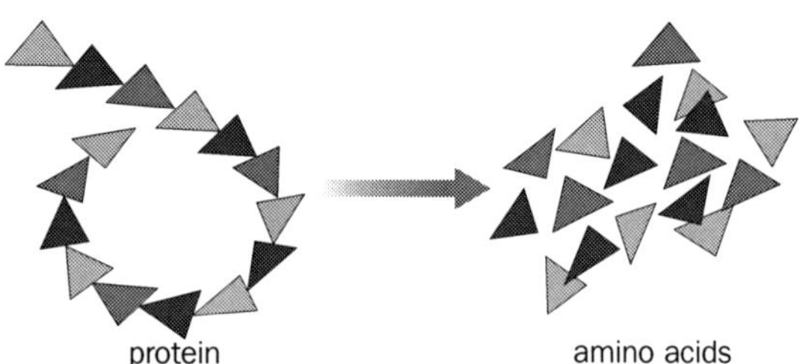

In the mouth, starch is broken down into glucose by amylase (also called carbohydrase.)

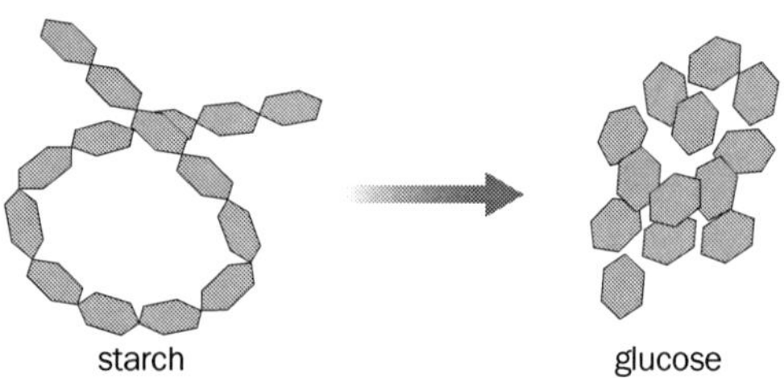

b Give one way in which a starch molecule and a protein molecule are **i** similar **ii** different.

c How do you think protease and carbohydrase got their names?

d Saliva is pH7, stomach juice is pH2.

i Explain why carbohydrase won't work in the stomach.

ii Suggest why stomach juice is neutralised by liver juices when it leaves the stomach.

10 marks

4 A student investigated the effect of protease on protein. She put some egg-white in a narrow tube and stood the tube in protease enzyme. Every minute she measured how much egg white had been digested by the enzyme.

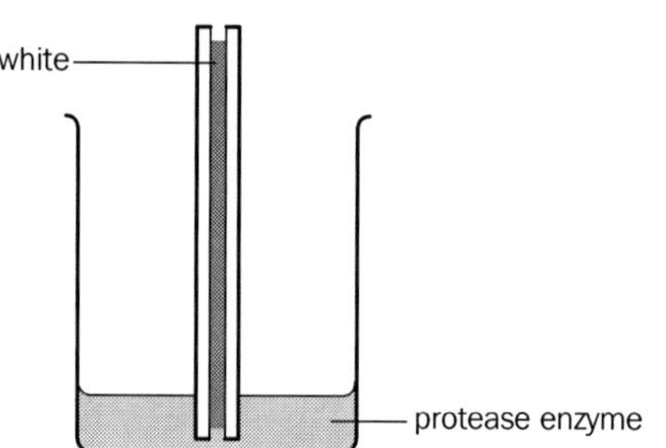

Temperature in C	mm of egg white digested									
	1 min	2 min	3 min	4 min	5 min	6 min	7 min	8 min	9 min	10 min
0	0		0	0	0	0	0	2	2	2
10	0	0	0	0	2	2	3	3	3	4
20	0	2	6	8	10	12	14	17	20	22
40	2	6	10	16	22	27	32	36	40	46
60	0	0	0	0	0	0	0	0	0	0

This was repeated at different temperatures. These are the student's results.

a What is the best temperature for the digestion of egg white by protease ?

b What evidence is there that this enzyme came from a human?

c Is there any evidence of enzymes slowing down as they get older?

d Suggest why this enzyme would not work very well in the stomach of a fish living in water at 15 °C?

e How could the student have got more accurate results?

5 marks

Chapter 13 ► Mark Scheme

White

Keeping the body working

Question	Answer	Marks	Level
1 a i	carbohydrates		
ii	protein	2	D
b i	appropriate answer, e.g. cakes, bread, etc.		
ii	appropriate answer, e.g. cheese, butter, etc.	2	D
c	gives intestine muscles something to push on	1	D
		5	
2 a	A mouth		
	B large intestine		
	C gullet		
	D stomach		
	E small intestine		
	F rectum		
	G anus	7	D
b i	D		
ii	A		
iii	F		
iv	E		
v	A, D, or E	5	D
		12	
3 a	enamel	1	
b	dentine	1	
c	pulp cavity	1	
d	plaque	1	
e	bacteria	1	D
		5	

Question	Answer		Marks	Level
4 a	**Breathing in**	**Breathing out**		
	rib muscles contract	rib muscles relax		
	diaphragm contracts	diaphragm relaxes		
	chest gets bigger	chest gets smaller		
	air sucked into lungs	air pushed out of lungs		
	Words in correct columns 1 mark (all words in correct order = 4 marks; delete 1 mark per error)		5	D
b	oxygen (1) carbon dioxide (1)		2	D
c	gases cross easily/ less distance to travel, etc.		1	D
			8	
			TOTAL 30 marks	

Suggested grade/level boundaries

C = 13/30

D = 25/30

Chapter 13 ► Mark scheme
Keeping the body working

Question	Answer	Marks	Level
1 a	**Breathing in** / **Breathing out**		
	rib muscles contract / rib muscles relax		
	diaphragm contracts / diaphragm relaxes		
	chest gets bigger / chest gets smaller		
	air sucked into lungs / air pushed out of lungs		
	(words in correct columns = 1 mark; all words in correct order = 4 marks; delete 1 mark per error)	5	D
b	oxygen and carbon dioxide	1	D
c	gases cross easily/ less distance to travel, etc.	1	D
		7	
2 a	carbon dioxide	1	D
b	more carbon dioxide in exhaled than inhaled air	1	F
c	mouse exhaling	1	F
d	water (1) carbon dioxide (1)	2	F
e	respiration is a chemical reaction (1) breathing is a means of exchanging gases (1)	2	F
f	prevent the mouse from getting stressed, etc.	1	D
		8	
3 a	A mouth B large intestine C gullet D stomach E small intestine F rectum G anus (all correct = 4 marks; delete 1 mark per error)	4	D
b i	both chains		
ii	different components in chain	2	F
c	named after the food type that they digest	1	F
d i	enzymes are pH specific (1) stomach too acid for carbohydrase (1)	2	F
ii	strong acid could damage intestines, etc.	1	F
		10	
4 a	40 °C	1	F
b	human body temp is 37 °C/ close to 40 °C	1	F
c	(rate of) reaction gets slower as temperature falls	1	F
d	reaction is too slow at this temperature/ not fast enough for fish	1	F
e	repeated the experiment/ greater range of temperatures	1	F
		5	
	TOTAL 30 marks		

Suggested grade/level boundaries

D = 7/30

E = 13/30

F = 25/30

14 Metals

This chapter begins by looking at the reactivity series before moving on to look at mining and extraction of metals from their ores. Costs relating to production and corrosion of metals lead naturally on to aspects of recycling. Following on from its introduction in Book 1, the Periodic Table is covered in more detail, this time linking it closely to atomic structure.

Assessment opportunities

Formative assessment opportunities are provided by worksheets, homework sheets, and an investigation.

The **worksheets** cover material at levels E and F for attainment targets for knowledge and understanding. Teachers may wish to use these worksheets not only as part of practical activities but also to provide evidence of pupil achievement.

Worksheet	Level
14.1a	E
14.1b	E
14.1c	E
14.2a	E
14.2b	E
14.4a	F
14.4b	F

The **homework sheets** cover material at levels E and F for attainment targets for knowledge and understanding. These homework sheets can be used individually as a follow-up to work done in class or assembled into a homework booklet allied closely to schemes of work.

Homework sheet	Level
14.2	E
14.3a	E
14.3b	F
14.3c	E

The **investigation** covers all three skill areas at levels C, D, E, and F. It is written in a way that allows for pupils to be assessed in all three skill areas at one level. Alternatively, customised assessments can be constructed enabling pupils to be assessed at different levels in all three skills. The latter approach is more time consuming, but it does provide the opportunity for pupils to show evidence of achievement at different levels in different skills in the same investigation. Teachers will need to use their professional judgement when deciding which level is appropriate to individual pupils. It is envisaged that pupils will show progression through the levels as they work through their science course.

A single **summative test** is provided at levels E and F only. This is because Chapter 14 only covers material at this level. The test has a total of 30 marks and will take about 30 minutes for pupils to complete, although this can be varied depending on pupil ability. A mark scheme is provided together with suggested grade/level boundaries. It is envisaged that this test will be given to pupils on completion of the material covered in Chapter 14.

ICT opportunities

The use of data loggers/remote sensors can extend the range, speed, and sensitivity of measurements in many of the worksheets for this chapter. Once downloaded onto a PC, data-handling programs can be used to analyse information gathered, data can be manipulated, and appropriate graphs etc. presented. The Internet provides pupils with access to a huge range of scientific information. A list of suitable websites is included in this Teacher's Guide.

Students' book chapter 14 contents and guide levels

14.1	What metals have in common	*Starting off*	E
	More or less reactive	*Going further*	E
	The reactivity series	*For the enthusiast*	E
14.2	Mining metals	*Starting off*	E
	Metals from their ores	*Going further*	E
14.3	Corrosion (1): coatings	*Starting off*	E
	Saving the metal	*Going further*	E
	Corrosion (2): alloys	*For the enthusiast*	E
14.4	It's all because of the atoms	*Starting off*	F
	Atomic shorthand	*Going further*	F
	Big ideas about atoms	*For the enthusiast 1*	F
	The periodic table – again!	*For the enthusiast 2*	F

14.1a Reacting metals with water W/S

Name: Date: Group:

What you need:

Distilled water, samples of metals, tweezers, test tubes, test tube rack, labels, splint, Bunsen burner, heatproof mat, safety goggles.

What to do:

1 Draw a table in your book using these headings:

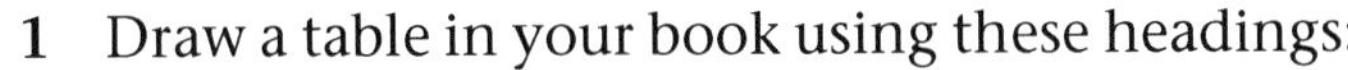

Metal	What happens when put in water	Reacted a lot (yes/no)	Reacted a little (yes/no)	No reaction (yes/no)

2 Put on safety goggles.

3 Put the test tubes in the rack and half fill them with distilled water.

4 Label the test tubes with the names of the metals you are going to use.

5 Use tweezers to put a different metal into each test tube and watch what happens.

(Get rid of any air bubbles on the surface of the metal by gently tapping the test tube.)

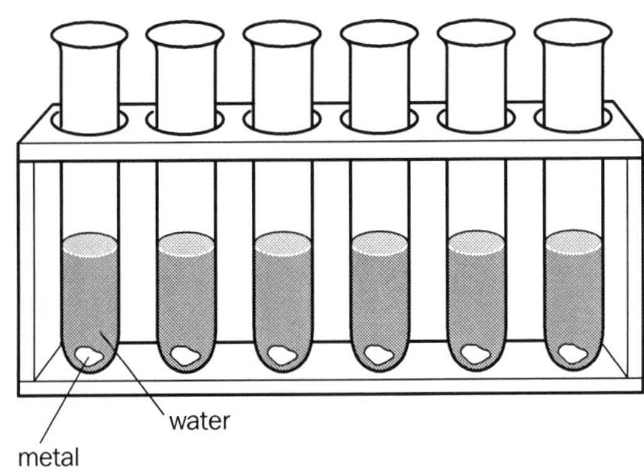

6 If a gas is produced, hold a lighted splint near the top of the test tube. A 'pop' shows that hydrogen is being produced.

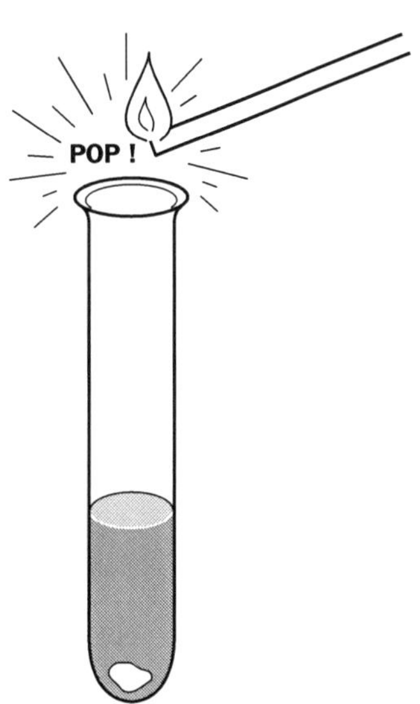

7 Write your results in the table.

- Which is the most reactive metal in water?
- Which is the least reactive metal in water?

SAFETY WARNING

Safety goggles to be worn at all times.

Keep calcium away from skin.

14.1a Practical notes

Reacting metals with water

This is a simple starter activity on reactivity of metals. Only calcium (and possibly magnesium if it is thoroughly cleaned beforehand) will show any visible reaction with water. A teacher demonstration of the reactions of sodium and potassium usually livens things up a bit. The activity provides a useful link with work on the Periodic Table, Group II, the alkali metals. Ensure all relevant safety precautions are observed.

Pupils must only be given a small amount of calcium, about the size of a rice grain. Make certain that safety goggles are worn at all times.

14.1a Technician's notes

Number of apparatus sets:

Number of pupils:

Number of groups:

Visual aids: ....

ICT resources: ....

Equipment/apparatus needed: ....

Reacting metals with water

Each group will need:

- a copy of worksheet 14.1a
- distilled water (metals may react with impurities in tap water)
- samples of copper, tin, iron, magnesium, and calcium (rice grain size only) (metals must be cleaned with fine abrasive paper)
- six test tubes
- test tube rack
- tweezers
- labels
- splints
- Bunsen burner
- heatproof mat
- safety goggles.

For teacher demonstration:

- sodium
- potassium
- tile
- knife
- tweezers
- glass bowl
- splints
- Bunsen burner
- heatproof mat
- safety goggles
- safety screen.

Safety notes

- **See HAZCARDS for calcium, sodium, and potassium.**
- **Pupils must wear safety goggles at all times.**
- **Only rice grain sized pieces of calcium to be used.**

CLEAPSS/SSERC SAFETY REFERENCE:

14.1b Reacting metals with acid

W/S

Name: **Date:** **Group:**

What you need:

Dilute hydrochloric acid, samples of metals, tweezers, test tubes, test tube rack, labels, splint, Bunsen burner, heatproof mat, safety goggles.

What to do:

1 Draw a table in your book using these headings:

Metal	What happens when put in acid	Fast reaction (yes/no)	Slow reaction (yes/no)	No reaction (yes/no)

2 Put on safety goggles.

3 Put the test tubes in the rack and half fill them with dilute hydrochloric acid.

4 Label the test tubes with the names of the metals you are going to use.

5 Add a different metal to each test tube and watch what happens.

(Get rid of any air bubbles on the surface of the metal by gently tapping the test tube.)

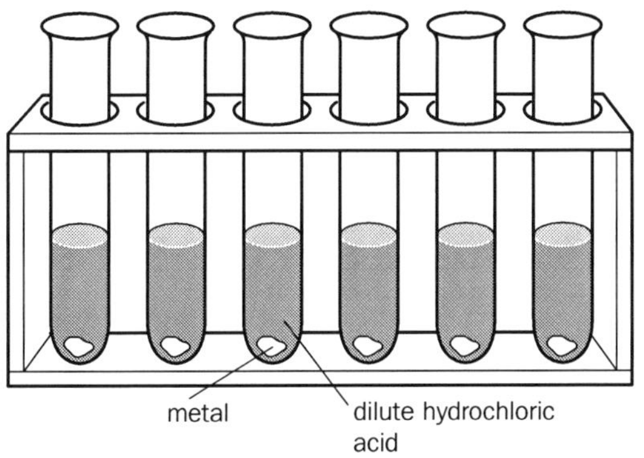

6 If a gas is produced hold a lighted splint near the top of the test tube. A 'pop' shows that hydrogen is being produced.

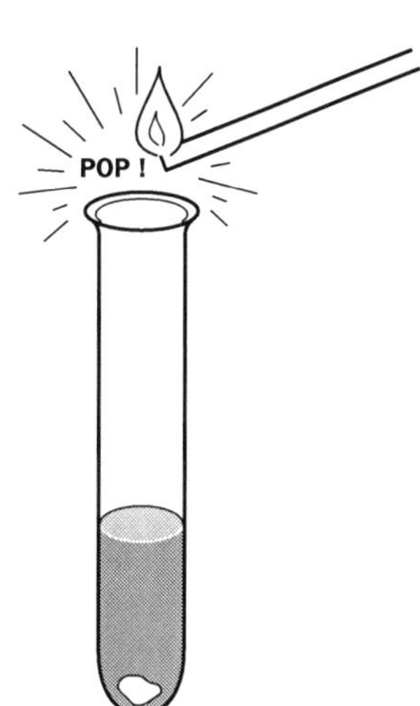

7 Write your results in the table.

- Which is the most reactive metal you have tested?
- Which is the least reactive metal you have tested?
- Is there a link between the way metals react with air, water, and acid? If so, what is it?

SAFETY WARNING

Hydrochloric acid may cause burns.

Keep calcium away from skin.

Wear safety goggles.

14.1b
Practical notes

Reacting metals with acid

This activity continues work on reactivity series. This time pupils will see a little more happening in their test tubes. Teachers may need to demonstrate the test for hydrogen beforehand. If calcium is used make sure that only rice grain sized pieces are used. Make sure that safety goggles are worn at all times.

14.1b
Technician's notes

Reacting metals with acid

Each group will need:

- a copy of worksheet 14.1b
- dilute hydrochloric acid
- samples of copper, tin, iron, magnesium, and calcium (rice grain size only) (metals must be cleaned with fine abrasive paper)
- six test tubes
- test tube rack
- tweezers
- labels
- splints
- Bunsen burner
- heatproof mat
- safety goggles.

Number of apparatus sets:

Number of pupils:

Number of groups:

Visual aids:

ICT resources:

Equipment/apparatus needed:

Safety notes

- **See HAZCARDS for calcium.**
- **Pupils must wear safety goggles at all times.**
- **Only rice sized pieces of calcium to be used.**

CLEAPSS/SSERC SAFETY REFERENCE:

14.1c Displacement reactions

W/S

Name: Date: Group:

What you need:

Zinc, tin, copper, copper sulphate solution, zinc nitrate solution, six test tubes, test tube rack, labels, safety goggles.

What to do:

1 Put on safety goggles.

2 Label the six test tubes and half fill them with solutions as shown in the diagram.

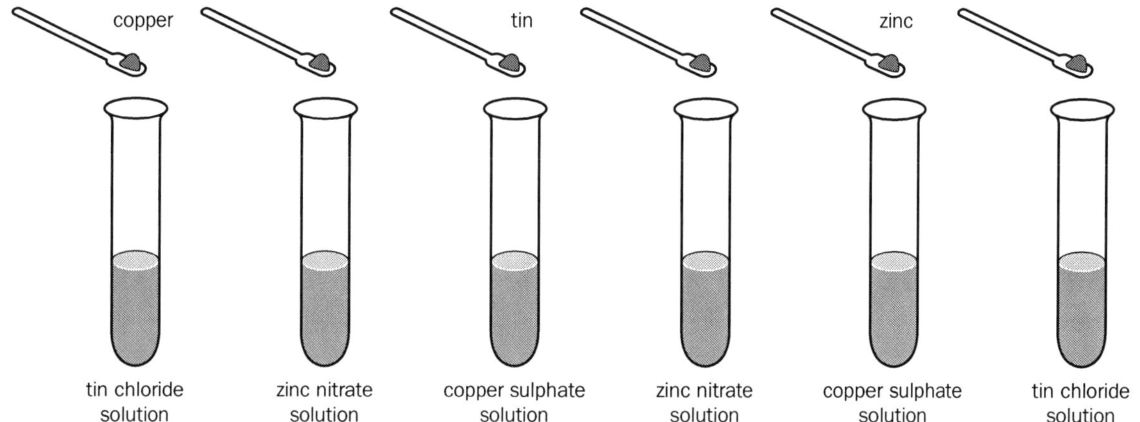

3 Add one spatula full of copper to the tin chloride and zinc nitrate solutions, one spatula full of tin to the copper sulphate and zinc nitrate solutions, and one spatula full of zinc to the copper sulphate and tin chloride solutions.

4 Look carefully to see if a reaction is happening. If it is, then one metal is displacing the other metal in the solution. You should be able to see the metal you added disappearing and the other metal appearing as a solid at the bottom of the test tube.

- Does copper displace tin?
- Does copper displace zinc?
- Does tin displace copper?
- Does tin displace zinc?
- Does zinc displace copper?
- Does zinc displace tin?
- Write the metals in order of reactivity starting with the **least** reactive.

SAFETY WARNING

Wear safety goggles.

14.1c
Practical notes

Displacement reactions

This is a straightforward activity developing further the idea of reactivity. As long as pupils get the mixtures right, they should get satisfactory results within around 10 minutes. Solid metal strips could be used but granules react quicker. Make sure the metals are clean. These are good examples of chemical reactions where new substances are formed. Make sure that safety goggles are worn at all times.

14.1c
Technician's notes

Displacement reactions

Each group will need:

Number of apparatus sets:

Number of pupils:

Number of groups:

Visual aids:

ICT resources:

Equipment/apparatus needed:

- a copy of worksheet 14.1c
- copper sulphate solution
- tin chloride solution
- zinc nitrate solution
- samples of copper, tin, and zinc granules (metals must be clean)
- six test tubes
- test tube rack
- spatula
- labels
- safety goggles.

Safety notes

- **See HAZCARDS for copper sulphate, tin chloride, and zinc nitrate.**
- **Pupils must wear safety goggles at all times.**

CLEAPSS/SSERC SAFETY REFERENCE:

14.2a Getting copper from copper ore 1

W/S

Name: Date: Group:

What you need:

Copper ore, beaker, dilute sulphuric acid, spatula, tweezers, Bunsen burner, heatproof mat, tripod, gauze, iron nail, safety goggles; thermometer.

What to do:

1 Put on safety goggles.

2 Put four spatulas full of copper ore into the beaker.

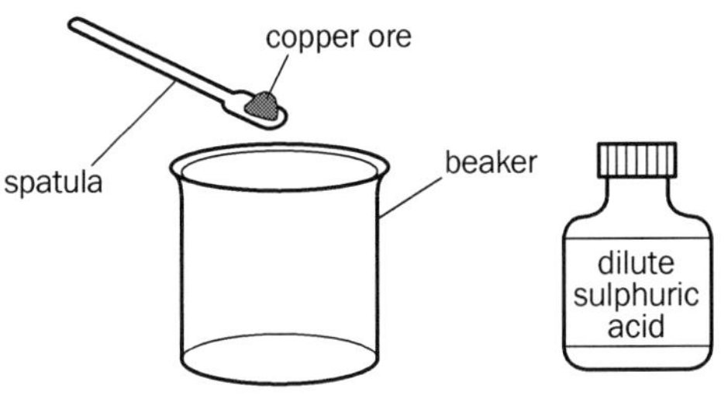

3 Carefully add dilute sulphuric acid until the beaker is half full. Gently warm the mixture over a low Bunsen flame. **Do not boil acids.** Stop heating the solution when it reaches 50 °C. The solution should be blue; this is copper sulphate solution.

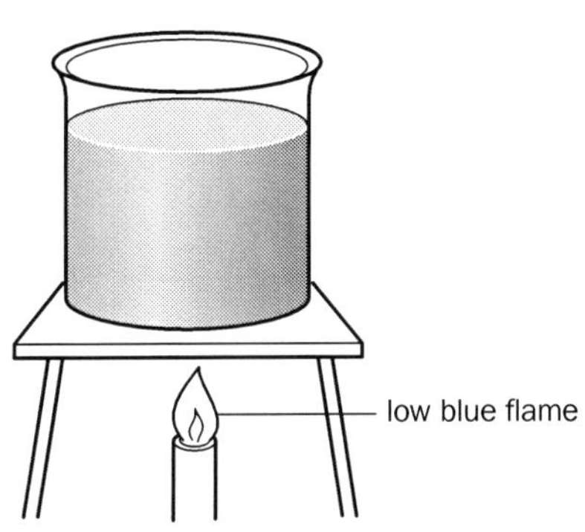

4 Use tweezers to carefully put the iron nail into the solution and leave it for about five minutes. Watch what happens.

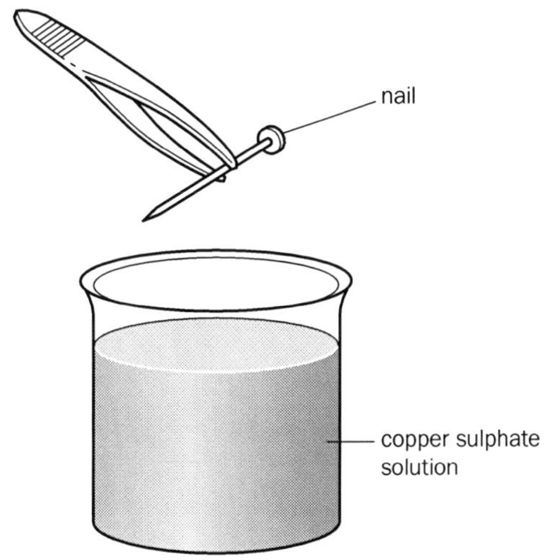

5 Use tweezers to remove the nail from the solution.

6 Try scraping some of the copper from the surface of the nail.

Which is more reactive, copper, or iron?

SAFETY WARNING

Sulphuric acid may cause burns.

Wear safety goggles.

14.2a
Practical notes

Getting copper from copper ore 1

This activity helps to demonstrate the usefulness of displacement reactions in industry. Copper extraction from poor quality ores is expensive by normal methods (e.g. electrolysis). Remind pupils to warm the acid/copper ore mixture gently. Acids should not be heated strongly. Make sure that safety goggles are worn at all times.

14.2a
Technician's notes

Getting copper from copper ore 1

Each group will need:

- a copy of worksheet 14.2a
- copper ore crushed (black copper oxide will do)
- 100 cm^3 beaker
- dilute sulphuric acid
- spatula
- tweezers
- Bunsen burner
- heatproof mat
- tripod
- gauze
- iron nail
- thermometer
- safety goggles.

Number of apparatus sets:

Number of pupils:

Number of groups:

Visual aids:

ICT resources:

Equipment/apparatus needed:

Safety notes

- **See HAZCARDS for dilute sulphuric acid and copper sulphate solution.**
- **Pupils must wear safety goggles at all times.**

CLEAPSS/SSERC SAFETY REFERENCE:

14.2b Getting copper from copper ore 2 W/S

Name: **Date:** **Group:**

What you need:

Copper ore, dilute sulphuric acid, two beakers, thermometer, spatula, Bunsen burner, heatproof mat, tripod, gauze, stirrer, filter funnel, filter paper, two carbon rods, two wires with crocodile clips, 6 V (dc) supply.

What to do:

1 Put on safety goggles.

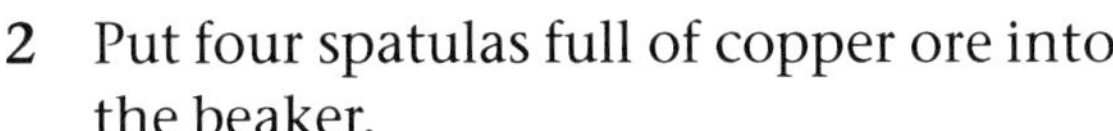

2 Put four spatulas full of copper ore into the beaker.

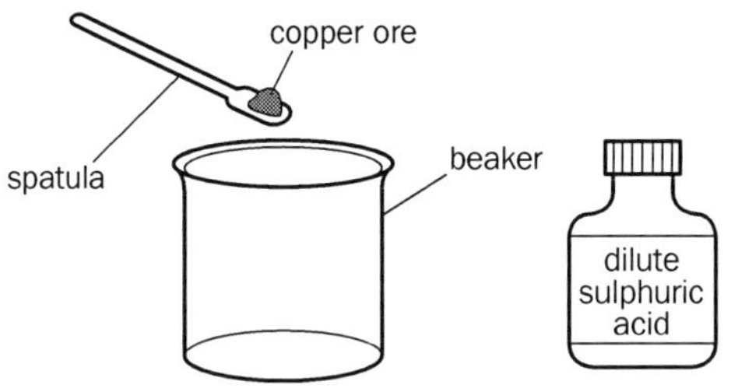

3 Carefully add dilute sulphuric acid until the beaker is half full. Gently warm the mixture over a low Bunsen flame. **Do not boil acids.** Stop heating the solution when it reaches 50 °C. The solution should be blue; this is copper sulphate solution.

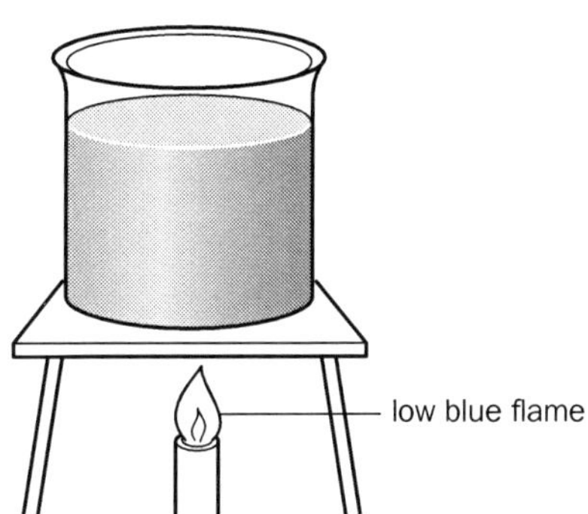

4 Add a little more copper ore and stir. Keep doing this until no more ore will dissolve.

(You will see a layer of black insoluble ore at the bottom of the beaker.)

5 Carefully filter the solution into a clean beaker.

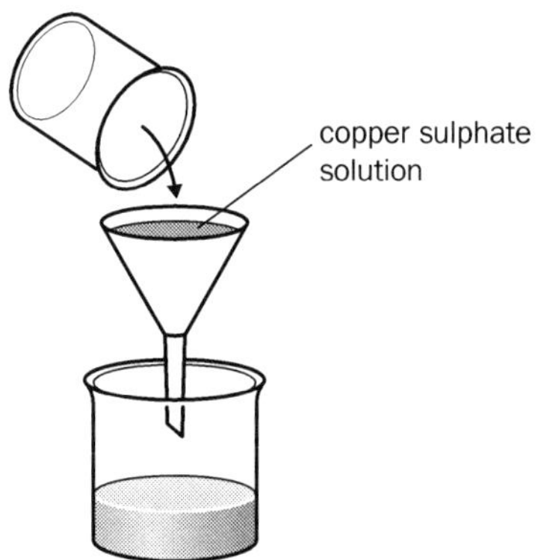

6 Make this circuit. Do not let the carbon rods touch.

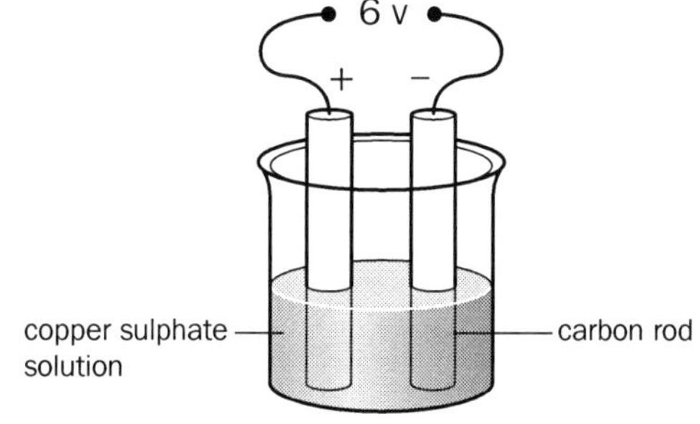

Switch it on and watch what happens.

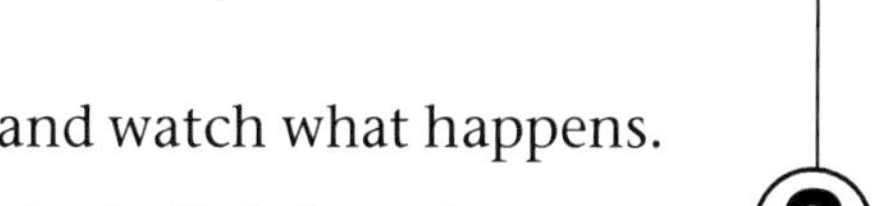

- Why does the bulb light up?
- Which carbon rod gets covered with copper?
- What happens at the other carbon rod?
- How do you think the process could be speeded up?

SAFETY WARNING

Sulphuric acid may cause burns.

Wear safety goggles.

14.2b
Practical notes

Getting copper from copper ore 2

This activity gives pupils the opportunity to see how electrolysis can be used as an alternative method of extracting copper from its ore. Useful economic comparisons between this and displacement reactions can be made. The activity could be split into two parts depending on the time available and level of pupil ability.

Remind pupils to warm the acid/copper ore mixture gently. Acids should not be heated strongly. Make sure that safety goggles are worn at all times.

14.2b
Technician's notes

Getting copper from copper ore 2

Each group will need:

Number of apparatus sets:

....

Number of pupils: Number of groups:

Visual aids:

....

....

ICT resources:

....

....

Equipment/apparatus needed:

....

....

- a copy of worksheet 14.2b
- copper ore crushed (black copper oxide will do)
- two 100 cm^3 beakers
- dilute sulphuric acid
- thermometer
- spatula
- Bunsen burner
- heatproof mat
- tripod
- gauze
- stirrer
- filter funnel
- filter paper
- two carbon rods/electrodes
- circle of card with two holes to support electrodes
- two wires with crocodile clips
- 6 V dc supply
- safety goggles
- access to oven gloves.

Safety notes

- **See HAZCARDS for dilute sulphuric acid and copper sulphate solution.**
- **Pupils must wear safety goggles at all times.**

CLEAPSS/SSERC SAFETY REFERENCE:

14.4a Word equation game

W/S

Name: **Date:** **Group:**

What to do:

This is a game for two players. The winner is the first to make three correct word equations.

1 Shuffle the cards. Deal each player nine cards.

2 Put the remaining cards in a pile face downwards. Turn the top card face upwards and put it beside the pile.

3 Take turns to take the top card from the pack or from the pile facing upwards.

4 Put one unwanted card, face upwards on the appropriate pile.

5 Write down all the word equations you and your opponent have made in your word equation game.

aluminium	iodine	aluminium iodide
calcium	chlorine	calcium chloride
carbon	oxygen	carbon dioxide
copper	oxygen	copper oxide
hydrogen	chlorine	hydrogen chloride
hydrogen	oxygen	hydrogen oxide
iron	chlorine	iron chloride
iron	sulphur	iron sulphide
lead	bromine	lead bromide
magnesium	oxygen	magnesium oxide
magnesium	sulphur	magnesium sulphide
sodium	chlorine	sodium chloride
sulphur	oxygen	sulphur dioxide
zinc	iodine	zinc iodide

14.4a
Practical notes

Word equation game

This is a light-hearted way of learning about word equations.

14.4a
Technician's notes

Word equation game

Each group will need:

- Set of word equation playing cards (cards can be made by photocopying Worksheet 14.4a, gluing them on to card and cutting them up. Sets of cards can be secured with elastic bands or put into envelopes for storage and later use).

Number of apparatus sets:

Number of pupils:

Number of groups:

Visual aids:

ICT resources:

Equipment/apparatus needed:

Safety notes

CLEAPSS/SSERC SAFETY REFERENCE:

Name: Date: Group:

KEY: Mass number A / Atomic number (proton number) Z — 1 H Hydrogen 1

1	2											3	4	5	6	7	0
																	4 He Helium 2
7 Li Lithium 3	9 Be Beryllium 4											11 B Boron 5	12 C Carbon 6	14 N Nitrogen 7	16 O Oxygen 8	19 F Fluorine 9	20 Ne Neon 10
23 Na Sodium 11	24 Mg Magnesium 12											27 Al Aluminium 13	28 Si Silicon 14	31 P Phosphorus 15	32 S Sulphur 16	35 Cl Chlorine 17	40 Ar Argon 18
39 K Potassium 19	40 Ca Calcium 20	45 Sc Scandium 21	48 Ti Titanium 22	51 V Vanadium 23	52 Cr Chromium 24	55 Mn Manganese 25	56 Fe Iron 26	59 Co Cobalt 27	59 Ni Nickel 28	63 Cu Copper 29	64 Zn Zinc 30	70 Ga Gallium 31	73 Ge Germanium 32	75 As Arsenic 33	79 Se Selenium 34	80 Br Bromine 35	84 Kr Krypton 36
85 Rb Rubicium 37	88 Sr Strontium 38	89 Y Yttrium 39	91 Zr Zirconium 40	93 Nb Niobium 41	96 Mo Molybdenum 42	Tc Technetium 43	101 Ru Ruthenium 44	103 Rh Rhodium 45	106 Pd Palladium 46	108 Ag Silver 47	112 Cd Cadmium 48	115 In Indium 49	119 Sn Tin 50	122 Sb Antimony 51	128 Te Tellurium 52	127 I Iodine 53	131 Xe Xenon 54
133 Cs Caesium 55	137 Ba Barium 56	139 La Lanthanum 57	178 Hf Hafnium 72	181 Ta Tantalum 73	184 W Tungsten 74	186 Re Rhenium 75	190 Os Osmium 76	192 Ir Iridium 77	195 Pt Platinum 78	197 Au Gold 79	201 Hg Mercury 80	204 Tl Thallium 81	207 Pb Lead 82	209 Bi Bismuth 83	Po Polonium 84	At Astatine 85	Rn Radon 86
Fr Francium 87	226 Ra Radium 88	227 Ac Actinium 89															

Elements 58–71 and 90–103 have been omitted.

14.4b
Practical notes

The periodic table

This is a photocopiable sheet for pupils to stick into their book.

14.4b
Technician's notes

The periodic table

Each pupil will need:

- a copy of worksheet 14.4b.

Number of apparatus sets:

Number of pupils:

Number of groups:

Visual aids:

ICT resources:

Equipment/apparatus needed:

Safety notes

CLEAPSS/SSERC SAFETY REFERENCE:

14.2 Getting metals from their ores

H/W

Name: Date: Group:

What you need to know …

Most of our metals come from rocks in the ground. They are usually in compounds called ores. The more reactive a metal is, the more difficult it is to separate from its ore. Iron can be extracted from iron ore by smelting in a blast furnace. Aluminium cannot be smelted. Aluminium compounds are melted, and then the metal is extracted by electrolysis.

What to do:

1 The diagram shows a blast furnace.

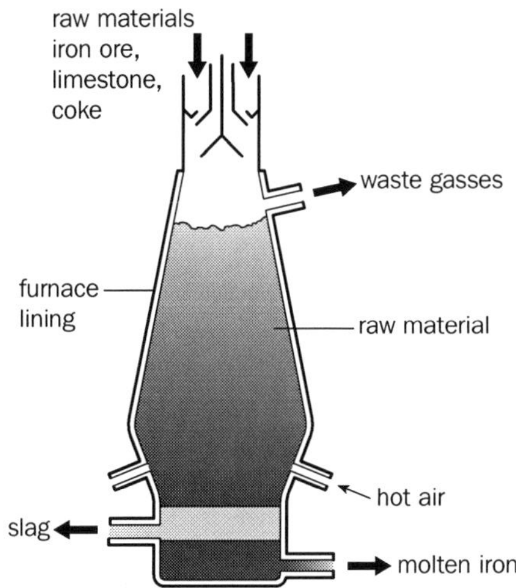

a Why is limestone put into the blast furnace?

b What must be removed from iron ore to leave iron behind?

c How is the iron collected from the furnace?

d Suggest two reasons why hot air is blown into the furnace.

e The temperature at the top of a blast furnace is about 400 °C. Suggest what temperature it might be at the bottom.

f Complete this word equation to summarise what happens in a blast furnace:

iron oxide + carbon =
____________ + ____________

g Give two uses for iron.

2 This diagram shows how aluminium is produced by electrolysis. Aluminium dissolves in molten cryolite and splits into positive (+) aluminium ions and negative (-) oxide ions.

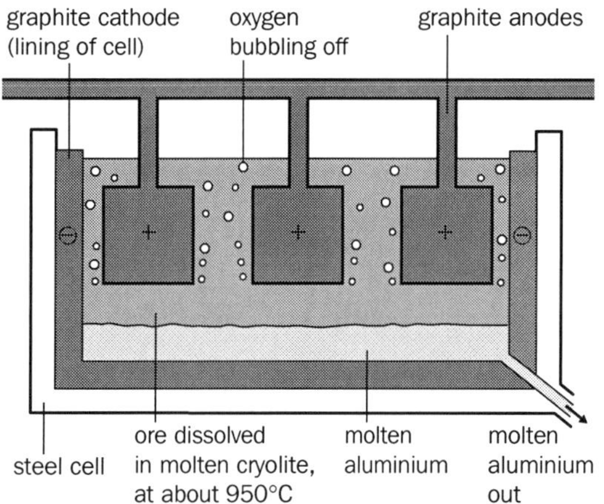

a Why is electrolysis used to extract aluminium from its ore instead of the blast furnace?

b What is
  i an anode
  ii a cathode?

c What are the
  i anodes and
  ii cathode made of?

d Explain why molten aluminium collects at the bottom of the tank.

e Why do you think the anodes have to be frequently replaced?

f Why do you think aluminium is expensive?

g Give two uses for aluminium.

14.3a Corrosion check

H/W

Name: Date: Group:

What you need to know ...

A metal corrodes whenever a chemical, such as water, air, or acid, attacks its surface. First the metal loses its shiny surface, then it is slowly eaten away and weakened.

What to do:

1 Draw a table into your book using these headings:

Object	How old is it?	What metal is it made of?	Is the surface dull or shiny?	Has the surface been eaten away?	What is it affected by?

a Look for ten things made of, or coated, with metal.

b Try to find out how old each object is and what metal it is made of. Write this information in the table.

c Look at the surface of the object. Is it dull or shiny? Has it been eaten away? Write this in the table.

d Try to decide what affects the metal. Is it the air? Is it water? Could it be something else? Write your answer in the table.

2 Name two metals that are used to make jewellery and explain why these metals are used.

3 Explain why you won't find any objects made of sodium or potassium in the home.

4 Explain why it is dangerous to drive a very rusty car.

HANDY HINTS

Metallic objects are everywhere, in and around the home, even on (and in) people. Look carefully for as many different metals as you can. Scratching the surface gives an idea about how much corrosion has happened.

14.3b A family called the halogens H/W

Name: **Date:** **Group:**

What you need to know ...

Knowing about atomic numbers and about electrons in their shells helps us to understand the Periodic Table better. Whenever an electron shell has 8 electrons in it (or 2 in the case of helium), it is full. If another electron is added, it will have to go into the next shell. The number of electrons in the outer shell has a big effect on the way an element reacts. All the elements in a family tend to react in the same way. The table gives some information about the family of elements called the Halogens.

Element	Fluorine	Chlorine	Bromine	Iodine
Formula	Fl_2	**a**	Br_2	I_2
Melting point	-220 °C	-101 °C	-7 °C	114 °C
Boiling point	-188 °C	-35 °C	59 °C	184 °C
Name of compound with iron	**b**	Iron (III) chloride	Iron (III) bromide	Iron (III) iodide
Formula of compound with iron	FeF_3	**c**	$FeBr_3$	FeI_3
Reaction with iron wool	Glows very brightly on gentle heating. Brown smoke and solid formed very quickly	Glows very brightly on gentle heating. Brown smoke and solid formed very quickly	**d**	Glows on heating. Brown solid formed slowly
Effect on indicator paper	Bleached very quickly	Bleached quickly	Bleached	**e**

What to do:

1 Complete the table by writing what should go in the boxes **a** – **e**.

2 At room temperature, what state is

a fluorine

b chlorine

c bromine

d iodine?

3 The Halogen family live in Group 7 of the Periodic Table.

a Draw a diagram showing the arrangement of electrons in a chlorine atom.

b How many electrons has
i fluorine
ii bromine
iii iodine in their outer shells?

c Explain why the Halogen family of elements are so reactive.

HANDY HINTS

Look for patterns in the information in the table.
Room temperature is about 20 °C.
Noble gases don't react because they have full outer electron shells.

14.3c Recycling survey

H/W

Name: **Date:** **Group:**

What you need to know …

Recycling means using waste materials to make new ones. Recycling makes sense because it protects the environment, saves the world's resources and saves the world's energy. The UK has one of the worst records in western Europe for recycling.

What to do:

1 Copy this tally chart into your book:

Name of household/ family										
Paper										
Plastic										
Cardboard										
Glass										
Metal										
None										

2 Find out how many of your relatives and family friends recycle the items listed in the table. Record your results by ticking the appropriate boxes.

3 Draw a bar graph of your results using these axes.

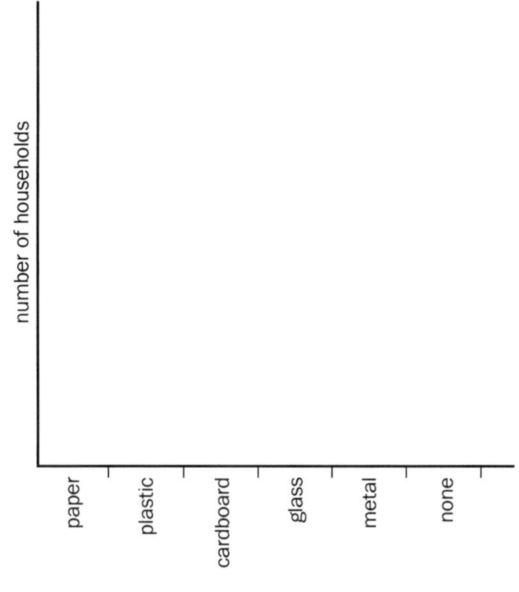

HANDY HINTS

Telephone your relatives or send them a simple questionnaire.
Gather as much information as you can.
Remember you are surveying households not individual people.

Chapter 14 ▶ Investigation 14F Metals and acid

In this investigation: you are going to find out if copper, iron, magnesium, and zinc react differently with hydrochloric acid.

Preparation: Predict

Finish the sentences in the box.

What I think will happen is...

I think this because...

Preparation: Plan

Write a short plan of your investigation.

Think about:

- the apparatus you are going to use
- how one variable depends upon another variable
- what you are going to measure and how you are going to measure it
- how many readings you are going to take
- how you are going to record your results
- how you are going to make your investigation fair
- how you are going to make your investigation safe.

Show your plan to your teacher before going on.

Carry out

Carry out your investigation and record your results.

Present your results in an appropriate way.

Report

Write a report on your investigation.

Here are some things you should include:

- what you did
- what happened
- explain your results
- if your prediction was correct or not
- how reliable your results were
- what you could have done if you had more time.

Chapter 14 ► Investigation 14E Metals and acid

In this investigation: you are going to find out if copper, iron, magnesium, and zinc react differently with hydrochloric acid.

Preparation: Predict

Finish the sentences in the box.

What I think will happen is...

I think this because...

Preparation: Plan

Write a short plan of your investigation.

Think about:

- the apparatus you are going to use
- what you are going to measure and how you are going to measure it
- how many readings you are going to take
- how you are going to record your results
- how you are going to make your investigation fair
- how you are going to make your investigation safe.

Show your plan to your teacher before going on.

Carry out

Carry out your investigation and record your results in a table.

Draw a bar graph of your results.

Report

Write a report on your investigation.

Here are some things you should include:

- what you did
- what happened
- explain your results
- if your prediction was correct or not
- what you could do to improve the investigation
- what you could have done if you had more time.

Chapter 14 ▶ Investigation 14D Metals and acid

In this investigation:	you are going to find out if copper, iron, magnesium, and zinc react differently with hydrochloric acid.

Preparation: Predict

Finish the sentence in the box.

I think that copper, iron, magnesium, and zinc (will/will not) react differently with hydrochloric acid because…

You are going to use this equipment to find out if copper, iron, magnesium, and zinc react differently with hydrochloric acid:

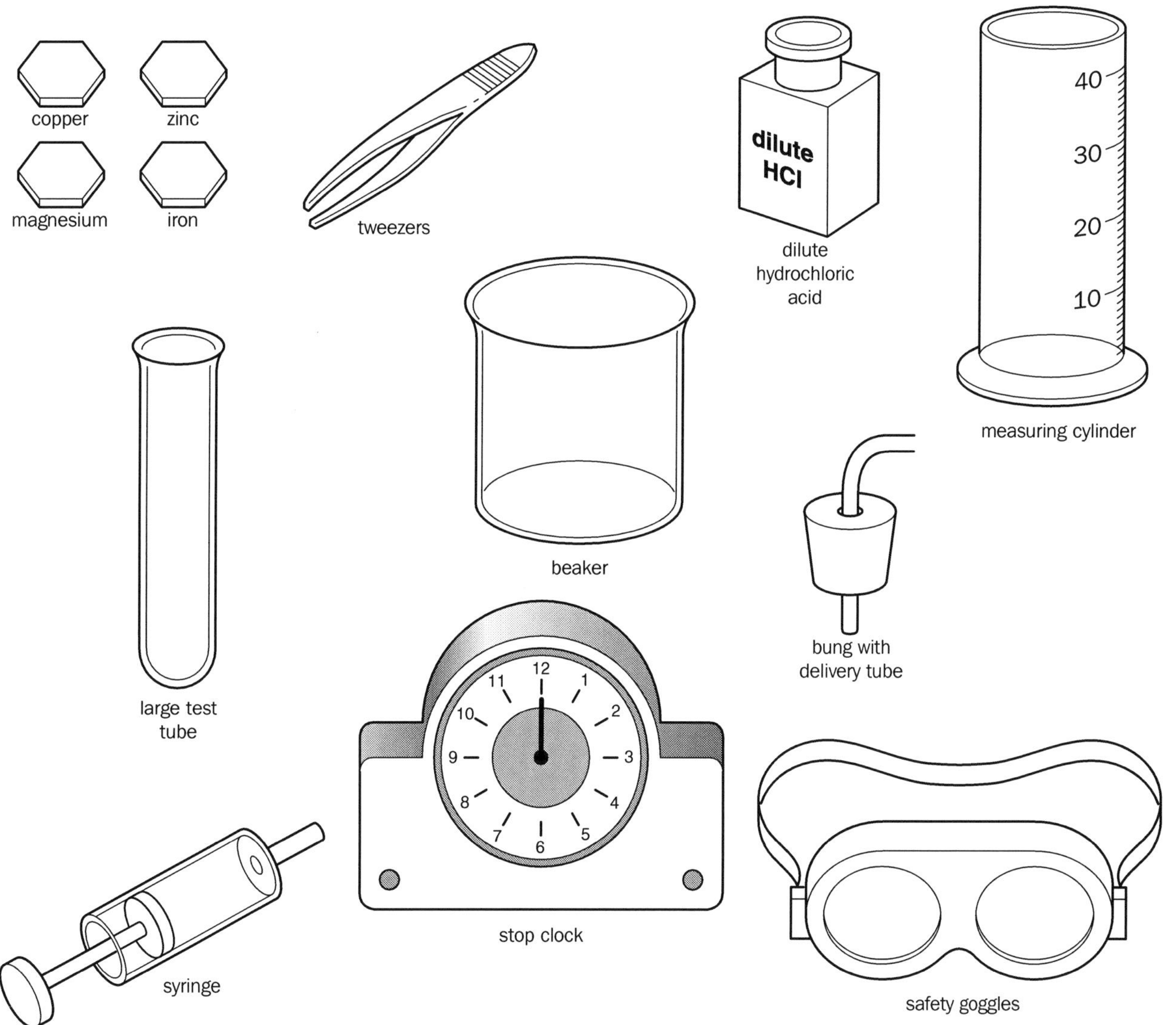

Chapter 14 ▶ Investigation 14D Metals and acid

Preparation: Plan

Finish the sentences in the box.

I will measure…

Things I will keep the same are…

My investigation will be fair because…

My investigation will be safe because…

Carry out

Set up your apparatus like this. See if there is a difference in the amount of gas produced by each metal in the same time.

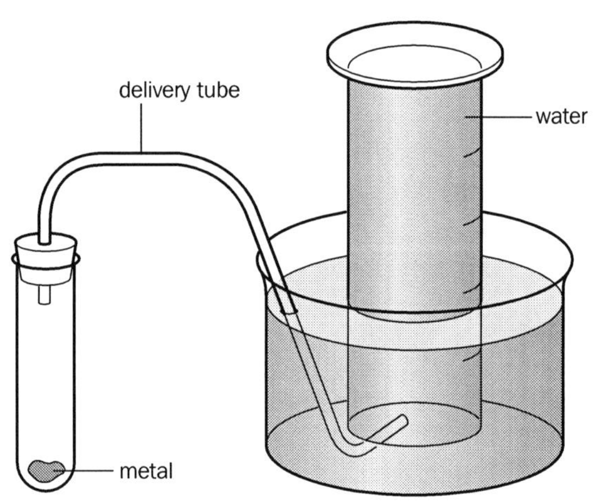

Put your results in a table like this:

Metal	Volume of gas produced in 2 minutes/cm^3
Copper	
Iron	
Magnesium	
Zinc	

Draw a bar graph of your results on a piece of graph paper.

Use different colours for each bar. Label the axes like this:

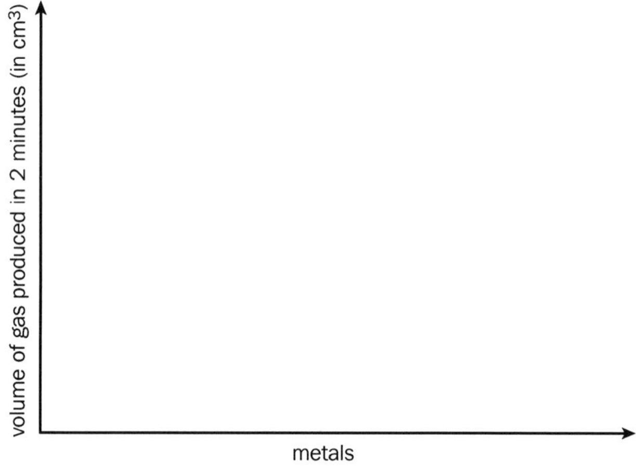

Report

Write a report on your investigation.

Here are some things you should include:

- what you did
- what happened
- explain your results
- if your prediction was correct or not
- what you could do to improve the investigation
- what you could have done if you had more time.

Chapter 14 ▶ Investigation 14C
Metals and acid

In this investigation:	you are going to find out if copper, iron, magnesium, and zinc react differently with hydrochloric acid.

Preparation: Predict

Finish the sentence in the box.

I think that copper, iron, magnesium, and zinc (will/will not) react differently with hydrochloric acid because...

You are going to use this equipment to find out if copper, iron, magnesium, and zinc react differently with hydrochloric acid:

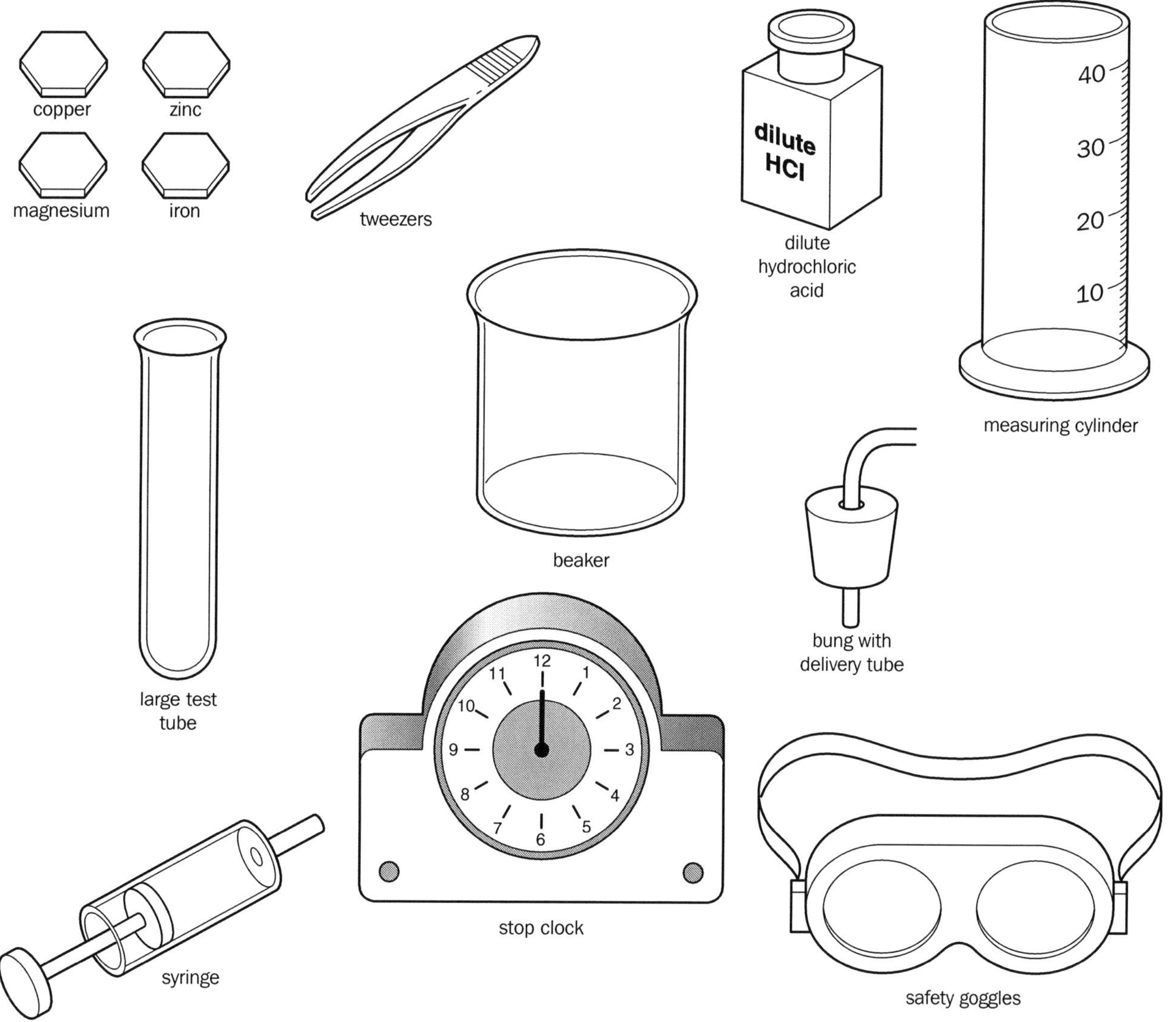

Chapter 14 ▶ Investigation 14C Metals and acid

Preparation: Plan

Finish the sentences in the box.

I will measure...

Things I will keep the same are...

My investigation will be fair because...

My investigation will be safe because...

Carry out

- Put on safety goggles.
- Set up your apparatus like this:

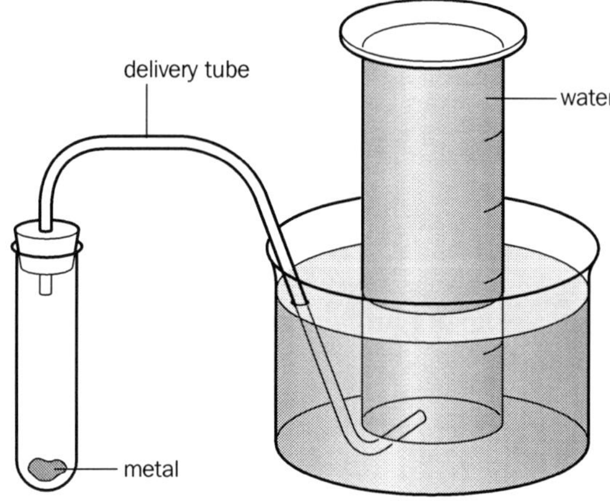

- Quickly but carefully use the syringe to add 10 cm^3 of hydrochloric acid to the metal in the test tube.
- Fit the bung and start timing.
- See how much gas collects in the measuring cylinder in 2 minutes.
- Do the same with the other pieces of metal using fresh acid each time.

Put your results in a table like this:

Metal	Volume of gas produced in 2 minutes/cm^3
Copper	
Iron	
Magnesium	
Zinc	

Chapter 14 ▶ Investigation 14C Metals and acid

Draw a graph of your results on this grid.

Use different colours for each bar.

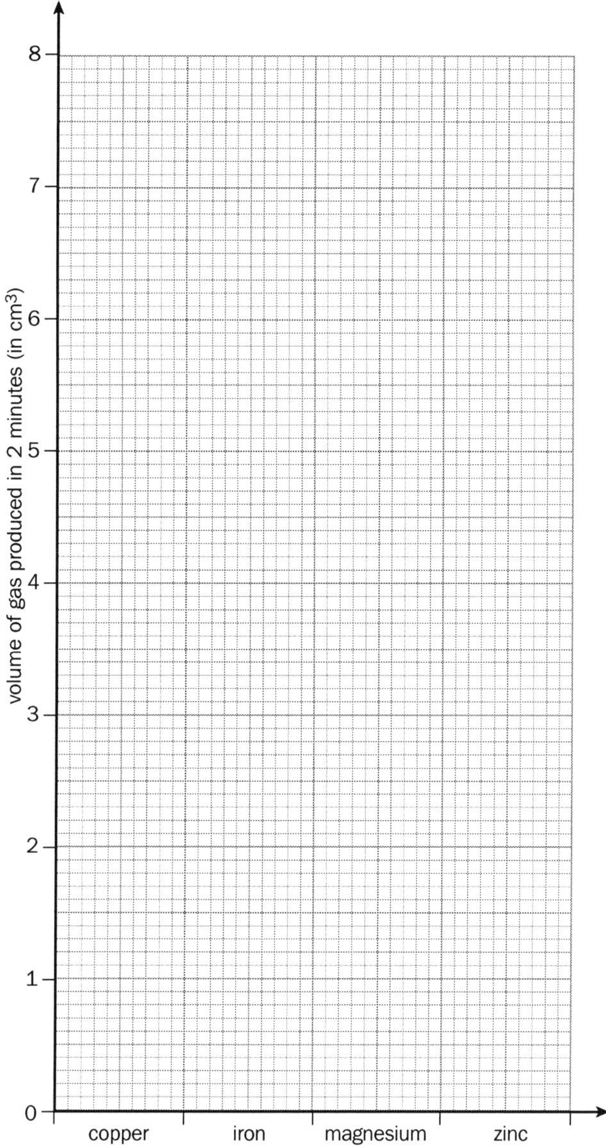

Report

Finish the sentences in the box.

What I did was…

What happened was…

From my results I found out that copper, iron, magnesium, and zinc (did/did not) react differently with hydrochloric acid. I know this because…

My prediction (was/wasn't) correct. If I could do the investigation again I would…

Investigation 14
Practical notes

Metals and acid

Investigation 14
Technician's notes

Metals and acid

Each group will need:

Number of apparatus sets:

Number of pupils:

Number of groups:

Visual aids:

ICT resources:

Equipment/apparatus needed:

- pieces of copper, iron, magnesium, and zinc (all same size and cleaned with fine abrasive paper)
- bottle of dilute hydrochloric acid
- large test tube
- bung with piece of rubber tubing attached/delivery tube
- 250 cm^3 beaker
- 10 cm^3 measuring cylinder
- tweezers
- syringe (10 cm^3)
- stop clock
- access to a large container for collection of waste acid and metal pieces
- safety goggles.

Safety notes

- **See HAZCARDS for dilute hydrochloric acid.**
- **Pupils must wear safety goggles.**
- **Watch for silly behaviour with syringes.**

CLEAPS/SSERC SAFETY REFERENCE:

Chapter 14 ► Test
Metals

1 Metals such as iron, copper, and tin have the same properties.

a Give three properties of metals.

b Give one use for
i iron
ii copper
iii tin.

5 marks

2 Some metals react more than others. Scientists have worked out a reactivity series for metals. Part of the reactivity series is shown below.

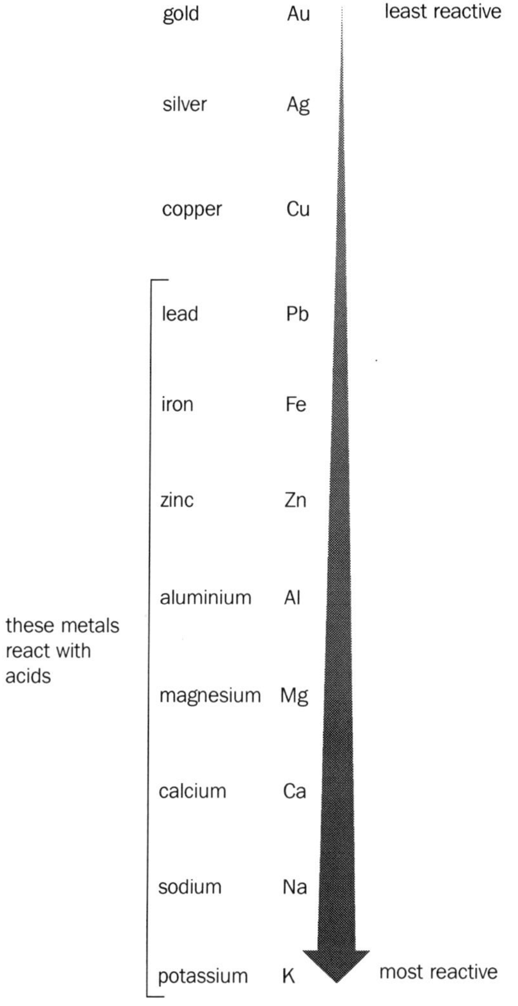

a Explain why gold can be dug from the ground as native metal whereas iron is only found as a compound in iron ore.

b Explain why metals such as gold and silver are used to make jewellery.

c Is aluminium found as a native metal? Explain your answer.

5 marks

3 Iron ore is iron oxide. Iron is extracted from iron ore in a blast furnace.

a What do we call the process of extracting iron in a blast furnace?

b Carbon is added to iron ore in the blast furnace to remove the oxygen from the iron oxide. Write a word equation for this chemical reaction.

c Explain why oxygen is blasted into the blast furnace.

5 marks

4 Three oxides OO, UO, and PO, are made when metals O, U, and P, react with oxygen. A sample of each oxide is put into a solution of U.

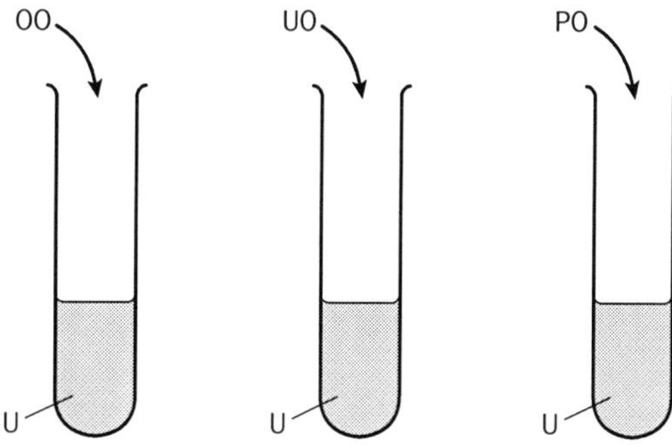

U displaces O in OO but it doesn't displace P in PO.

Write the reactivity series for these three metals starting with the most reactive.

2 marks

5 The diagram shows iron nails in three different test tubes. They have been left for a few days.

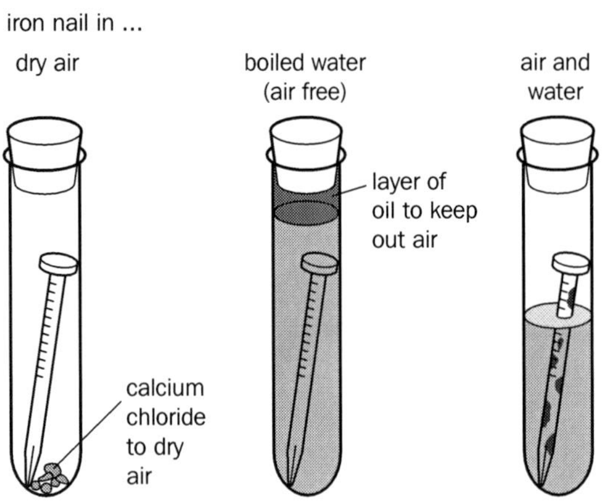

a Describe what you would expect to see in each test tube after a few days.

b Explain why this happens.

c Lots of things are made of iron. They are usually protected to stop them going rusty. Give three ways that iron can be treated to prevent rusting.

7 marks

6 The diagram shows a model of an atom.

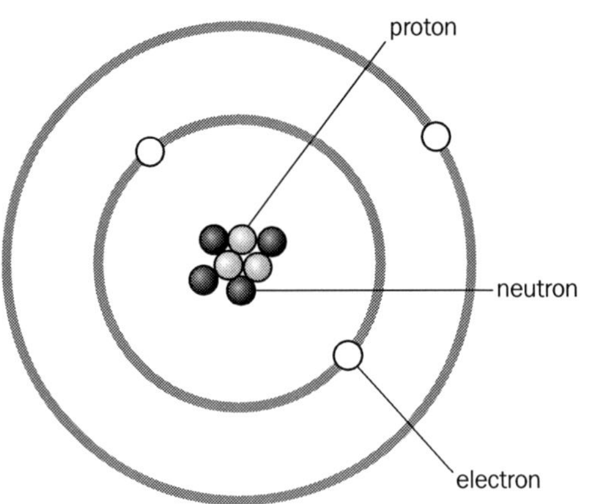

a Name one famous scientist who made an important discovery about atoms.

b What charge does
   i a proton
   ii an electron have?

c The atom shown in the diagram is reactive. Explain why.

d Which group in the periodic table does the atom belong to?

6 marks

Chapter 14 ▶ Mark scheme Metals

White/Blue

Question	Answer	Marks	Level
1 a	three from: reflect light, shiny, good electrical conductor, malleable, ductile, etc. (all three = 2 marks; two = 1 mark; one = 0 marks)	2	E
b i	nails		
ii	water pipes		
iii	food cans	3	E
		5	
2 a	gold is low/iron high in reactivity series	1	E
b	not reactive (1) keeps shine/ decorative (1)	2	E
c	no (1) it is reactive/ will form compounds (1)	2	E
		5	
3 a	smelting	1	E
b	carbon + iron oxide (1) ⟶ iron + carbon dioxide (1)	2	F
c	increase temperature (1) increased rate of reaction (1)	2	F
		5	
4	PUO (all in correct order = 2 marks; two in correct order = 1 mark)	2	E
		2	

Question	Answer	Marks	Level
5 a	rusting in test tube 3 (1) no rust in tubes 1 and 2 (1)	2	E
b	oxygen (1) and water (1) needed for rusting	2	E
c	paint, grease, oil, alloys, electroplating, etc. (any three)	3	E
		7	
6 a	Dalton, Thomson, Rutherford, Bohr (any one)	1	F
b i	positive		
ii	negative	2	F
c	outer shell not full of electrons (1) only atoms with full outer shells don't react (1)	2	F
d	group 2	1	F
		6	
	TOTAL 30 marks		

Suggested grade/level boundaries

E = 18/30

F = 25/30

(This page is intentionally blank.)

15 Light and sound

Pupils are introduced to the properties of light and sound in this chapter. Reflection of light by plane and curved mirrors is followed by refraction in glass blocks and lenses. Refraction of white light by glass prisms leads on to reflection and absorption of colours by different surfaces including colour filters. Sound production, its transmission and absorption leads finally to the relationship between pitch and frequency, and between loudness and amplitude.

Assessment opportunities

Formative assessment opportunities are provided by worksheets, homework sheets, and an investigation.

The **worksheets** cover material at levels C, D, E, and F for attainment targets for knowledge and understanding. Teachers may wish to use these worksheets not only as part of practical activities but also to provide evidence of pupil achievement.

Worksheet	Level
15.1a	C
15.1b	C
15.2a	E
15.2b	D
15.3a	E
15.3b	F
15.4a	C/D
15.4b	E
15.4c	F

The **homework sheets** cover material at levels C, E, and F for attainment targets for knowledge and understanding. These homework sheets can be used individually as a follow-up to work done in class or assembled into a homework booklet allied closely to schemes of work.

Homework sheet	Level
15.1	C
15.2	E
15.3	F
15.4	C/F

The **investigation** covers all three skill areas at levels C, D, E, and F. It is written in a way that allows for pupils to be assessed in all three skill areas at one level. Alternatively, customised assessments can be constructed enabling pupils to be assessed at different levels in all three skills. The latter approach is more time consuming, but it does provide the opportunity for pupils to show evidence of achievement at different levels in different skills in the same investigation. Teachers will need to use their professional judgement when deciding which level is appropriate to individual pupils. It is envisaged that pupils will show progression through the levels as they work through their science course.

Summative tests are provided at two levels, white and blue. The white test contains questions covering attainment target levels C and D. The blue test contains questions covering attainment target levels E and F. Each test has a total of 30 marks and will take about 30 minutes for pupils to complete, although this can be varied depending on pupil ability. Mark schemes are provided together with suggested grade/level boundaries.

It is envisaged that these tests will be given to pupils on completion of the material covered in Chapter 15.

ICT opportunities

The use of data loggers/remote sensors can extend the range, speed, and sensitivity of measurements in many of the worksheets for this chapter. Once downloaded onto a PC, data-handling programs can be used to analyse information gathered, data can be manipulated, and appropriate graphs etc. presented. The Internet provides pupils with access to a huge range of scientific information. A list of suitable websites is included in this Teacher's Guide.

Students' book chapter 15 contents and guide levels

15.1	Light travels in straight lines	*Starting off 1*	C
	Shining lights on surfaces	*Starting off 2*	C
	Shining lights on curved mirrors	*Going further*	C
15.2	Light through lenses (1)	*Starting off*	D
	Light through lenses (2)	*Going further*	D
	Shining light through materials (1)	*For the enthusiast*	E
15.3	Prisms	*Starting off*	E
	Shining light through materials (2)	*Going further*	E
	Why is grass green?	*For the enthusiast*	F
15.4	Making sound – making vibrations	*Starting off*	C/D
	Travelling sound	*Going further 1*	D/E
	Too much or too little	*Going further 2*	D/E
	High and low, loud and soft	*For the enthusiast*	F

15.1a Reflection in a plane mirror W/S

Name: **Date:** **Group:**

What you need:

Piece of plain paper, mirror, ray box with single slit, power supply, ruler, protractor.

What to do:

1 Draw up a table in your book using these headings:

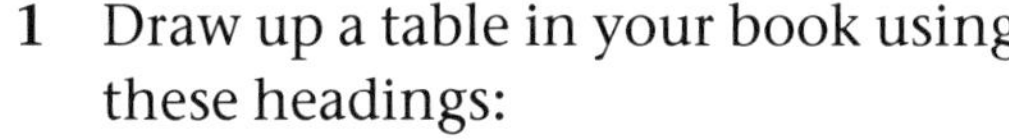

Angle of incident ray/ angle of incidence	Angle of reflected ray/angle of reflection

2 Draw a straight line on a piece of paper. Set up the mirror so that the back of the mirror is on the line you have drawn.

3 Point a ray of light from the ray box at the mirror at an angle.

4 Put two crosses on the paper to mark the path of the incident ray. Do the same for the reflected ray.

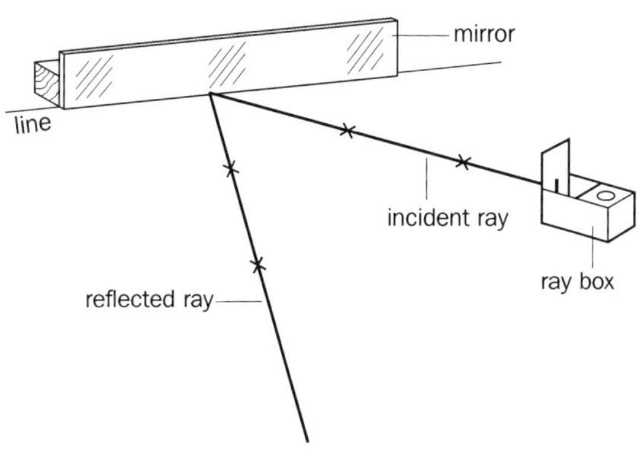

5 Switch off the ray box and move it and the mirror off the paper. Join the crosses so the lines touch the line where the back of the mirror was.

6 Draw a line at right angles to the mirror line at the point where the lines meet.

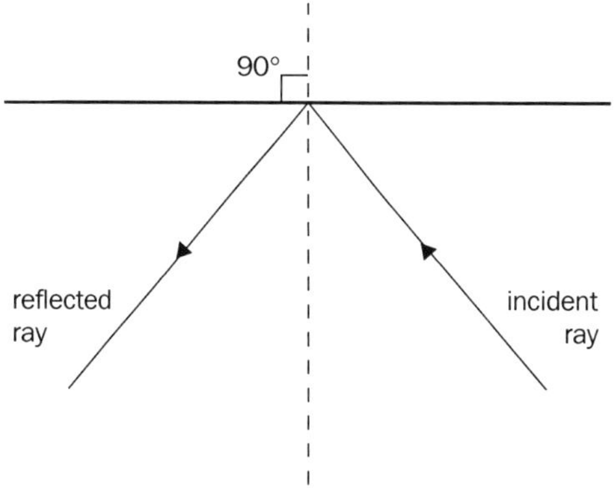

7 Measure the angle of incidence and the angle of reflection and write these in the table.

8 Try doing the activity again with different angles of incidence. Put your results in the table.

What do you notice about the angles of incidence and the angles of reflection?

9 Look at your image in the mirror.

Describe the size of the image, is it bigger, smaller or the same size as you, the object?

Is the image the right way up or upside down?

Why isn't the image a true picture of you?

SAFETY WARNING

Check electrical equipment before use.

Report any faulty equipment.

15.1a
Practical notes

Reflection in a plane mirror

This is a simple activity designed to enable pupils to investigate the relationship between angles of incidence and reflection. In practice, the results will be a few degrees out, but the pattern should be clear to see. It is important to emphasise that the back (reflecting surface) of the mirror sits on the line drawn on the paper.

Warn pupils that the ray box will get hot during the activity.

15.1a
Technician's notes

Reflection in a plane mirror

Each group will need:

- a copy of worksheet 15.1a
- access to A4 plain paper
- plane mirror mounted on block
- ray box with single slit
- power supply
- ruler
- protractor.

Number of apparatus sets:

Number of pupils:

Number of groups:

Visual aids:

ICT resources:

Equipment/apparatus needed:

Safety notes

- **Check electrical equipment before use.**
- **Report any faulty equipment.**
- **Warn pupils that the ray box will get hot during the activity.**

CLEAPSS/SSERC SAFETY REFERENCE:

15.1b Reflections in curved mirrors

W/S

Name: Date: Group:

What you need:

Concave mirror, convex mirror, ray box with three slits, power supply, two pieces of plain paper, ruler.

What to do:

1 Put the convex mirror on a piece of paper and draw a line round the back of the mirror.

2 Point the three rays of light at the mirror.

3 Put two crosses on each ray to mark its path.

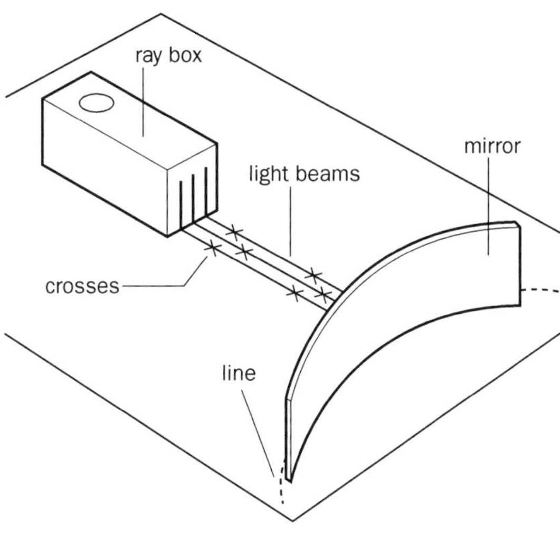

4 Mark each reflected ray with crosses to mark their path.

5 Switch off the ray box and move it and the mirror off the paper. Join the crosses so the lines touch the line where the back of the mirror was.

6 Do the same with a concave mirror using a separate piece of paper.

- What happens to the rays of light when they are reflected by a convex mirror?
- Explain why you can see more of your surroundings if you look into a convex mirror. When might this be useful?
- What happens to the rays of light when they are reflected by a concave mirror?
- Explain why near images are magnified in a concave mirror. When might this be useful?

SAFETY WARNING

Check electrical equipment before use.

Report any faulty equipment.

15.1b
Practical notes

Reflections in curved mirrors

It is important to emphasise that the line is drawn around the back (reflecting surface) of the mirror in both cases. Accurate drawing will enable pupils to see light rays brought to a focus from a concave mirror. Warn pupils that the ray box will get hot during the activity.

15.1b
Technician's notes

Reflections in curved mirrors

Each group will need:

- a copy of worksheet 15.1b
- access to A4 plain paper
- concave mirror
- convex mirror
- ray box with three slits
- power supply
- ruler.

Number of apparatus sets:

Number of pupils:

Number of groups:

Visual aids:

ICT resources:

Equipment/apparatus needed:

Safety notes

- **Check electrical equipment before use.**
- **Report any faulty equipment.**
- **Warn pupils that the ray box will get hot during the activity.**

CLEAPSS/SSERC SAFETY REFERENCE:

15.2a Refraction 1

W/S

Name: **Date:** **Group:**

What you need:

Piece of plain paper, glass block, ray box with single slit, power supply, ruler, protractor.

What to do:

1 Draw up a table in your book using these headings:

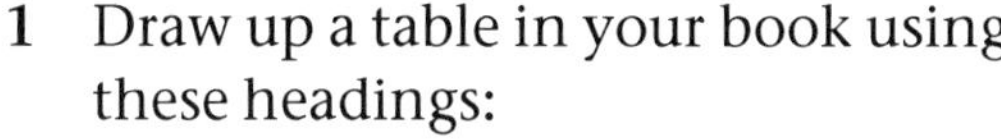

Angle of incoming ray	Angle of emerging ray

2 Put the glass block on a piece of paper.

3 Draw round the block to mark its position.

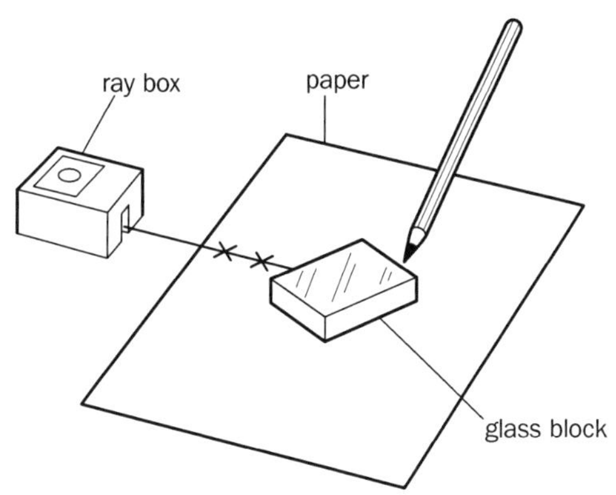

4 Point a ray of light from the ray box at an angle to one side of the block.

5 Put two crosses on the paper to mark the path of the incoming ray. Do the same for the emerging ray.

6 Switch off the ray box and move it and the glass block off the paper. Join the crosses so the lines touch the outline of the glass block.

7 Draw a line through the 'block' joining the incoming and emerging rays.

8 Draw a line at right angles to the block outline at the point where the incoming ray entered the block. Do the same where the emerging ray left the block.

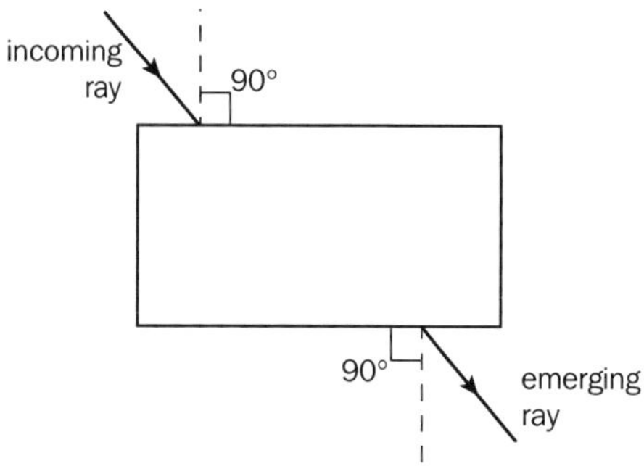

9 Measure the angle of the incoming ray and the angle of the emerging ray and write these in the table.

10 Try doing the activity again with different angles for the incoming ray. Put your results in the table.

What do you notice about the angles of the incoming rays and the angles of the emerging rays?

11 Put a spot on a piece of paper and look at it through the glass block. What do you notice about the position of the spot? Explain why this happens.

SAFETY WARNING

Check electrical equipment before use.

Report any faulty equipment.

15.2a
Practical notes

Refraction 1

This activity enables pupils to see that light is refracted as it passes into and out of glass. The investigation lends itself to measurement of angles and the collection of quantitative data. Warn pupils that the ray box will get hot during the activity.

15.2a
Technician's notes

Refraction 1

Each group will need:

- a copy of worksheet 15.2a
- access to A4 plain paper
- rectangular glass block
- ray box with single slit
- power supply
- ruler
- protractor.

Number of apparatus sets:

Number of pupils:

Number of groups:

Visual aids:

ICT resources:

Equipment/apparatus needed:

Safety notes

- **Check electrical equipment before use.**
- **Report any faulty equipment.**
- **Warn pupils that the ray box will get hot during the activity.**

CLEAPSS/SSERC SAFETY REFERENCE:

15.2b Refraction 2

W/S

Name: **Date:** **Group:**

What you need:

Two pieces of plain paper, concave lens, convex lens, ray box with three slits, power supply, ruler.

What to do:

1 Put the convex lens on a piece of paper.

2 Draw round the lens to mark its position.

3 Point the three rays of light from the ray box at the lens.

4 Put two crosses on each incoming ray to mark its path.

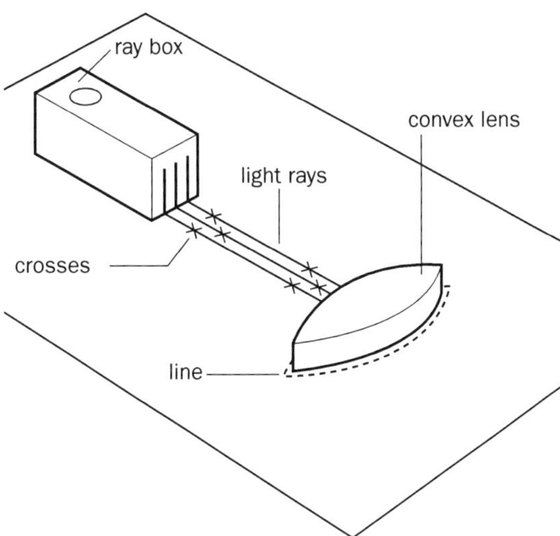

5 Mark each emerging ray with crosses to mark their paths.

6 Switch off the ray box and move it and the lens off the paper. Join the crosses so the lines touch the outline of the lens.

7 Draw lines through the 'lens' joining the incoming and emerging rays.

8 Do the same with a concave lens using a separate piece of paper.

?

- What happens to the rays of light when they are refracted by a convex lens?
- Explain why a convex lens held at arm's length produces an image which is upside down and smaller than the object. When might this be useful?
- What happens to the rays of light when they are refracted by a concave mirror?
- Explain why you can see a lot of your surroundings when you look through a concave lens. When might this be useful?

SAFETY WARNING

Check electrical equipment before use.

Report any faulty equipment.

15.2b
Practical notes

Refraction 2

This activity enables pupils to see how light is refracted as it passes through concave and convex lenses. Extension work could include investigating focal lengths of various lenses, introducing the concept of 'focus'.

Warn pupils that the ray box will get hot during the activity.

15.2b
Technician's notes

Refraction 2

Each group will need:

Number of apparatus sets:

Number of pupils:

Number of groups:

Visual aids:

ICT resources:

Equipment/apparatus needed:

- a copy of worksheet 15.2b
- access to A4 plain paper
- concave glass/plastic lens (2D to lie flat on bench)
- convex glass/plastic lens (2D to lie flat on bench)
- ray box with three slits
- power supply
- ruler.

Safety notes

- **Check electrical equipment before use.**
- **Report any faulty equipment.**
- **Warn pupils that the ray box will get hot during the activity.**

CLEAPSS/SSERC SAFETY REFERENCE:

15.3a Refraction in a prism

W/S

Name: Date: Group:

What you need:

Triangular prism, ray box with single slit, power supply, white screen and stand, coloured pencils.

What to do:

1 Stand the prism about 10 cm in front of the screen.

2 Point a ray of light from the ray box at an angle to one side of the prism.

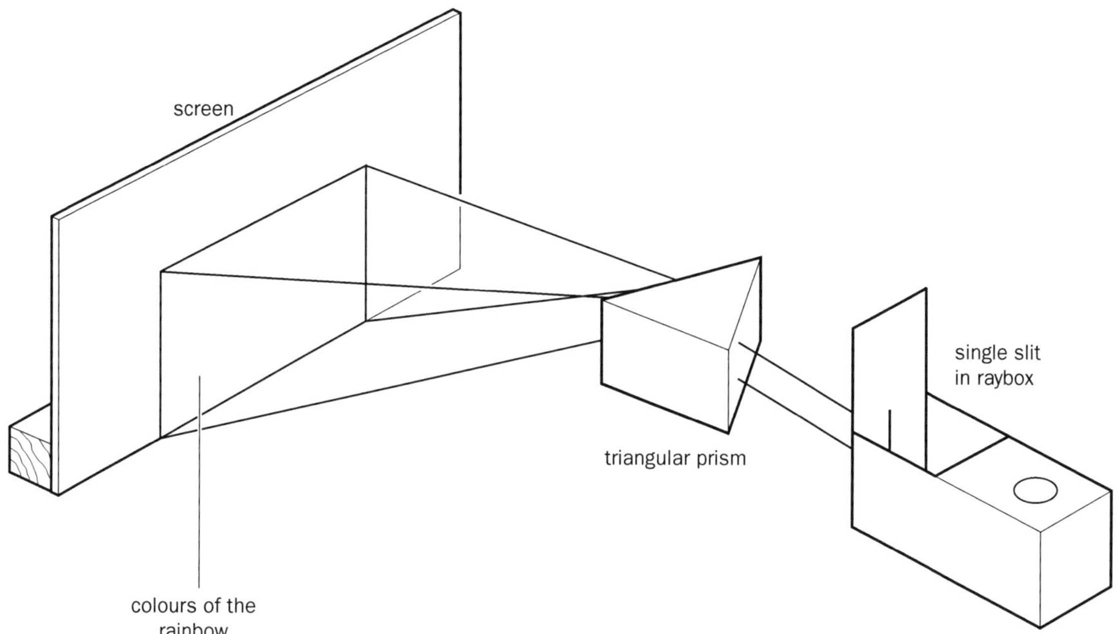

3 Adjust the prism by turning it slightly so that you get colours on the screen.

4 Draw a diagram like the one below in your book and put the colours, in order, in the boxes.

An easy way to remember the colours of the spectrum is to use a saying such as:

Richard **O**f **Y**ork **G**ave **B**attle **I**n **V**ain

The funnier the saying the better you will remember it. Try to make up your own saying to help you remember the colours of the spectrum.

SAFETY WARNING

Check electrical equipment before use.

Report any faulty equipment.

15.3a
Practical notes

Refraction in a prism

This activity is best done in a darkened room otherwise the colours are hard to distinguish. Pupils may need help adjusting their prisms to get a clear spectrum. A convex lens could be added to focus the spectrum onto the screen although perfectly satisfactory results can be obtained without.

Warn pupils that the ray box will get hot during the activity.

15.3a
Technician's notes

Refraction in a prism

Each group will need:

- a copy of worksheet 15.3a
- glass/plastic triangular prism
- ray box with single slit
- power supply
- screen made from white card supported vertically in wooden block.

Number of apparatus sets:

Number of pupils:

Number of groups:

Visual aids:

ICT resources:

Equipment/apparatus needed:

Safety notes

- **Check electrical equipment before use.**
- **Report any faulty equipment.**
- **Warn pupils that the ray box will get hot during the activity.**

CLEAPSS/SSERC SAFETY REFERENCE:

15.3b Coloured lights

W/S

Name: Date: Group:

What you need:

Three ray boxes or torches, power supply, white screen and stand, red filter, blue filter, green filter, yellow card, magenta card, cyan card.

Part 1

What to do:

1 Copy this diagram into your book:

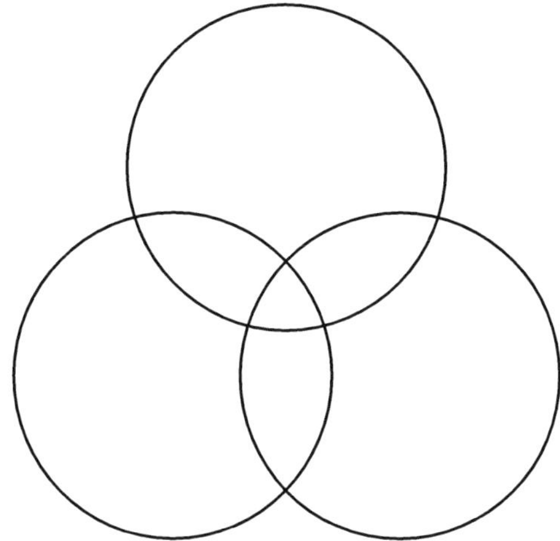

2 Put a different coloured filter into each of the ray boxes or torches.

3 Shine the lights on to the screen to find out what happens when the lights are mixed.

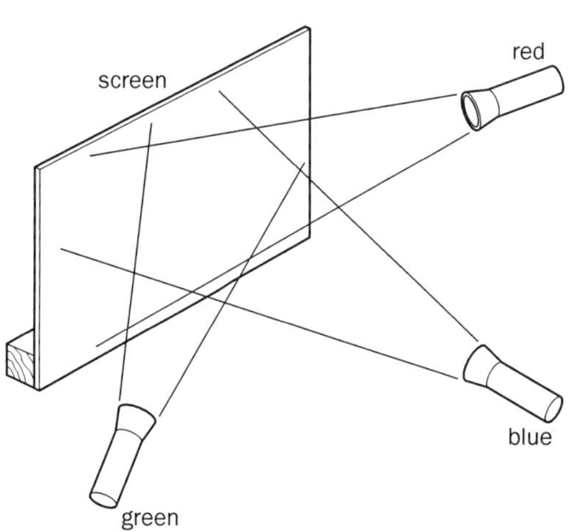

4 Put your results in the diagram.

Part 2

What to do:

1 Copy this table into your book.

Colour of light	Colour of surface	Appearance under coloured light
red	yellow	
red	magenta	
red	cyan	
blue	yellow	
blue	magenta	
blue	cyan	
green	yellow	
green	magenta	
green	cyan	

2 Put a red filter in a ray box and shine it onto yellow, magenta, and cyan cards.

See what colour they turn. Write the colours in the table.

3 Do the same with a blue filter then a green filter. Write the colours in the table.

SAFETY WARNING

Check electrical equipment before use.

Report any faulty equipment.

15.3b
Practical notes

Coloured lights

This simple activity is best carried out in a darkened room. It does have the potential to cause total confusion, but with guidance pupils should be able to grasp the basic concepts. The biggest problem will be getting over the confusion in some pupil's minds between mixing coloured pigments and mixing coloured light. The range of coloured cards could be extended or reduced as necessary; it is the concept that is important. Warn pupils that the ray boxes (if used) will get hot during the activity.

15.3b
Technician's notes

Coloured lights

Each group will need:

- a copy of worksheet 15.3b
- three ray boxes or torches with red, blue, and green filters
- power supply
- screen made from white card supported vertically in a wooden block
- one yellow, one magenta, and one cyan card.

Number of apparatus sets:

Number of pupils:

Number of groups:

Visual aids:

ICT resources:

Equipment/apparatus needed:

Safety notes

- **Check electrical equipment before use.**
- **Report any faulty equipment.**
- **Warn pupils that the ray boxes (if used) will get hot during the activity.**

CLEAPSS/SSERC SAFETY REFERENCE:

15.4a Making sounds

W/S

What you need:

Wooden ruler, G clamp.

What to do:

1 Clamp the ruler to the bench with 10 cm sticking out from the edge.

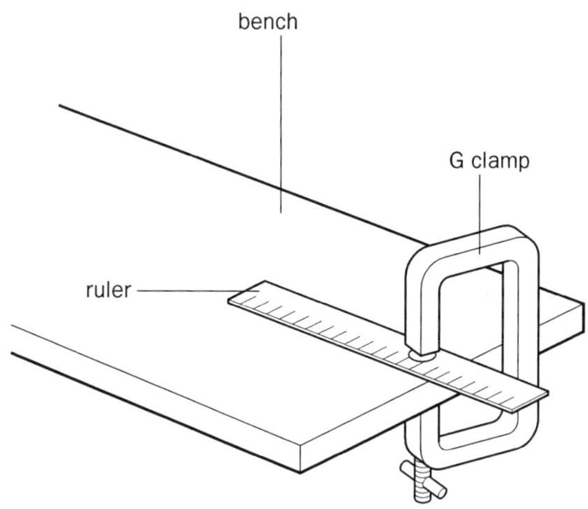

2 Gently pluck the ruler to make it vibrate. Listen to the sound it makes.

3 Pluck the ruler a second time; harder.

What happens to the sound?

4 Loosen the clamp and move the ruler so that 20 cm sticks out from the edge of the desk.

Pluck the ruler again.

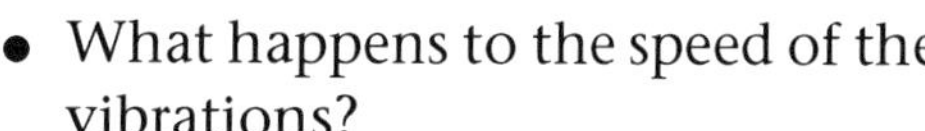

- What happens to the speed of the vibrations?

- What happens to the sound this time?

5 Do the activity several times more. Try different lengths of the ruler over the edge of the bench. Pluck the ruler gently and then harder.

What is the connection between the speed of the vibrations and the pitch of the note produced?

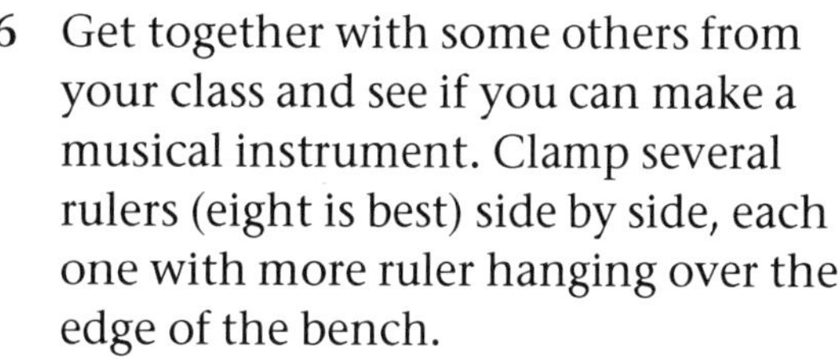

6 Get together with some others from your class and see if you can make a musical instrument. Clamp several rulers (eight is best) side by side, each one with more ruler hanging over the edge of the bench.

Tune your instrument by adjusting the lengths of ruler hanging over the edge of the bench.

15.4a
Practical notes

Making sounds

Pupils at last have an opportunity to legitimately 'twang' rulers. Plastic rulers do not work as well as wooden or metal varieties. Have a few spares in case of breakage. Pupils with a musical background will have a head start over the others. It is sensible to make sure that at least one member of the group has some musical knowledge. Extension work could include making a musical instrument from stretched rubber bands or a row of test tubes each filled with different amounts of water.

15.4a
Technician's notes

Making sounds

Each pupil will need:

- a copy of worksheet 15.4a
- wooden (or metal) ruler
- G clamp.

Number of apparatus sets:

Number of pupils:

Number of groups:

Visual aids:

ICT resources:

Equipment/apparatus needed:

Safety notes

CLEAPSS/SSERC SAFETY REFERENCE:

15.4b Travelling sound

Name: **Date:** **Group:**

What you need:

Wooden rod, watch, squeaky toy in plastic bag, bowl of water, two polystyrene cups, piece of string.

Part 1

What to do:

1. Put a ticking watch on the bench. Listen carefully to see if you can hear the tick.
2. Put one end of the solid wooden rod on the watch and the other at the entrance to your ear. Listen carefully.

- What difference does the wooden rod make?
- What does this tell you about the way sound travels through solids?

Part 2

What to do:

1. Make sure the plastic bag holding the squeaky toy is tied tightly. Give the toy a squeeze and listen to the sound it makes.
2. Put the toy into the bowl of water and squeeze it again.

- What difference is there in the sound made by the toy underwater?
- Does the depth of water make any difference to the sound?
- What does this tell you about the way sound travels through liquids?

Part 3

What to do:

1. Carefully make a hole in the bottom of two polystyrene cups large enough to pass string through.
2. Push string through the hole and tie a large knot on the end of the string inside the cup. Do the same with the other end of the string and the other cup.

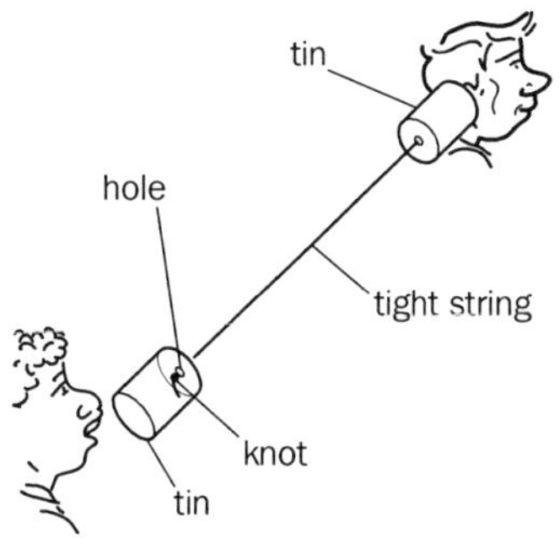

3. Move away from your partner so the string is tight. The idea is that you hold the cup to your ear while your partner speaks and vice versa. Do not try speaking at the same time as each other.
4. Slacken the string slightly.

   What difference does this make to the sound?

5. Try shortening the string but remember to keep it tight.

   What difference does this make?

15.4b
Practical notes

Travelling sound

This activity gives pupils the opportunity to investigate the way sound travels in different media. The first two parts are straightforward examples of sound travelling through a solid and a liquid. The third part still interests pupils even if it has been around a long time. Try to avoid calling it a string telephone though, it causes unnecessary confusion.

15.4b
Technician's notes

Travelling sound

Each group will need:

- a copy of worksheet 15.4b
- wooden rod about 30 cm long
- watch
- squeaky toy sealed securely in waterproof plastic bag
- bowl of water
- two polystyrene cups
- piece of string (the longer the better).

Number of apparatus sets:

Number of pupils:

Number of groups:

Visual aids:

ICT resources:

Equipment/apparatus needed:

Safety notes

- **If pupils stick the wooden rods in their ears, the rod must be sterilised before being reused.**

CLEAPSS/SSERC SAFETY REFERENCE:

15.4c High and low, loud and soft

W/S

Name: **Date:** **Group:**

What you need:

Oscilloscope with microphone, low pitch tuning fork, high pitch tuning fork.

Part 1

What to do:

1 Draw these circles in your book:

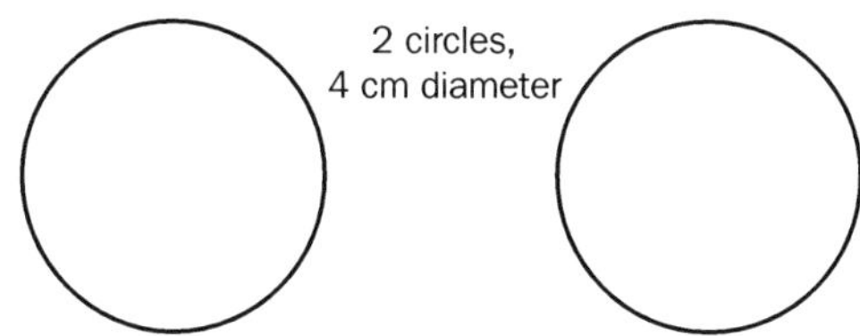

2 Make sure the oscilloscope is switched on and has a straight, horizontal line on the screen.

3 Tap the low pitch tuning fork on the bench and stand it on the bench next to the microphone.

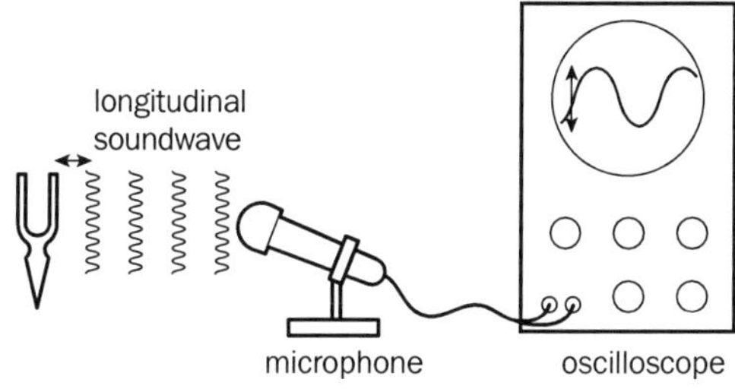

4 Draw the shape of the wave produced by the low pitched tuning fork in the first circle.

5 Repeat the activity this time using the high pitch tuning fork.

6 Draw the shape of the wave produced by the high pitch tuning fork in the second circle.

What difference is there in the shapes of the waves for a high pitch note and a low pitch note?

Part 2

What to do:

1 Draw these circles in your book:

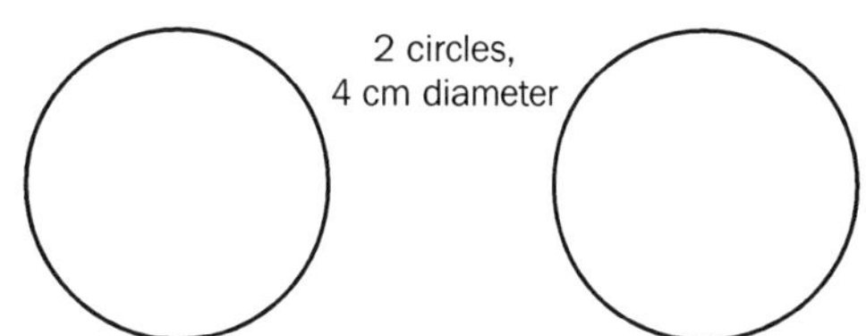

2 Make sure the oscilloscope is switched on and has a straight, horizontal line on the screen.

3 Whistle a soft note into the microphone. Look at the wave you have produced.

4 Draw the shape of the wave produced by a soft note in the first circle.

5 Whistle the same note but this time make it louder. Look at the wave you have produced.

6 Draw the shape of the wave produced by a loud note in the second circle.

What difference is there in the shapes of the waves for a soft note and a loud note?

SAFETY WARNING

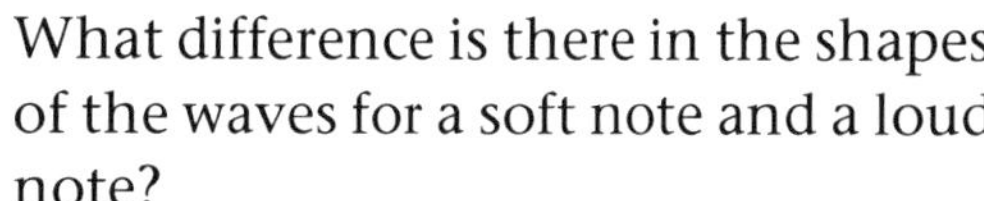

Check electrical equipment before use.

Report any faulty equipment.

15.4c
Practical notes

High and low, loud and soft

Since it is unlikely that many schools will have a class set of oscilloscopes, this activity is probably best incorporated into other classwork on this topic. Make sure the oscilloscope(s) is/are correctly adjusted before letting pupils near it/them. Sticky tape over the knobs might deter fiddling fingers. Pupils should be in no doubt of the sensitivity (and cost) of this equipment.

15.4c
Technician's notes

High and low, loud and soft

Each group will need:

- a copy of worksheet 15.4c
- oscilloscope with microphone connected to input terminals. Set oscilloscope so horizontal line appears on the screen
- low pitch tuning fork
- high pitch tuning fork.

Number of apparatus sets:

Number of pupils:

Number of groups:

Visual aids:

ICT resources:

Equipment/apparatus needed:

Safety notes

Check electrical equipment before use.

CLEAPSS/SSERC SAFETY REFERENCE:

15.1 Reflections H/W

Name: Date: Group:

What you need to know ...

If you shine a ray of light at a flat or plane mirror it is reflected in a definite direction. The angle of incoming (incident) light is always equal to the angle of the reflected light. When light falls on a rough surface, light is reflected in many different directions.
The rougher the surface the duller it looks.

What to do:

1 Imagine you live on a farm, and the farm road joins the main road at a blind junction. You want to fix a big plane mirror on the fence opposite the farm road entrance to help you drive on to the main road safely.

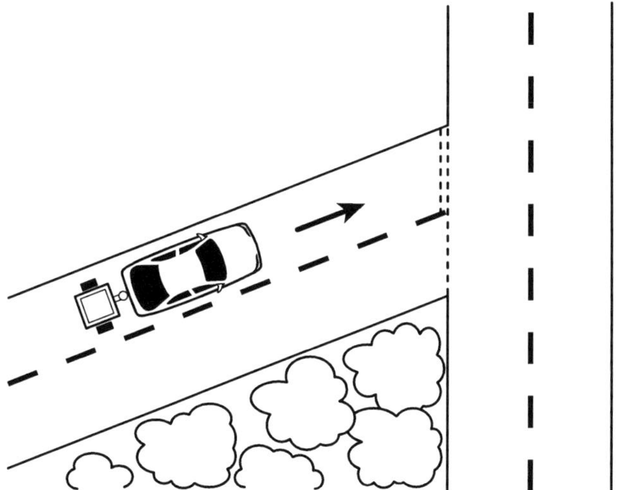

mirror goes here

At what angle should you fix the mirror to give you a good view of the road to your right?

2 A room has only one small window. Even on a sunny day the room is dark. Describe what sort of things you could do to brighten the room up.

3 Explain why you cannot see an image of yourself on this paper.

HANDY HINTS

In Question 1, copy the diagram and draw lines representing a ray of light coming from a car approaching from the blind side on the main road and being reflected by the mirror into your eye in the farm road.
In Question 2, think about wallpaper, mirrors, furnishings, etc.

Refraction

Name: **Date:** **Group:**

What you need to know ...

When light passes from one transparent material to another it is bent or refracted. Refraction makes things seem to be in a different place to where they really are. When a ray of white light passes through a glass prism it is dispersed into the colours of the spectrum. Red light is refracted least. Violet light is refracted most.

Part 1

What to do:

1. Put a coin in a mug and put the mug on the table in front of you. Look at the coin in the bottom of the mug, then move backwards until the coin just disappears from view. Ask someone to slowly pour some water into the mug. Watch what happens.

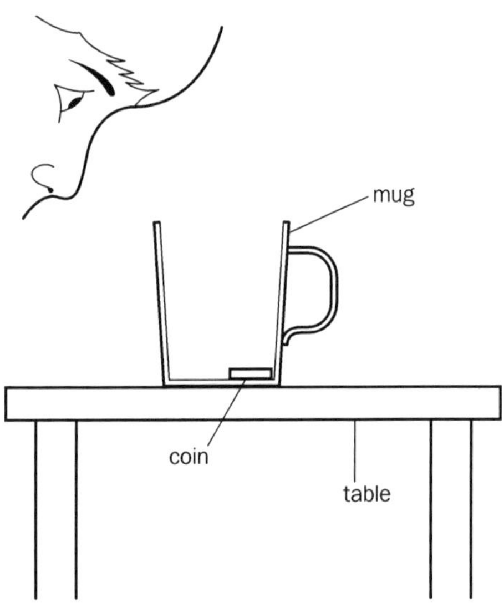

2. Copy the diagram and use it to explain why the coin cannot be seen before the water is poured into the mug.
3. Draw another diagram of the mug, this time with water in. Use this diagram to explain why the coin can be seen.

Part 2

What to do:

1. Carefully copy this diagram into your book.

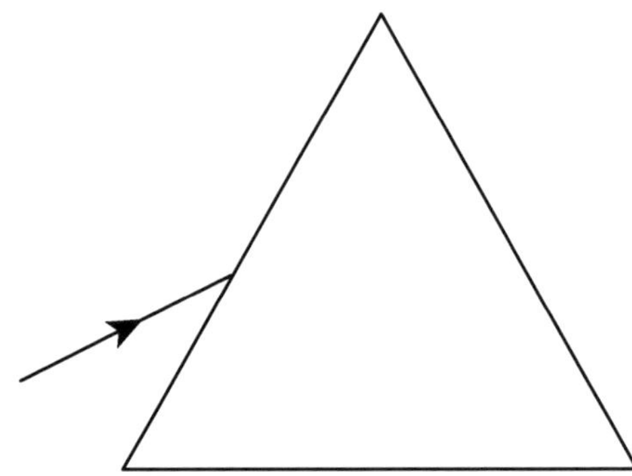

2. Complete the diagram to show how a ray of light passing through a prism is split into the colours of the spectrum.
3. Label the colours of the spectrum on your diagram.

HANDY HINTS

Draw the diagrams as accurately as you can and use a ruler to draw lines representing rays of light.
Do not forget the arrows to show which way the light is going.

15.3 Colour filters H/W

Name: **Date:** **Group:**

What you need to know ...

The diagram shows what happens when red, blue, and green lights are mixed.

Colour filters and coloured objects remove colours from the light that shines on them.

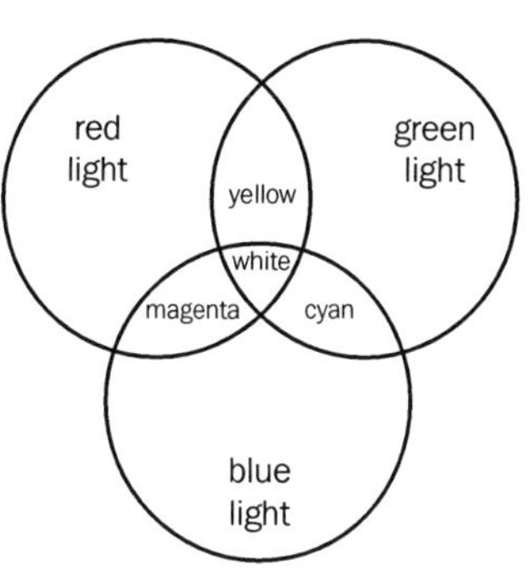

What to do:

1 Copy the passage below filling in the gaps using these words:

- absorb
- blue
- green
- reflect
- red

Leaves are green because they __________ green light and __________ all the other colours. Yellow flowers appear red because they absorb __________ light but reflect __________ and __________ light.

2 A coloured plastic sweet wrapper can be used as a colour filter.

a What does a colour filter do?

b Use a diagram to explain how a blue filter allows only blue light to pass through.

c Copy and complete this diagram to show what happens to white light when a green filter is put in front of a blue filter.

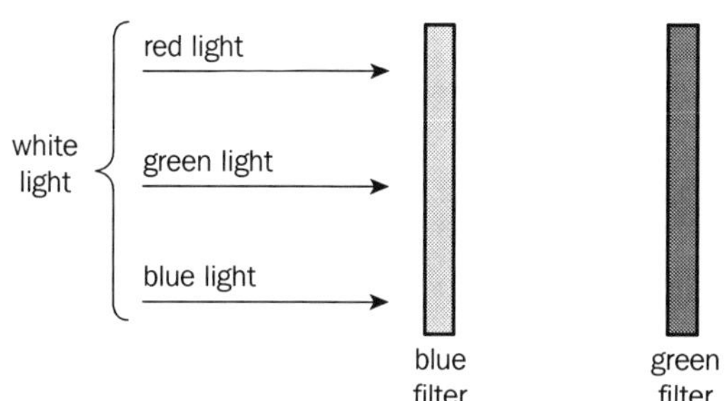

3 The diagram shows yellow light from a street lamp shining on a red car.

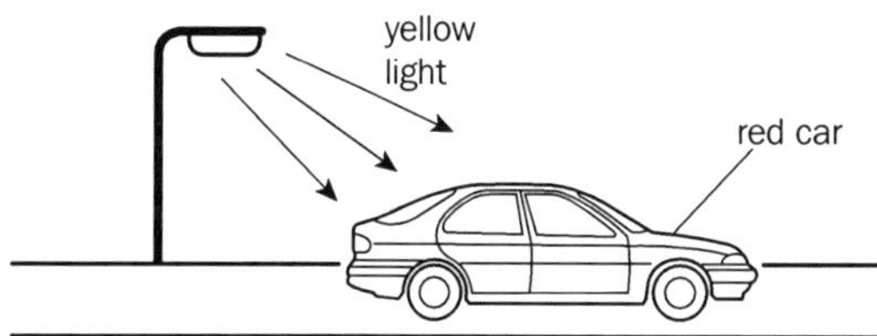

a What colour does the car appear to be?

b Draw simple diagrams to show what happens when
- **i** magenta light shines on a blue car
- **ii** cyan light shines on a red car
- **iii** yellow light shines on a green car.

HANDY HINTS

Read the information at the top of the page carefully.
Keep your diagrams neat and tidy, and use a ruler.

15.4 Sounds, high and low, loud and soft H/W

Name: Date: Group:

What you need to know …

Sounds are made when something vibrates. The pitch of a sound depends on the frequency of the vibrations produced by the 'sound producer'. The loudness of a sound depends on the amplitude or size of the vibrations.

What to do:

1 Look at these pictures of musical instruments. They produce sounds in different ways.

a Sort the instruments into **three** groups. Each group should contain instruments which are played in the same way. List the three groups in your book under separate headings.

b Choose one instrument from each group and think carefully about how the sound is produced.

2 The diagram shows a graph on the screen of an oscilloscope. The wave has been produced by playing a note on a piano next to a microphone connected to the oscilloscope.

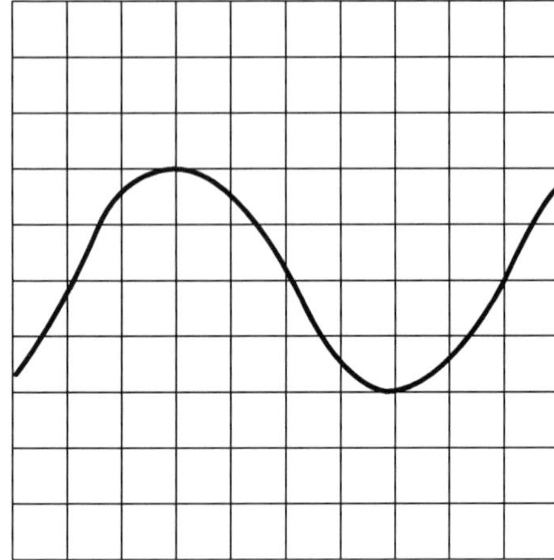

a Copy the outline of the graph paper into your book.

b Draw the shape of the wave as it would appear when a
  i higher note
  ii lower note was played on the piano.

c Draw the shape of the wave as it would appear when the same note on the piano was played
  i harder
  ii softer.

HANDY HINTS

In Question 1, think carefully about each stage in the production of the sound. For example, the harp strings are plucked making them vibrate so producing a sound.

Chapter 15 ► Investigation 15F Sound insulation

Excessive noise can damage the environment and our health. Soundproofing is one way of overcoming noise pollution.

In this investigation: you are going to find out which is the best soundproofing material; polystyrene foam, egg boxes, or newspaper.

Preparation: Predict

Finish the sentences in the box.

What I think will happen is…

I think this because…

Preparation: Plan

Write a short plan of your investigation.

Think about:

- the apparatus you are going to use
- how one variable depends upon another variable
- what you are going to measure and how you are going to measure it
- how many readings you are going to take
- how you are going to record your results
- how you are going to make your investigation fair
- how you are going to make your investigation safe.

Show your plan to your teacher before going on.

Carry out

Carry out your investigation and record your results.

Present your results in an appropriate way.

Report

Write a report on your investigation.

Here are some things you should include:

- what you did
- what happened
- explain your results
- if your prediction was correct or not
- how reliable your results were
- what you could have done if you had more time.

Chapter 15 ► Investigation 15E Sound insulation

Excessive noise can damage the environment and our health. Soundproofing is one way of overcoming noise pollution.

In this investigation: you are going to find out which is the best soundproofing material; polystyrene foam, egg boxes, or newspaper.

Preparation: Predict

Finish the sentences in the box.

What I think will happen is…

I think this because…

Preparation: Plan

Write a short plan of your investigation.

Think about:

- the apparatus you are going to use
- what you are going to measure and how you are going to measure it
- how many readings you are going to take
- how you are going to record your results
- how you are going to make your investigation fair
- how you are going to make your investigation safe.

Show your plan to your teacher before going on.

Carry out

Carry out your investigation and record your results in a table.

Draw a bar graph of your results.

Report

Write a report on your investigation.

Here are some things you should include:

- what you did
- what happened
- explain your results
- if your prediction was correct or not
- what you could do to improve the investigation
- what you could have done if you had more time.

Chapter 15 ► Investigation 15D Sound insulation

Excessive noise can damage the environment and our health. Soundproofing is one way of overcoming noise pollution.

In this investigation: you are going to find out which is the best soundproofing material; polystyrene foam, egg boxes, or newspaper.

Preparation: Predict

Finish the sentence in the box.

I think the best soundproofing material out of polystyrene foam, egg boxes, or newspaper is ______________ because...

You are going to use this equipment to find out which is the best soundproofing material; polystyrene foam, egg boxes, or newspaper:

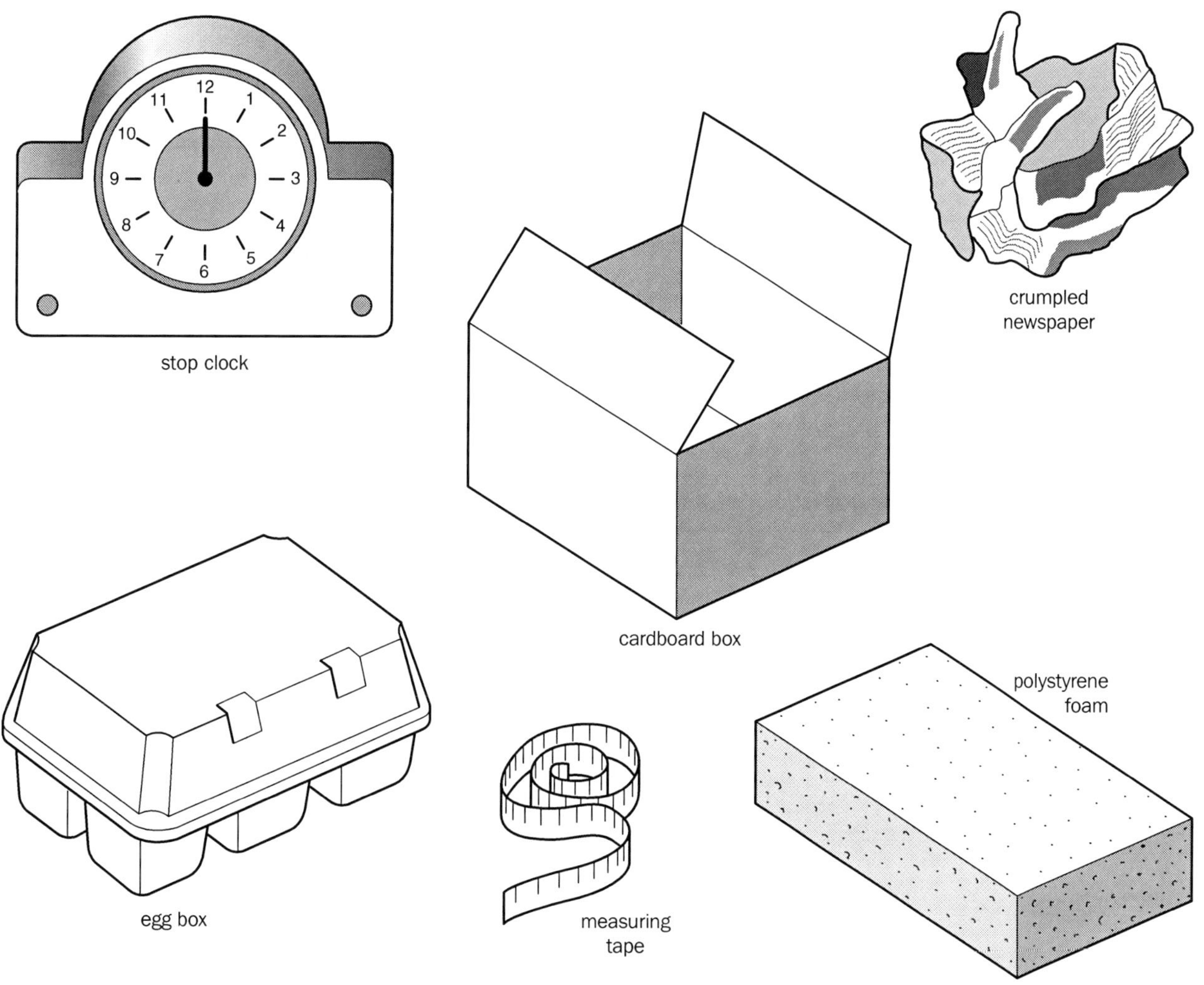

Chapter 15 ▶ Investigation 15D Sound insulation

Preparation: Plan

Finish the sentences in the box.

I will measure...

Things I will keep the same are...

My investigation will be fair because...

My investigation will be safe because...

Carry out

- Cut a piece of polythene sheet about 40 cm × 40 cm.
- Do your experiment first with no insulating material. Then try insulating the clock with polystyrene foam.
- Next, use the egg box and finally the crumpled newspaper to insulate the clock.
- Decide which is the best soundproofing material by measuring how far away you have to go before you stop hearing the clock ticking.

  Put your results in a table like this:

Material	Distance for no sound to be heard in m
Polystyrene foam	
Egg box	
Crumpled newspaper	

Draw a bar graph of your results on a piece of graph paper.

Use a key with different colours for each result. Label the axes like this:

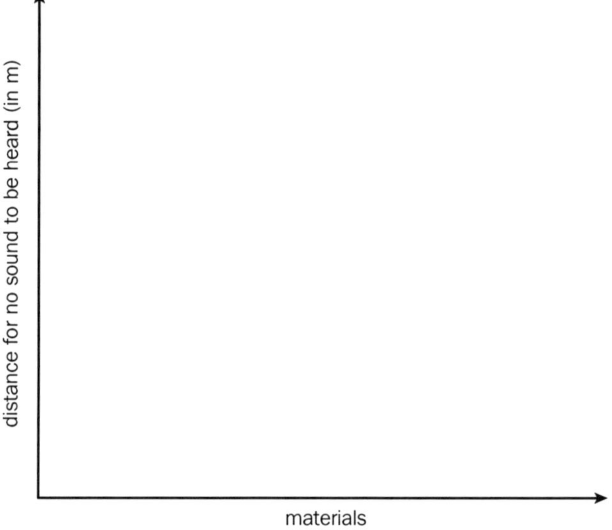

Report

Write a report on your investigation.

Here are some things you should include:

- what you did
- what happened
- explain your results
- if your prediction was correct or not
- what you could do to improve the investigation
- what you could have done if you had more time.

Chapter 15 ► Investigation 15C Sound insulation

Excessive noise can damage the environment and our health. Soundproofing is one way of overcoming noise pollution.

In this investigation: you are going to find out which is the best soundproofing material; polystyrene foam, egg boxes, or newspaper.

Preparation: Predict

Finish the sentence in the box.

I think the best soundproofing material out of polystyrene foam, egg boxes, or newspaper is ____________ because...

You are going to use this equipment to find out which is the best soundproofing material; polystyrene foam, egg boxes, or newspaper:

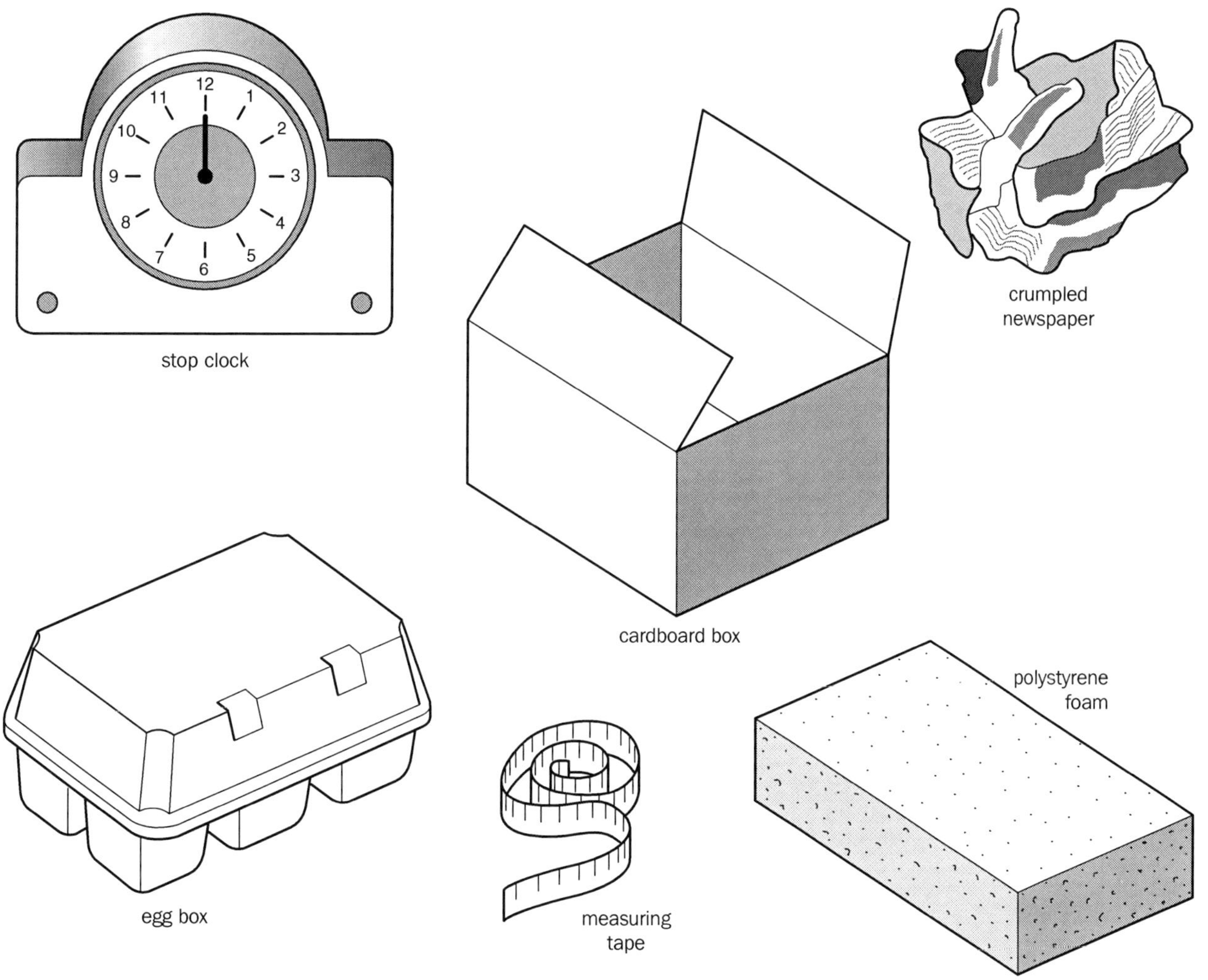

Chapter 15 ► Investigation 15C Sound insulation

Preparation: Plan

Finish the sentences in the box.

I will measure…

Things I will keep the same are…

My investigation will be fair because…

My investigation will be safe because…

Carry out

- Cut a piece of polythene sheet about 40 cm × 40 cm.
- Wind the stop clock and make sure it is ticking.
- Put the stop clock into the box and close the lid.
- Move slowly away from the box and stop when you can no longer hear the ticking.
- Measure the distance between you and the box.
- Open the box and put some polystyrene foam around the stop clock.
- Replace the lid and see how far you have to go before you can no longer hear the ticking.
- Do this again using the egg box and again with the crumpled newspaper.

  Put your results in a table like this:

Material	Distance for no sound to be heard in m
Polystyrene foam	
Egg box	
Crumpled newspaper	

Chapter 15 ► Investigation 15C Sound insulation

Draw a graph of your results on this grid.

Use different colours for each bar.

Report

Finish the sentences in the box.

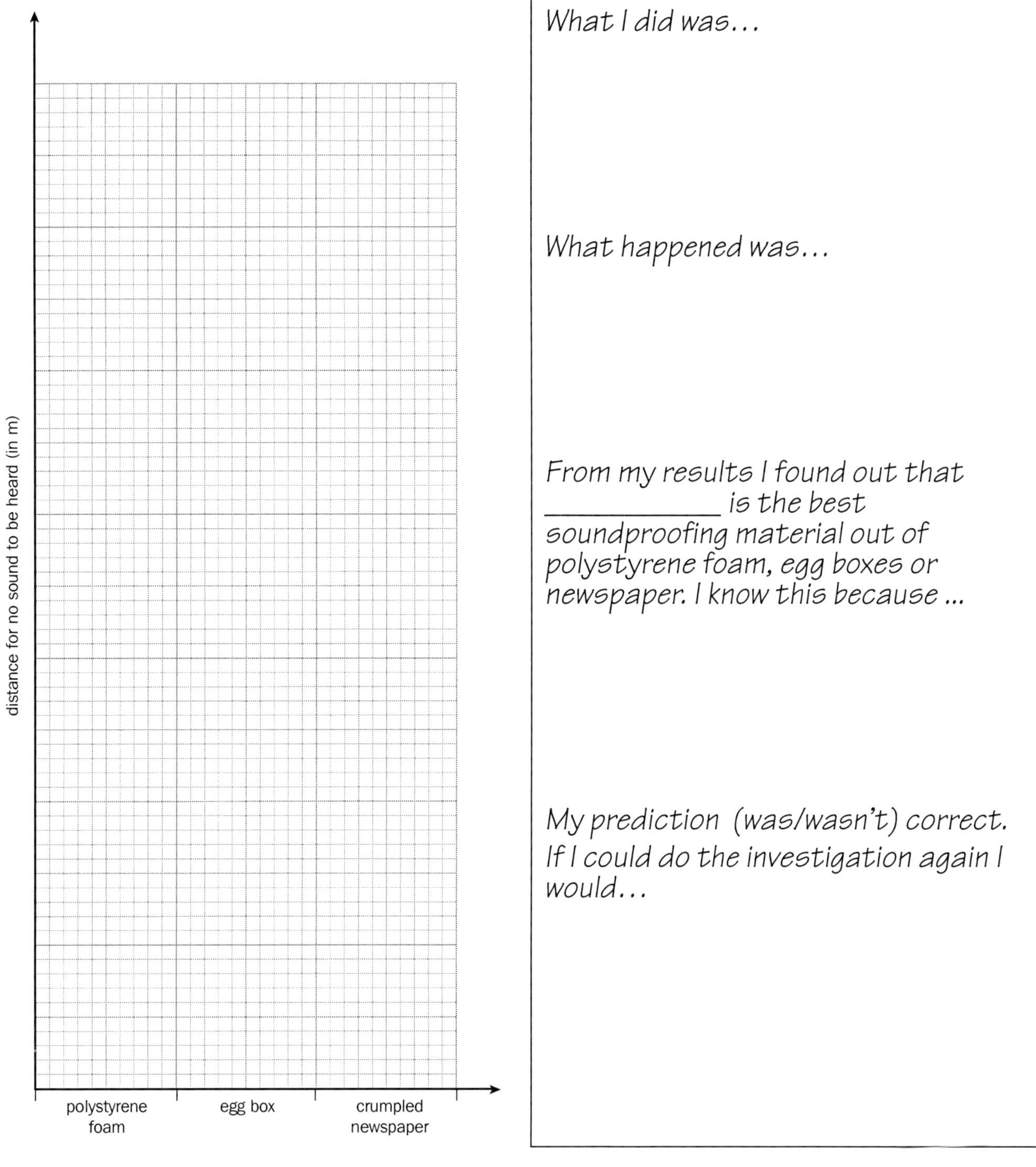
distance for no sound to be heard (in m)
polystyrene foam
egg box
crumpled newspaper
What I did was…
What happened was…
From my results I found out that ____________ is the best soundproofing material out of polystyrene foam, egg boxes or newspaper. I know this because ...
My prediction (was/wasn't) correct.
If I could do the investigation again I would…

Investigation 15
Practical notes

Sound insulation

Investigation 15
Technician's notes

Sound insulation

Each group will need:

- stop clock
- piece of polystyrene foam (can be broken up into chunks)
- crumpled newspaper (enough to pack loosely around clock)
- egg box or pieces of egg trays (enough to pack loosely around clock)
- cardboard box (big enough to hold clock and material, e.g. shoe box)
- tape measure.

Number of apparatus sets:

Number of pupils:

Number of groups:

Visual aids:

ICT resources:

Equipment/apparatus needed:

Safety notes

CLEAPSS/SSERC SAFETY REFERENCE:

Chapter 15 ► Test

White

Light and sound

1 The diagram shows a girl sitting by a window reading a book.

a Explain how the girl is able to see the book.

b Explain why the girl cannot see her reflection in the pages of the book.

c Which of these words best describes the pages of the book?

non-reflective   opaque
translucent   transparent

5 marks

2 The diagram shows a single ray of light being reflected from a plane mirror.

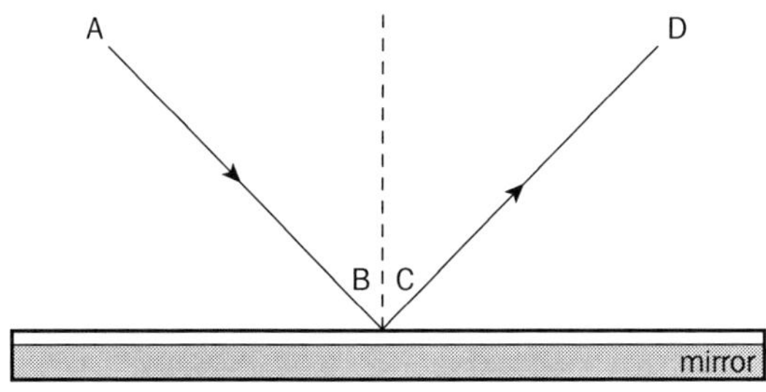

These labels are missing from the diagram:

angle of incidence   angle of reflection
incident ray   reflected ray

a Which label goes at:
i A
ii B
iii C
iv D?

b If the angle of incidence is 55°, what is the angle of reflection?

5 marks

3 Copy these diagrams.

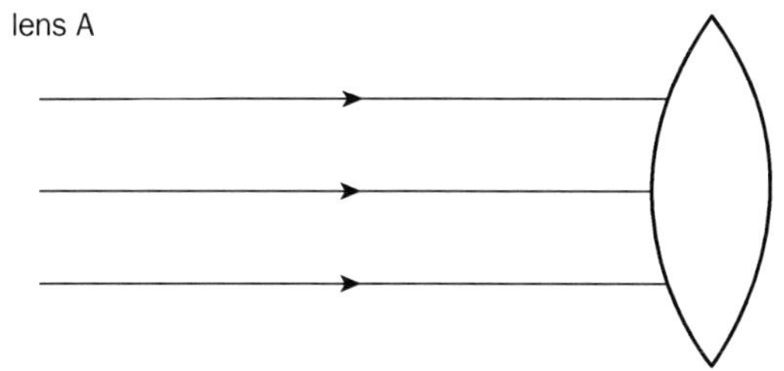

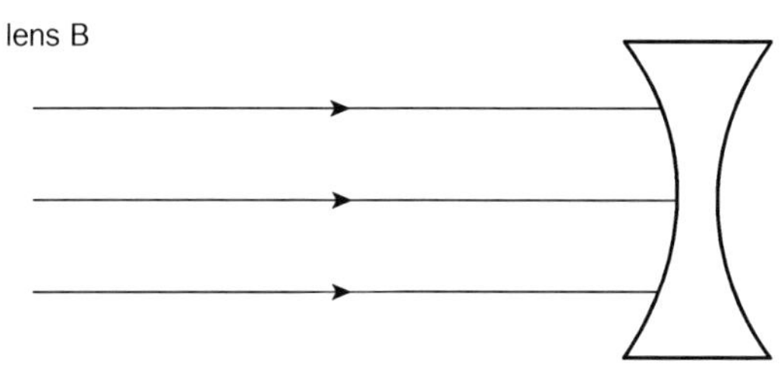

a Draw lines through the lenses to finish the diagrams.

b Imagine you are holding a paper clip close to each of these lenses. How would a paper clip look if you saw it through
i lens A
ii lens B?

c Give one example of where lens B might be used.

d Lens A can be used to focus an image on a screen. Give two ways in which the image on the screen is different from the object.

9 marks

Light and sound

4 Sound is produced when something vibrates.

a What vibrates when these musical instruments are played:
i a triangle
ii a guitar
iii a piano?

b Give one way in which you can change the volume of a triangle.

c Give two ways in which you can change the pitch of a guitar string.

d Explain why the sound from a piano is louder and clearer when the lid is open.

8 marks

5 The following sentences explain how the sound from a drum travels to your ears. They are in the wrong order. Write the letter for each sentence in the correct order starting with **C**.

A The vibrating drum skin makes air molecules vibrate backwards and forwards.

B Moving air molecules enter the ear and make the ear drum vibrate.

C The drummer strikes the drum with a drum stick.

D The sound spreads out as moving air molecules affect the molecules next to them.

E The drum skin vibrates up and down.

3 marks

Chapter 15 ► Test

Blue

Light and sound

1 The diagram shows a single ray of light passing through a glass block.

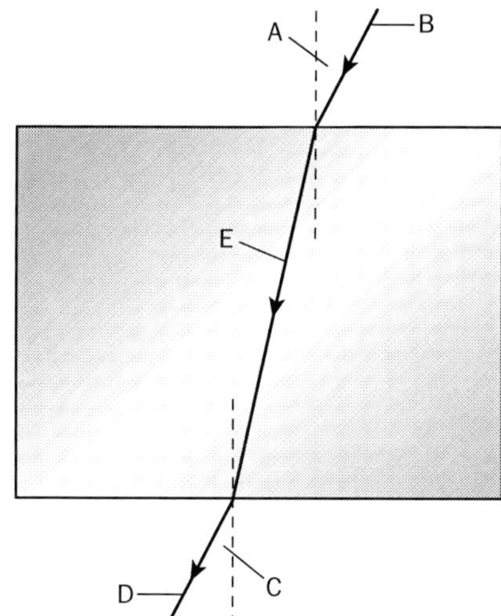

These labels are missing from the diagram:

angle of incidence
angle of refraction
emerging ray
incoming ray
refracted ray

a Which label goes at:
 i A
 ii B
 iii C
 iv D
 v E?

b What do the incoming ray and emerging ray have in common?

6 marks

2 The diagram shows what happens to a ray of white light when it passes through a triangular prism and on to a screen.

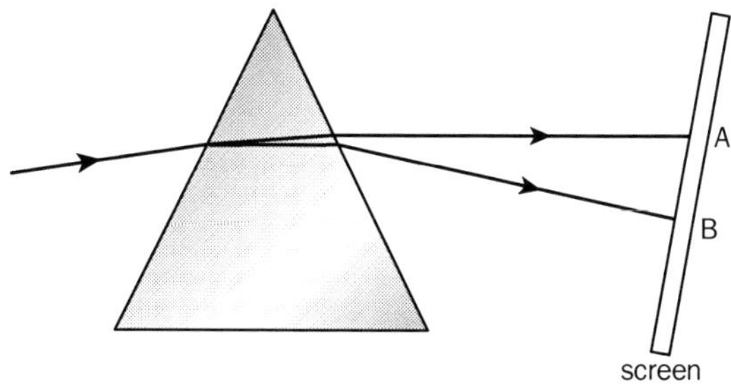

a Name the seven colours seen on the screen.

b What colour is at
 i A
 ii B?

5 marks

3 Coloured filters absorb some colours and allow other colours to pass through. The diagram shows rays of red, blue, and green light shining on two coloured filters.

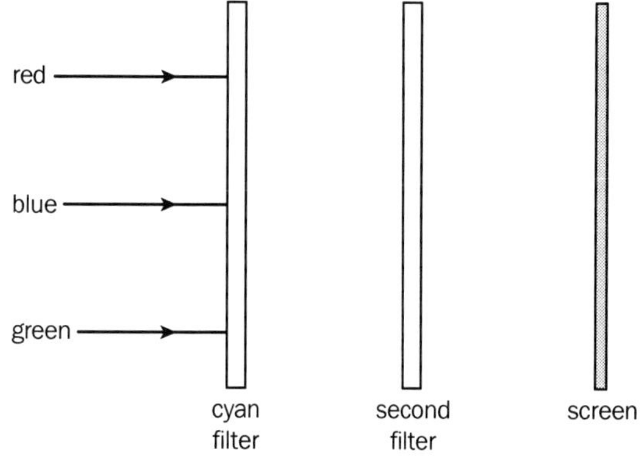

a Which rays of light pass through the cyan filter?

b If the second filter is red, what colours appear on the screen?

3 marks

4 What colours do the following surfaces appear:

a a red surface in white light?

b a red surface in red light?

c a red bus in yellow street lights?

d a blue car in yellow street lights?

4 marks

Chapter 15 ▶ Test
Light and sound

5 The table shows the level of noise produced in various situations.

Situation	Noise level in dB
A quiet library	20
An English lesson in a classroom	40
A washing machine working in a kitchen	60
Mowing a lawn with a motor mower	80
A Technology lesson in a workshop	100
A rock concert	120

a What does dB stand for?

b Explain why a Technology lesson is noisier than an English lesson.

c Which situations could cause deafness?

d **i** How could someone working in a noisy environment avoid permanent damage to their ears?
ii Explain how this would work.

e Modern hearing aids can be of great help to deaf people. What do hearing aids do to the sounds entering the ear of a deaf person?

6 marks

6 The diagrams show four traces of sound waves on an oscilloscope.

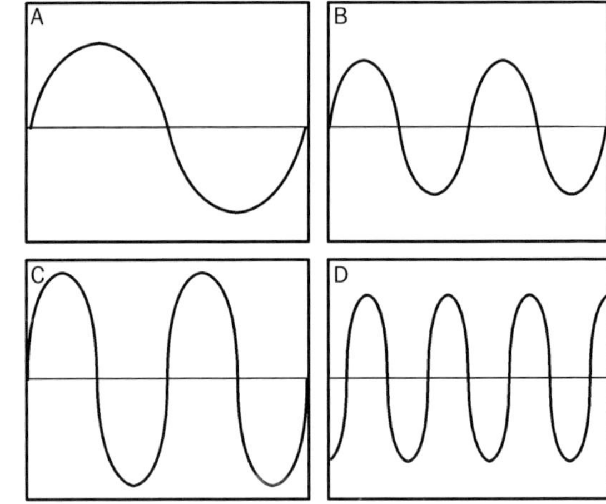

a Which sound is the loudest?

b Which sound has the highest pitch?

c Which sound has the smallest amplitude?

d Which sound has the highest frequency?

e Which trace is most likely to have been produced by
i gently plucking a thin, tight guitar string
ii banging a bass drum hard?

6 marks

Chapter 15 ► Mark scheme

White

Light and sound

Question	Answer	Marks	Level
1 a	light reflected (1) from book (to eye) (1)	2	C
b	paper surface rough (1) light reflected in all directions (1)	2	C
c	opaque	1	C
		5	
2a i	incident ray	1	
ii	angle of incidence	1	
iii	angle of reflection	1	
iv	reflected ray	1	C
b	55°	1	C
		5	
3 a	Lens A: continuous line through centre of lens (1) rays converging after passing through (1)	2	
	Lens B: continuous line through centre of lens (1) rays diverging after passing through (1)	2	D
b i	magnified/bigger	1	
ii	smaller/reduced	1	D
c	any correct use, e.g. spectacles	1	D
d	upside down (1) smaller (1)	2	D
		9	

Question	Answer	Marks	Level
4 a i	metal	1	
ii	string	1	
iii	string	1	
b	hit it harder	1	D
c	tighten string (1) shorten string (1)	2	D
d	vibrations travel better (1) through air than solid (1)	2	D
		8	
5	C E A D B (all in correct order = 3, one wrong = 2, two wrong = 1)	3	C
		3	
	TOTAL 30 marks		

Suggested grade/level boundaries

C = 14/30

D = 23/30

Chapter 15 ► Mark scheme

Light and sound

Question	Answer	Marks	Level
1 a i	angle of incidence	1	
ii	incoming ray	1	
iii	angle of refraction	1	
iv	emerging ray	1	
v	refracted ray	1	E
b	they are parallel	1	E
		6	
2 a	red, orange, yellow, green, blue, indigo, violet	3	E
	(all correct = 3 marks; deduct 1 mark for each error)		
b i	red	1	
ii	violet (allow blue or indigo)	1	E
		5	
3 a	blue (1) and green (1)	2	F
b	nothing	1	F
		3	
4 a	red	1	F
b	red	1	F
c	red	1	F
d	black	1	F
		4	

Question	Answer	Marks	Level
5 a	decibels	1	E
b	any sensible reason, e.g. no carpet/noisy equipment	1	E
c	any over 80–85 dB, i.e. bottom three in table	1	E
d i	wear ear protectors/muffs /plugs, etc.	1	
ii	(material) absorbs sound	1	E
e	amplify sound/make sounds louder	1	E
		6	
6 a	C	1	F
b	D	1	F
c	B	1	F
d	D	1	F
e i	D	1	
ii	A	1	F
		6	
	TOTAL 30 marks		

Suggested grade/level boundaries

E = 15/30

F = 24/30

16 Microorganisms/biotechnology

This chapter introduces pupils to the world of microorganisms. Useful and harmful microbes are considered together with their impact on individuals and the world as a whole. The principles of modern biotechnology are outlined. Work on genetics is developed from Book 1 with examples of the roles of genes and chromosomes in monohybrid inheritance. The nature of genes and chromosomes leads on briefly to the structure of DNA.

Assessment opportunities

Formative assessment opportunities are provided by worksheets, homework sheets, and an investigation.

The **worksheets** cover material at levels E and F for attainment targets for knowledge and understanding. Teachers may wish to use these worksheets not only as part of practical activities but also to provide evidence of pupil achievement.

Worksheet	Level
16.1a	E
16.1b	E
16.2a	E/F
16.2b	E/F
16.2c	F
16.3	F

The **homework sheets** cover material at levels E and F for attainment targets for knowledge and understanding. These homework sheets can be used individually as a follow-up to work done in class or assembled into a homework booklet allied closely to schemes of work.

Homework sheet	Level
16.1	E
16.2	F
16.3a	F
16.3b	F

The **investigation** covers all three skill areas at levels C, D, E, and F. It is written in a way that allows for pupils to be assessed in all three skill areas at one level. Alternatively, customised assessments can be constructed enabling pupils to be assessed at different levels in all three skills. The latter approach is more time consuming, but it does provide the opportunity for pupils to show evidence of achievement at different levels in different skills in the same investigation. Teachers will need to use their professional judgement when deciding which level is appropriate to individual pupils. It is envisaged that pupils will show progression through the levels as they work through their science course.

A single **summative test** is provided at levels E and F only. This is because Chapter 16 only covers material at this level. The test has a total of 30 marks and will take about 30 minutes for pupils to complete, although this can be varied depending on pupil ability. A mark scheme is provided together with suggested grade/level boundaries. It is envisaged that this test will be given to pupils on completion of the material covered in Chapter 16.

ICT opportunities

The use of data loggers/remote sensors can extend the range, speed, and sensitivity of measurements in many of the worksheets for this chapter. Once downloaded onto a PC, data-handling programs can be used to analyse information gathered, data can be manipulated, and appropriate graphs etc. presented. The Internet provides pupils with access to a huge range of scientific information. A list of suitable websites is included in this Teacher's Guide.

Students' book chapter 16 contents and guide levels

16.1	What are microorganisms?	*Starting off 1*	F
	Useful microbes (1)	*Starting off 2*	F
	Useful microbes (2)	*Going further*	F
	Microbes and biotechnology	*For the enthusiast*	F
16.2	Harmful microbes	*Starting off*	F
	Defences against disease	*Going further*	F
16.3	Chromosomes and genes	*Starting off*	F
	Selective breeding	*Going further 1*	F
	Cloning	*Going further 2*	F
	DNA	*For the enthusiast*	F

16.1a Making yoghurt

W/S

Name: Date: Group:

What you need:

Clean beaker, teaspoon, fresh natural yoghurt, fresh milk, plastic pot with lid, thermometer, Bunsen burner, tripod, gauze, heatproof mat.

What to do:

1 Put 150 cm^3 of milk into the beaker and heat the milk until it boils. Boiling kills bacteria already in the milk.

2 Let the milk cool to about 37 °C. While you are waiting for the milk to cool, rinse the plastic container and the spoon with boiling water. Take care not to get burned by the hot water. Dry them with a clean tissue or towel.

3 Put the warm milk into the empty plastic pot.

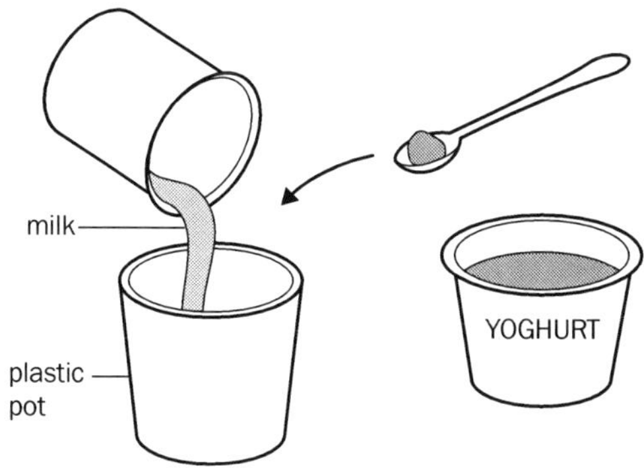

4 Add 1 teaspoonful of yoghurt and stir it into the warm milk. This adds yoghurt bacteria to the milk.

5 Put the lid on the plastic pot and put it in a warm place for 6 – 8 hours.

6 If you have been careful and used only clean equipment you may be able to taste your home-made yoghurt. Check with your teacher first.

Try adding some fruit, nuts or chocolate to flavour your yoghurt.

SAFETY WARNING

Take care using hot water.

Use only sterile equipment.

Taste nothing until you have checked with your teacher first.

16.1a
Practical notes

Making yoghurt

Sterilised milk could be used as an alternative to boiling fresh milk, although pupils will then miss a key stage of the process, i.e. killing the naturally occurring bacteria in milk. If the pupils are to be allowed to taste the yoghurt, teachers must check health and safety guidance. It may be that the activity will have to be carried out in a food preparation area, e.g. the food technology room.

16.1a
Technician's notes

Making yoghurt

Each group will need:

- a copy of worksheet 16.1a
- clean 250 cm^3 beaker
- teaspoon
- fresh natural yoghurt; organic usually gives good results
- fresh milk
- thermometer
- Bunsen burner
- tripod
- gauze
- heatproof mat
- access to kettle of boiling water
- oven gloves
- access to oven on low setting (about 40 °C) or warm cupboard, e.g. drying cabinet.

Number of apparatus sets:

Number of pupils:

Number of groups:

Visual aids:

ICT resources:

Equipment/apparatus needed:

Safety notes

- **Pupils must be told to wash their hands before starting the activity and reminded of the importance of working in a hygienic manner.**
- **Tasting of yoghurt is at the teacher's discretion after he/she has checked appropriate regulations.**

CLEAPSS/SSERC SAFETY REFERENCE:

16.1b Making bread

W/S

Name: **Date:** **Group:**

What you need:

Beaker, clean bowl, plain flour, sugar, dried yeast, warm water, tablespoon, teaspoon, plastic bag, cooking oil, baking tray, access to an oven, oven gloves.

What to do:

1 Put about 25 cm^3 of warm water into the beaker and stir in one teaspoon of sugar until it dissolves. Add half a teaspoon of dried yeast to the sugar solution and leave it for about 10 minutes.

2 Put four tablespoons of plain flour in the bowl. When the yeast/sugar mixture has become frothy, pour it into the flour and stir it in. Use your (clean) fingers to kneed the dough until it is well mixed.

3 Rub a few drops of cooking oil inside the plastic bag and put the dough inside. Tie the neck of the bag and put it in a warm place so the dough will rise.

4 When the dough has risen to about twice its original volume, remove it from the bag and put it on to a baking tray. (Some flour on the tray will help stop the bread from sticking.)

5 Put the dough into a hot oven (220 °C or Gas mark 7) for about 15 minutes.

6 Use oven gloves and carefully remove the bread from the oven. Put it on a heatproof surface and leave it to cool down.

7 When it is cool, tear your bread roll open and look at its texture. Look for the holes where the yeast cells produced carbon dioxide gas.

Check with your teacher to see if you are allowed to taste your bread.

What else could you have added to your bread to improve the flavour?

SAFETY WARNING

Use only sterile equipment.

Take care using a hot oven, use oven gloves.

Taste nothing until you have checked with your teacher first.

16.1b
Practical notes

Making bread

There are numerous 'simple' methods for making bread, this one has worked every time so far! If the pupils are to be allowed to taste the bread, teachers must check health and safety guidance. It may be that the activity will have to be carried out in a food preparation area, e.g. the food technology room. Remind pupils to take care when putting bread into and taking it out of a hot oven, use oven gloves.

16.1b
Technician's notes

Making bread

Each group will need:

- a copy of worksheet 16.1b
- clean 250 cm^3 beaker
- clean mixing bowl
- access to plain flour or issue each group with four tablespoonfuls
- sugar
- access to dried yeast or issue each group with half a teaspoonful
- access to kettle of warm water
- tablespoon
- teaspoon
- clean plastic bag (sandwich bags with ties are good for this)
- access to cooking oil (groups need only a few drops to oil the inside of the bag)
- baking tray
- access to oven set at 220 °C
- oven gloves.

Number of apparatus sets:

Number of pupils:

Number of groups:

Visual aids:

ICT resources:

Equipment/apparatus needed:

Safety notes

- **Pupils must be told to wash their hands before starting the activity and reminded of the importance of working in a hygienic manner.**
- **Tasting of bread is at the teacher's discretion after he/she has checked appropriate regulations.**

CLEAPSS/SSERC SAFETY REFERENCE:

16.2a Growing fungi

W/S

Name: Date: Group:

What you need:

Dish, beaker or jar, stale bread, water, label.

What to do:

1 Put the piece of stale bread on to the dish and moisten it with water. Be careful not to make the bread too wet.

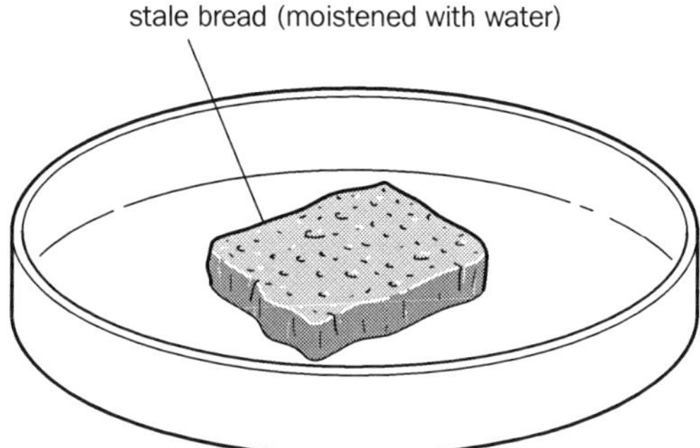

2 Label the dish with your name and the date and put it in an open, well-ventilated place. Leave the dish for about 24 hours.

3 Moisten the bread again if it has dried out and then cover it with the jar. Put the equipment in a safe place for a few days.

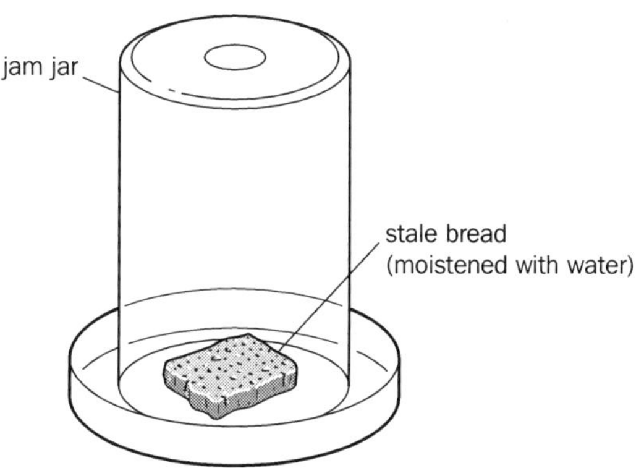

4 After a few days have a look at your bread, you should see colonies of mould fungi growing on it.

Try looking at the mould using a hand lens or a microscope set on lower power.

What does the mould look like?

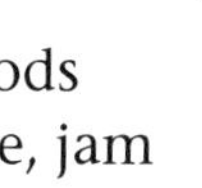

5 Try growing mould on different foods such as squashed banana, stale cake, jam or apple sauce.

- Which food does mould fungi like best?
- How can you tell?

SAFETY WARNING

***Wash your hands after handling dishes of fungi.**

***Taste nothing during this activity.**

16.2a
Practical notes

Growing fungi

This is a simple yet effective activity that enables pupils to see the structure of a mould fungus. While appearing 'furry', closer inspection will show the mould consists of hundreds of microscopic fruiting bodies (they look a bit like pins sticking out of a pin cushion). These fruiting bodies produce the millions of tiny spores by which fungi reproduce. Check with health and safety advice for safe disposal of fungi. Watch out for silliness with food samples especially the soggy ones.

16.2a
Technician's notes

Growing fungi

Each group will need:

- a copy of worksheet 16.2a
- dish (old plastic petri dish lid or base)
- beaker or glass jar to cover bread on dish
- supply of stale bread/cake, squashed banana, jam, etc.
- label
- access to warm water, soap and tissues/towel.

Number of apparatus sets:

Number of pupils:

Number of groups:

Visual aids:

ICT resources:

Equipment/apparatus needed:

Safety notes

- **Remind pupils to taste nothing during the activity.**
- **Make sure pupils wash their hands after the activity.**
- **Check health and safety advice for safe disposal of fungi**

CLEAPSS/SSERC SAFETY REFERENCE:

16.2b Growing bacteria

W/S

Name: **Date:** **Group:**

What you need:

Dish of agar, sticky tape, label.

What to do:

1 Wash your hands.

2 Agar is a jelly containing food that bacteria like.

Collect a dish of agar.
Do not take the lid off.

Take the dish of agar and put it in one of the following places:

- on a window sill in the classroom
- outside in the open air
- on the floor
- in a cupboard
- in a drawer.

Check with your teacher that your choice is OK before continuing.

3 Remove the lid from the dish. Try not to touch the jelly or breathe on it.

Leave the lid off the dish for a few minutes. Keep a close watch on it to make sure no one else goes near it.

4 After a few minutes put the lid back on the dish and seal it with sticky tape. Write your name, the date, and where you put the dish on a label. Stick the label on the bottom of the dish.

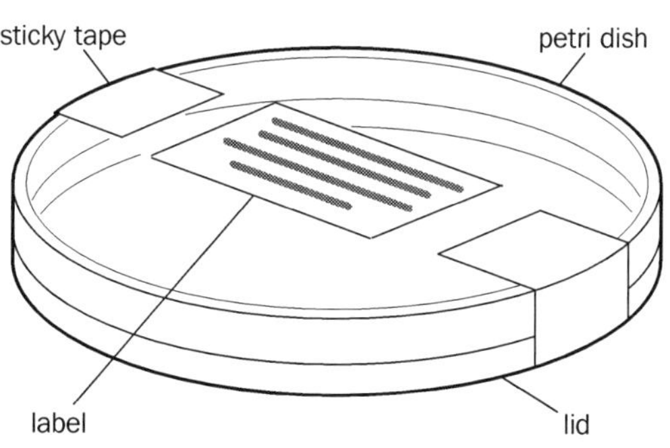

5 Put the dish in a warm place and leave it for a few days.

6 After a few days, look at your dish. Use a hand lens to get a closer look.

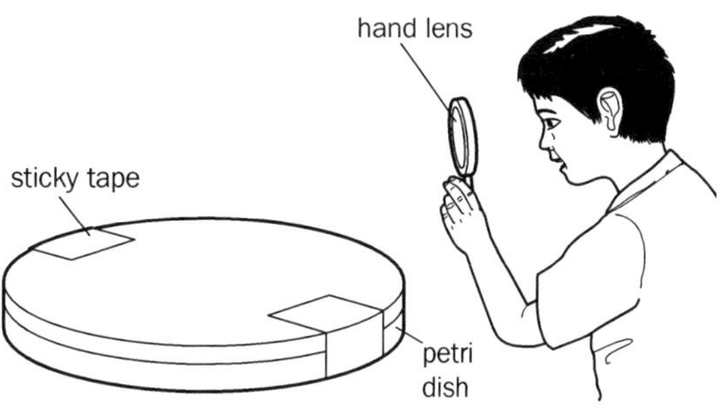

What do you see?

Colonies of bacteria look like smooth, round blobs (fungi have furry edges). The colonies have grown from single bacteria that landed on your dish when the lid was off. Different bacteria are different colours.

- How many colonies have you got on your dish?
- How many different colonies have you got on your dish?

?

7 Compare your dish with others in your class.

Which place produced the most bacteria?

SAFETY WARNING

Do not touch or breathe on the jelly.

Once sealed, do not open the dish again, some of the bacteria may be very dangerous.

Follow instructions for the disposal of dishes.

16.2b
Practical notes

Growing bacteria

Despite the inherent risks involved with growing bacteria, if pupils follow instructions and take the recommended precautions, there is no reason why they should not attempt this activity. The culture method can give rise to dense colonies of potentially harmful bacteria, hence the warning not to open the dishes once they have been sealed. Put the dishes upside-down in an incubator/oven. This stops any condensation dropping onto the agar. Arrangements for the safe disposal of dishes must be made and adhered to by all concerned.

16.2b
Technician's notes

Growing bacteria

Each group will need:

Number of apparatus sets:

Number of pupils:

Number of groups:

Visual aids:

ICT resources:

Equipment/apparatus needed:

- a copy of worksheet 16.2b
- dish of nutrient agar, mixed as per instructions and using new sterile dishes
- sticky tape
- label
- access to warm water, soap and tissues/towel.

Safety notes

- **Make sure pupils wash their hands before and after the activity.**
- **Make sure dishes are sealed before incubation and are kept sealed up to and during disposal.**
- **Check health and safety advice for safe disposal of bacterial plates.**

CLEAPSS/SSERC SAFETY REFERENCE:

16.2c What does mould fungus feed on? W/S

Name: Date: Group:

What you need:

Dish of starch agar, bit of mould fungus, tweezers, iodine solution, label.

What to do:

1 Wash your hands

2 Starch agar is a jelly that contains starch.

Collect a dish of starch agar. **Do not take the lid off.** Write your name and the date on a label and stick it on to the lid of the dish.

3 Use tweezers to remove a small piece of mould fungi from a fungal colony. When you are ready, lift the lid of your dish just enough to let you smear the piece of mould over the jelly.

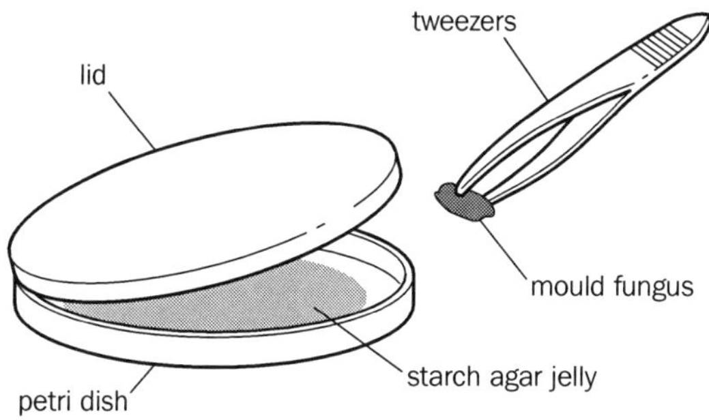

Replace the lid and leave the dish for a few days.

4 After a few days you should be able to see fungus growing along the smear lines.

Remove the lid and pour iodine solution over the starch agar; make sure it is all covered.

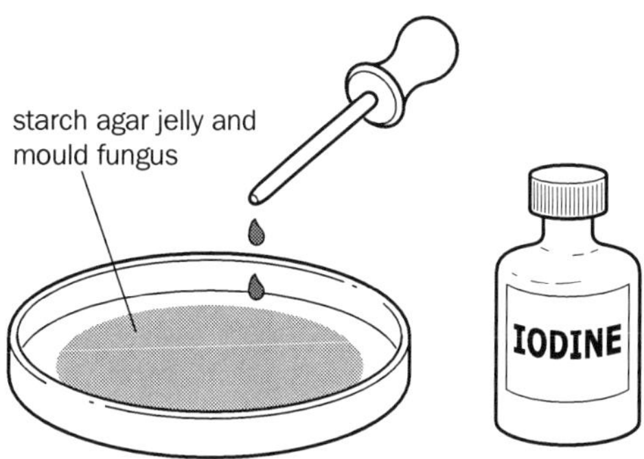

Leave the iodine for a few minutes then rinse it off with water.

5 Iodine solution turns starch blue/black. Brown areas on the agar have no starch in them.

Where do you think the starch has gone?

6 Wash your hands.

SAFETY WARNING

Wash your hands before and after this activity.

16.2c
Practical notes

What does mould fungus feed on?

This activity is a useful follow on to 'Growing fungi' and also links nicely with work on enzymes. Pupils may need some help with inoculating their plates, as it is very easy to cut into the surface of the agar when smearing with tweezers. Check with health and safety advice for safe disposal of fungi.

16.2c
Technician's notes

What does mould fungus feed on?

Number of apparatus sets:

....

Number of pupils:

Number of groups:

Visual aids:

....

....

ICT resources:

....

....

Equipment/apparatus needed:

....

....

Each group will need:

- a copy of worksheet 16.2c
- dish of starch agar; make a paste using 0.3 g starch and a little cold water. Add this to 1 g of agar and then to 100 cm of water and boil, stirring until mixed. Cover until cool then pour into sterile petri dishes
- access to mould fungus
- tweezers
- iodine solution
- label
- access to warm water, soap and tissues/towel.

Safety notes

- **Make sure pupils wash their hands before and after the activity.**
- **Check health and safety advice for safe disposal of fungi.**

CLEAPSS/SSERC SAFETY REFERENCE:

16.3 Hair colour, genes and beads W/S

Name: Date: Group:

What you need:

Two bags, three beakers labelled A, B and C, 50 black beads, 50 white beads.

What to do:

The black beads represent the gene allele for black hair.

The white beads represent the gene allele for blond hair.

1 Put 25 black beads in one bag. This bag represents a woman who only has the gene allele for black hair.

2 Put 25 white beads in the other bag. This bag represents a man who only has the gene allele for blond hair.

3 Pour the beads from one bag into the other and mix the beads up well. This bag represents a child produced by the woman with black hair and the man with blond hair.

If the gene allele for black hair is dominant, what colour hair will the child have?

4 Mix the remaining 25 black beads and 25 white beads in the empty bag. You now have two 'people' with the same mixture of gene alleles.

Let's see what happens when these two 'people' produce children.

5 Close your eyes and pick one bead from each bag.

- If you pick two black beads, put them in beaker A.
- If you pick one black bead and one white bead, put them in beaker B.
- If you pick two white beads, put them in beaker C.

Carry on doing this until the bags are empty.

6 Count the number of beads in each of the beakers. Write your results in your book.

- Which beaker has most beads?
- Explain why there is a greater chance of this couple having a child with black hair rather than blond hair.

16.3
Practical notes

Hair colour, genes and beads

This activity aims to help pupils understand the mechanism of monohybrid inheritance by using coloured beads to represent gene alleles. The example of a cross between a woman with black hair and a man with blond hair is the same as that shown in *Starting Science for Scotland Book 2* (p75). The beads are pulled at random from two bags representing 'parents' to show the possible genetic outcomes in their offspring. The activity shows how models can be used to predict statistically the results of genetic crosses. Depending on ability, class results could be collated and percentages calculated. The more data collected, the closer the results will be to the theoretical prediction of 25 per cent white beads, 50 per cent black and white beads and 25 per cent black beads.

16.3
Technician's notes

Hair colour, genes and beads

Each group will need:

- a copy of worksheet 16.3
- two bags, preferably cloth or other opaque material
- three beakers or plastic containers labelled A, B, and C
- 50 black beads
- 50 white beads, or any suitable alternative.

Number of apparatus sets:

Number of pupils:

Number of groups:

Visual aids:

ICT resources:

Equipment/apparatus needed:

Safety notes

- **Watch for silly behaviour with beads. Count them out and count them in.**

CLEAPSS/SSERC SAFETY REFERENCE

16.1 Useful microbes H/W

Name: Date: Group:

What you need to know ...

Microbes are very tiny living things. The three main types of microbes are bacteria, viruses, and microscopic fungi. Microbes can be useful. Some microbes are used to make food. Other microbes feed on the remains of dead animals and plants causing them to decompose. Materials that can be decomposed by microbes are called biodegradable materials.

What to do:

1 The diagrams show some microbes. The diagrams are labelled A, B, C, and D.

A

B

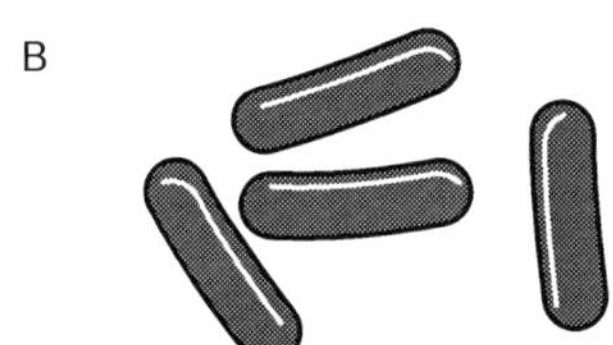

C

D

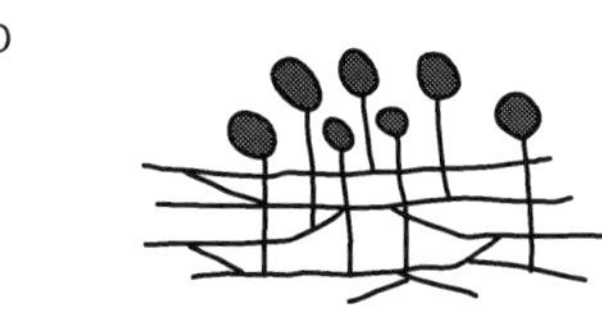

Which diagram shows

- bacteria
- mould fungus
- a virus
- yeast cells?

2 Name **two** foods that are made using bacteria and **two** that are made using fungi.

3 Which of these things are biodegradable and which are non-biodegradable?

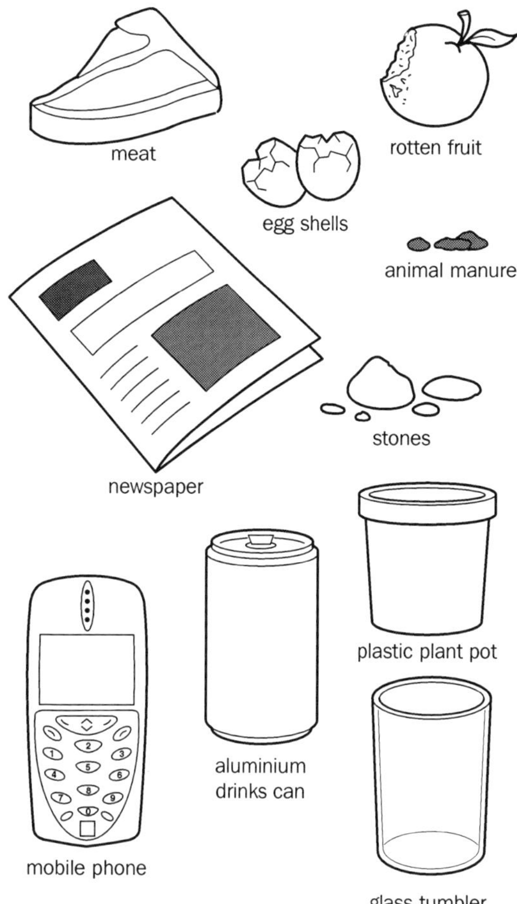

Make two lists in your book. Head one list 'Biodegradable' and the other 'Non-biodegradable'.

HANDY HINTS

Remember, biodegradable things come from living things.

16.2 Harmful microbes

H/W

Name: Date: Group:

What you need to know …

Microbes are very tiny living things. The three main types of microbes are bacteria, viruses, and microscopic fungi. Microbes can be harmful, they can cause diseases.

What to do:

1 Copy and complete this table about microbes and diseases:

Microbe	Disease
fungus	
	cold
virus	
	pneumonia
	thrush
bacteria	

2 Give two ways that microbes can get into the body.

3 Explain how your body defends itself against microbes that get inside you.

4 Suggest why make-up should be washed off every night.

HANDY HINTS

Make-up is usually oil-based to make it stick to the skin.

16.3a Matching chromosomes

H/W

Name: Date: Group:

What you need to know ...

There are 46 chromosomes inside the nucleus of every one of your body cells. The chromosomes are paired off, making 23 pairs altogether. The diagram shows the chromosomes from a human body cell. They are jumbled up, just as they would be inside a cell nucleus.

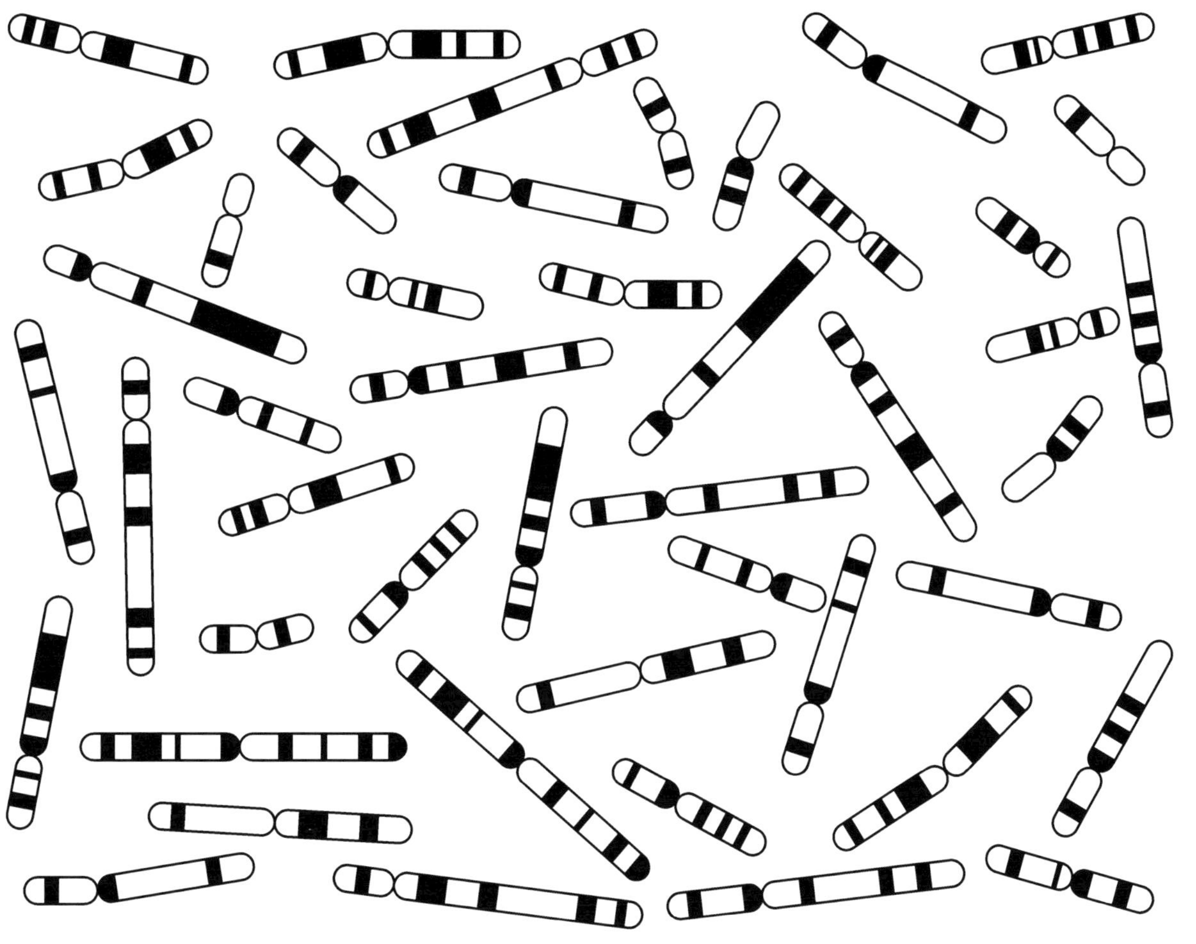

What to do:

1 Cut out the chromosomes, check to see you haven't lost any.

2 Match the pairs of chromosomes.

3 Stick the pairs of chromosomes into your book. Start with the largest pair and end with the smallest.

HANDY HINTS

Look carefully at the length of each chromosome and at their patterns.

16.3b Tall peas and short peas

H/W

Name: **Date:** **Group:**

What you need to know ...

Genes are always in pairs, one coming from each parent. The genes in a pair may carry the same message or one gene may carry a different message from the other. These are called gene alleles; different forms of the same gene. Humans have been selectively breeding animals and plants long before we knew about chromosomes and genes.

What to do:

1 Copy the diagram below. It shows a cross between two tall pea plants. **T** stands for the gene allele for tall pea plants and **t** stands for the gene allele for short pea plants.

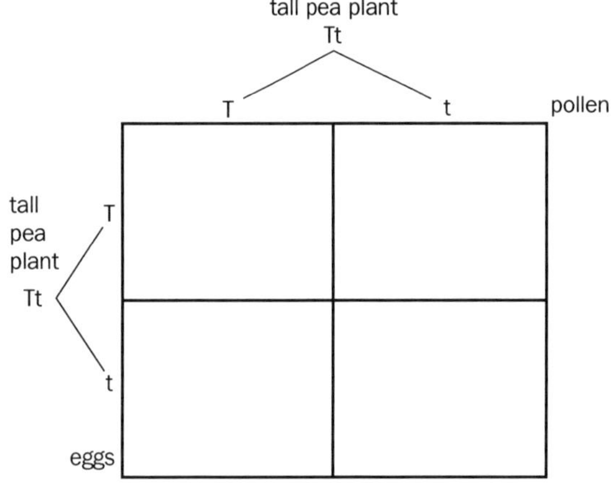

a Finish the diagram by filling in the boxes.

b Which is the dominant gene allele?

c What do we call the other gene allele?

d Which offspring have two gene alleles the same?

e Which offspring could eventually produce
  i more tall pea plants
  ii more short pea plants?

f Explain why you cannot be sure exactly what gene alleles a tall pea plant has.

2 A farmer wants to grow all tall pea plants because he can harvest the crop more easily. He knows nothing about chromosomes and genes. Explain how the farmer can use selective breeding to produce tall pea plants.

HANDY HINTS

Make sure you follow the lines and fill in the boxes correctly.

Chapter 16 ► Investigation 16F
Sour milk

Milk contains microbes. These microbes turn milk sour.
The speed at which microbes turn milk sour depends on where the milk is kept.

In this investigation: you are going to find out which conditions cause milk to go sour fastest.

Preparation: Predict

Finish the sentences in the box.

What I think will happen is...

I think this because...

Preparation: Plan

Write a short plan of your investigation.

Think about:

- the apparatus you are going to use
- how one variable depends upon another variable
- what you are going to measure and how you are going to measure it
- how many readings you are going to take
- how you are going to record your results
- how you are going to make your investigation fair
- how you are going to make your investigation safe.

Show your plan to your teacher before going on.

Carry out

Carry out your investigation and record your results.

Present your results in an appropriate way.

Report

Write a report on your investigation.

Here are some things you should include:

- what you did
- what happened
- explain your results
- if your prediction was correct or not
- how reliable your results were
- what you could have done if you had more time.

Chapter 16 ► Investigation 16E Sour milk

Milk contains microbes. These microbes turn milk sour.
The speed at which microbes turn milk sour depends on where the milk is kept.

In this investigation: you are going to find out which conditions cause milk to go sour fastest.

Preparation: Predict

Finish the sentences in the box.

What I think will happen is…

I think this because…

Preparation: Plan

Write a short plan of your investigation.

Think about:

- the apparatus you are going to use
- what you are going to measure and how you are going to measure it
- how many readings you are going to take
- how you are going to record your results
- how you are going to make your investigation fair
- how you are going to make your investigation safe.

Show your plan to your teacher before going on.

Carry out

Carry out your investigation and record your results in a table.

Draw a bar graph of your results.

Report

Write a report on your investigation.

Here are some things you should include:

- what you did
- what happened
- explain your results
- if your prediction was correct or not
- what you could do to improve the investigation
- what you could have done if you had more time.

Chapter 16 ► Investigation 16D Sour milk

Milk contains microbes. These microbes turn milk sour.
The speed at which microbes turn milk sour depends on where the milk is kept.

In this investigation: you are going to find out which conditions cause milk to go sour fastest.

Preparation: Predict

Finish the sentence in the box.

I think milk will go sour fastest when it is kept in a (refrigerator/classroom/warm oven/hot oven) because…

You are going to use this equipment to find out which conditions cause milk to go sour fastest:

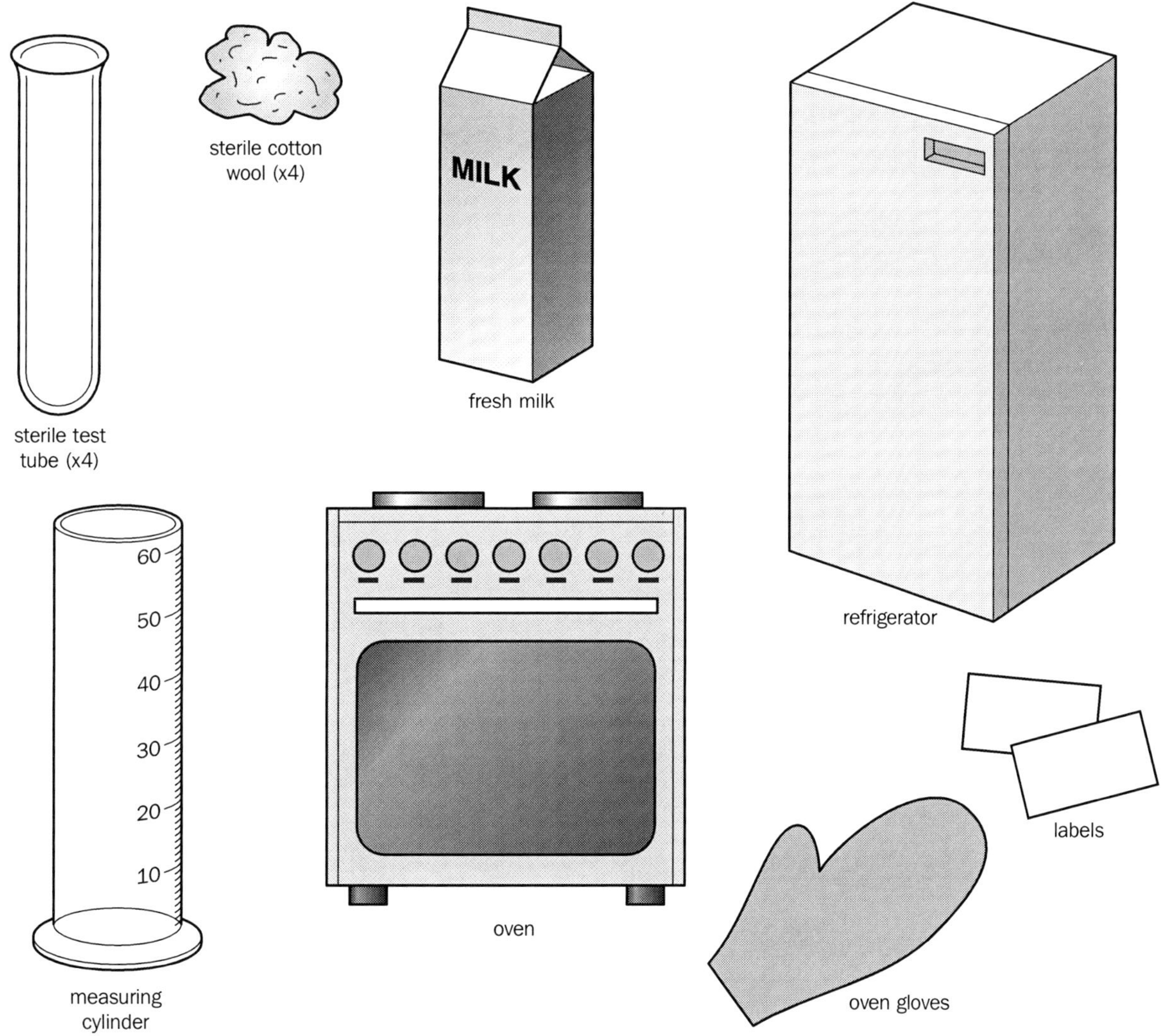

Chapter 16 ► Investigation 16D Sour milk

Preparation: Plan

Finish the sentences in the box.

I will measure…

Things I will keep the same are…

My investigation will be fair because…

My investigation will be safe because…

Carry out

See if milk goes sour in different conditions after two days and four days.

Put your results in a table like this:

Test tube	Appearance of milk: fresh/sour/ very sour		
	At start	After two days	After four days
1 (refrigerator)			
2 (classroom)			
3 (warm oven)			
4 (hot oven)			

Draw a bar graph of your results on a piece of graph paper.

Use different colours for each of the four conditions. Label the axes like this:

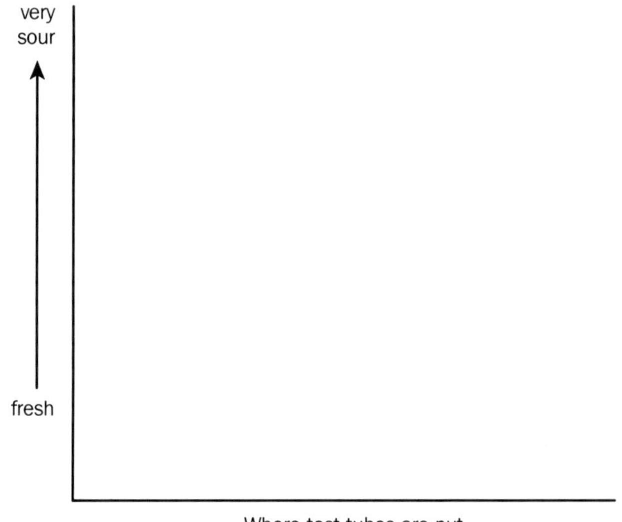

Report

Write a report on your investigation.

Here are some things you should include:

- what you did
- what happened
- explain your results
- if your prediction was correct or not
- what you could do to improve the investigation
- what you could have done if you had more time.

Chapter 16 ► Investigation 16C Sour milk

Milk contains microbes. These microbes turn milk sour.
The speed at which microbes turn milk sour depends on where the milk is kept.

In this investigation: you are going to find out which conditions cause milk to go sour fastest.

Preparation: Predict

Finish the sentence in the box.

I think milk will go sour fastest when it is kept in a (refrigerator/classroom/warm oven/hot oven) because...

You are going to use this equipment to find out which conditions cause milk to go sour fastest:

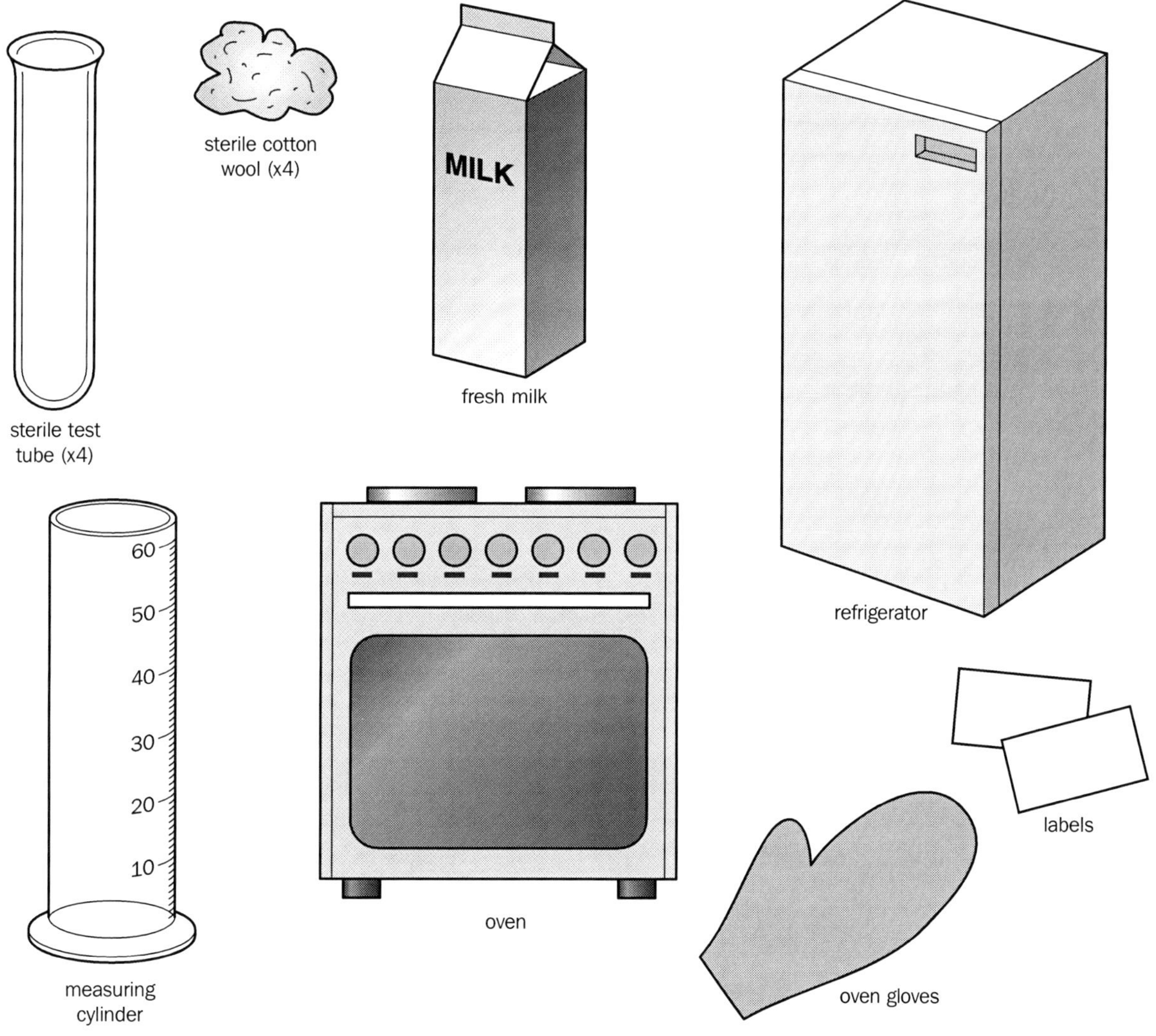

Chapter 16 ▶ Investigation 16C Sour milk

Preparation: Plan

Finish the sentences in the box.

I will measure...

Things I will keep the same are...

My investigation will be fair because...

My investigation will be safe because...

Carry out

- Wash your hands.
- Label four test tubes 1, 2, 3, and 4.
- Put 5 cm^3 of fresh milk in each test tube and bung each test tube with cotton wool.
- Put test tube 1 in a refrigerator, test tube 2 on a shelf in the classroom, test tube 3 in a warm oven set at about 35 °C, and test tube 4 in a hot oven set at about 100 °C.
- After two days look at the milk in your test tubes to see if it has it gone lumpy. Carefully remove the cotton wool and smell the milk.
- Bung each test tube with cotton wool again and put them back where they were.
- After two more days look at the milk again. Decide if the milk is fresh, sour, or very sour.
- Put your results in a table like this:

Test tube	Appearance of milk: fresh/sour/ very sour		
	At start	After two days	After four days
1 (refrigerator)			
2 (classroom)			
3 (warm oven)			
4 (hot oven)			

Chapter 16 ► Investigation 16C Sour milk

Draw a bar graph of your results on this grid.

Use different colours for each of the four conditions.

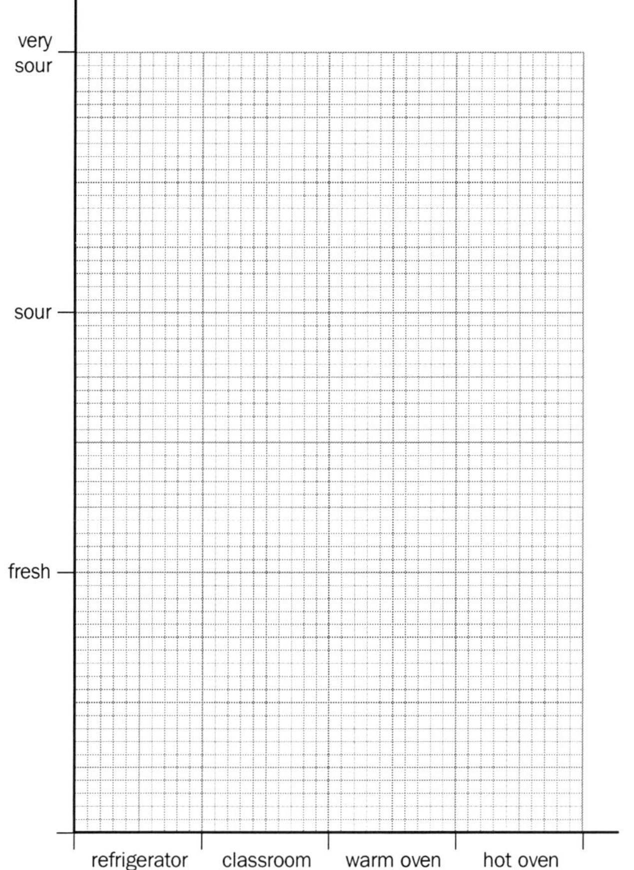

Report

Finish the sentences in the box.

What I did was…

What happened was…

From my results I found out that milk goes sour fastest when it is kept in a (refrigerator/classroom/warm oven/hot oven). I know this because…

My prediction (was/wasn't) correct.
If I could do the investigation again I would…

Investigation 16
Practical notes

Sour milk

Investigation 16
Technician's notes

Sour milk

Each group will need:

Number of apparatus sets:

Number of pupils:

Number of groups:

Visual aids:

ICT resources:

Equipment/apparatus needed:

- four sterile test tubes
- four pieces of sterile cotton wool (to bung test tubes)
- fresh milk
- 10 cm^3 measuring cylinder
- access to refrigerator
- access to incubator/oven set to 35 °C
- access to incubator/oven set to 100 °C
- labels
- oven gloves.

Safety notes

CLEAPSS/SSERC SAFETY REFERENCE:

Chapter 16 ▶ Test

White/Blue

Microorganisms/biotechnology

1 **a** What are microorganisms?

b Read these three descriptions of microorganisms:

A Very simple and cannot be seen under an ordinary microscope. They are not cells and can only reproduce inside other cells.

B Reproduce by releasing spores into the air which land and grow on the surface of food. They like warm, damp places.

C Very small cells which do not have a nucleus. They reproduce about once every 20 minutes.

Which best describes a
i bacteria
ii virus
iii microscopic fungi?

4 marks

2 The diagram shows a compost bin.

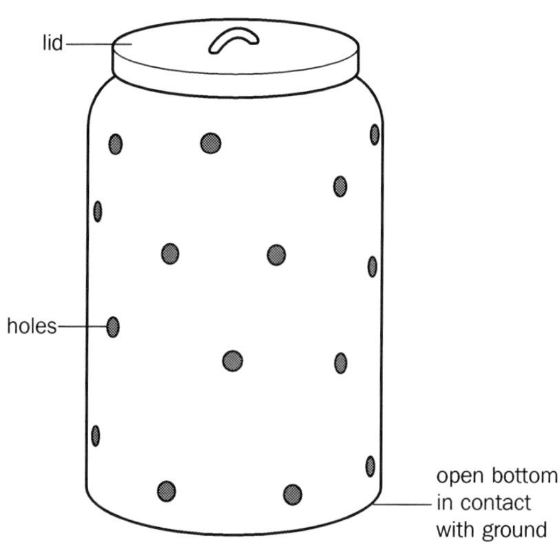

a Explain why each of the labelled features is important.

b Suggest why compost takes longer to make in winter than in summer.

c **i** Name two things you can make compost from.
ii What do scientists call materials that decompose?

7 marks

3 Some microorganisms cause diseases.

a Explain why each of the following is a good way of stopping the spread of disease:
i covering your mouth when you cough or sneeze
ii cooking food thoroughly
iii washing your hands before eating
iv showering or bathing regularly
v putting a plaster over a cut until it heals.

b Explain how the body defends itself against microbes that get into the body.

c What does immune mean?

8 marks

4 Copy the diagram below. It shows a cross between a woman with brown eyes and a man with brown eyes. B stands for the gene allele for brown eyes and b stands for the gene allele for blue eyes.

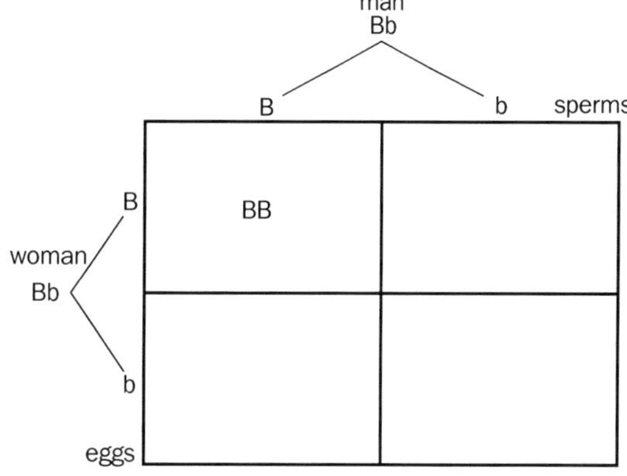

a Finish the diagram by filling in the boxes.

b Which allele is dominant?

c What do we call the other allele?

d Which offspring could eventually produce children with
i brown eyes
ii blue eyes.

6 marks

5 The drawings show a cart horse, used for pulling heavy loads, and a racehorse, used for sprinting.

Both horses are varieties of the same species. They have been bred using selective breeding to enable them to do different jobs.

a How is the cart horse built to do its job?

b Describe how selective breeding has been used to produce the cart horse.

c Modern racehorses have been produced by breeding mares (females) with stallions (males) that can run fast. Explain why a person should check the pedigree (family tree) of a racehorse before buying it.

5 marks

Chapter 16 ► Mark scheme

White/Blue

Microorganisms/biotechnology

Question	Answer	Marks	Level
1 a	very tiny living things/ seen only with microscope	1	E
b	C	1	E
	A	1	E
	B	1	E
		4	
2 a	suitable temperature/ warmth		
	oxygen/air		
	microbes from soil do the decomposing		
	not too much water		
	(any three)	3	F
b	lower temperature (1) microbes like it warm (1)	2	F
c i	any two suitable materials, e.g. fruit, vegetable peelings etc.	1	F
ii	biodegradable	1	F
		7	
3 a i	stops microbes going into air	1	F
ii	kills any microbes in food	1	F
iii	removes any microbes from skin	1	F
iv	stops growth of microbes on body	1	F
v	stops microbes getting into cut	1	F
b	white blood cells eat microbes (1) produce antibodies (1)	2	F
c	protection against disease	1	F
		8	

Question	Answer	Marks	Level
4 a	(three boxes filled correctly = 2 marks		
	two boxes filled correctly = 1 mark)	2	F
b	brown/B	1	F
c	recessive	1	F
d i	BB and/or Bb		
ii	Bb and or bb	2	F
		6	
5 a	big/heavy/strong/ muscular	1	F
b	select animals with desired characteristics (1)		
	breed them (1) choose offspring/repeat process (1)	3	F
c	history of good/poor characteristics	1	F
		5	
	TOTAL 30 marks		

Suggested grade/level boundaries

E = 11/30

F = 22/30

(This page is intentionally blank.)

17 Changing materials

This chapter deals with three short interrelated topics. In the first part pupils learn more about water and its properties. The processes of evaporation and condensation are considered together with solubility and crystal formation. In the second part pupils are introduced to acids, alkalis and neutralisation. Finally, in the third part, chemical reactions and the factors that affect them are introduced.

Assessment opportunities

Formative assessment opportunities are provided by worksheets, homework sheets, and an investigation.

The **worksheets** cover material at levels C, D, and E for attainment targets for knowledge and understanding. Teachers may wish to use these worksheets not only as part of practical activities but also to provide evidence of pupil achievement.

Worksheet	Level
17.1	C
17.2a	D
17.2b	D
17.2c	E
17.2d	D
17.3a	E
17.3b	E

The **homework sheets** cover material at levels C, E, and F for attainment targets for knowledge and understanding. These homework sheets can be used individually as a follow-up to work done in class or assembled into a homework booklet allied closely to schemes of work.

Homework sheet	Level
17.2a	C
17.2b	E
17.3	E
17.4	F

The **investigation** covers all three skill areas at levels C, D, E, and F. It is written in a way that allows for pupils to be assessed in all three skill areas at one level. Alternatively, customised assessments can be constructed enabling pupils to be assessed at different levels in all three skills. The latter approach is more time consuming, but it does provide the opportunity for pupils to show evidence of achievement at different levels in different skills in the same investigation. Teachers will need to use their professional judgement when deciding which level is appropriate to individual pupils. It is envisaged that pupils will show progression through the levels as they work through their science course.

Summative tests are provided at two levels, white and blue. The white test contains questions covering attainment target levels C and D. The blue test contains questions covering attainment target levels E and F. Each test has a total of 30 marks and will take about 30 minutes for pupils to complete, although this can be varied depending on pupil ability. Mark schemes are provided together with suggested grade/level boundaries.

It is envisaged that these tests will be given to pupils on completion of the material covered in Chapter 17.

ICT opportunities

The use of data loggers/remote sensors can extend the range, speed, and sensitivity of measurements in many of the worksheets for this chapter. Once downloaded onto a PC, data-handling programs can be used to analyse information gathered, data can be manipulated, and appropriate graphs etc. presented. The Internet provides pupils with access to a huge range of scientific information. A list of suitable websites is included in this Teacher's Guide.

Students' book chapter 17 contents and guide levels

17.1	Water – some reminders	*Starting off*	C
	What happens to molecules?	*Going further*	C
	Looking at the water cycle	*For the enthusiast*	C
17.2	Soluble or insoluble?	*Starting off*	C
	Temperature vs dissolving?	*Going further*	E
	Growing crystals	*For the enthusiast*	E
17.3	Acids and alkalis	*Starting off*	E
	More about neutralisation	*Going further*	E
	Colours and flavours	*For the enthusiast*	E
17.4	Speeding up reactions	*Starting off*	F
	Everyday rates of reaction	*Going further*	F
	Catalysts and clean air	*For the enthusiast 1*	F
	What kind of change?	*For the enthusiast 2*	F

17.1 Evaporation

Name: **Date:** **Group:**

What you need:

Watch glass, beaker, syringe, test tube, stop clock, methylated spirits (meths), cold tile, hot water, safety goggles, access to a fridge.

What to do:

1 Copy this table into your book. Put your results in the table when you get them.

Position	Start time	Finish time	Time for meths to evaporate
Watch glass in still air/cold place			
Watch glass in cold moving air			
Watch glass in a warm place			
Test tube in a cold place			

2 Put on safety goggles.

3 Time how long it takes for 1 cm^3 of methylated spirits to evaporate when it is placed in these positions:

On a watch glass on a cold surface in still air.

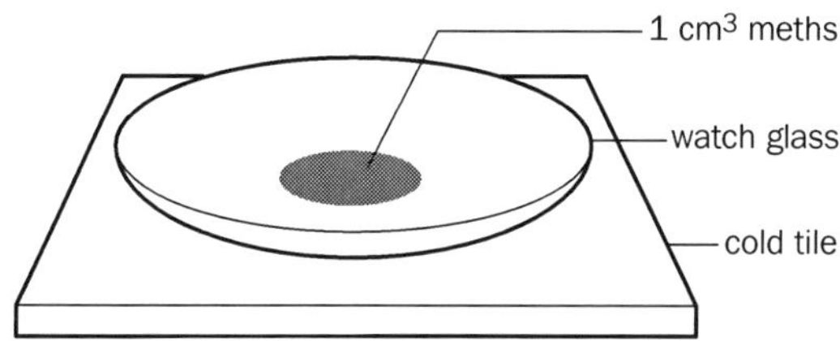

On a watch glass in cold moving air.

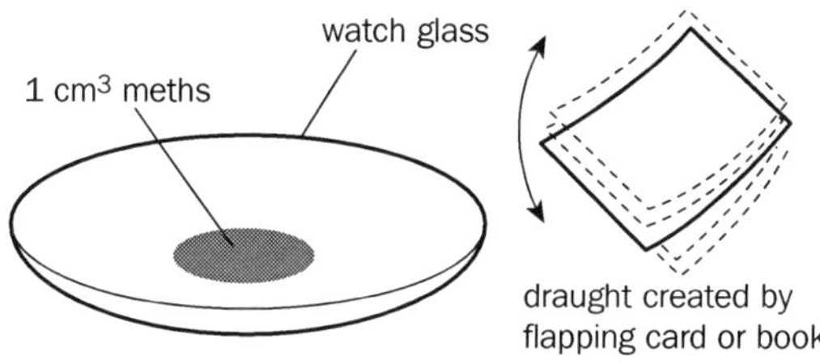

In a test tube in a cold place.

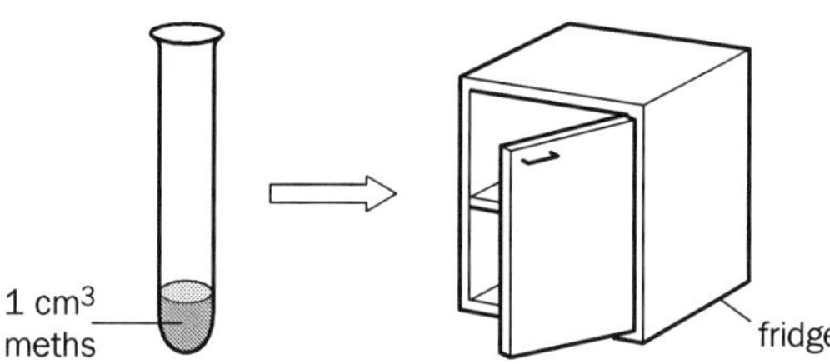

On a watch glass heated over a beaker of hot water.

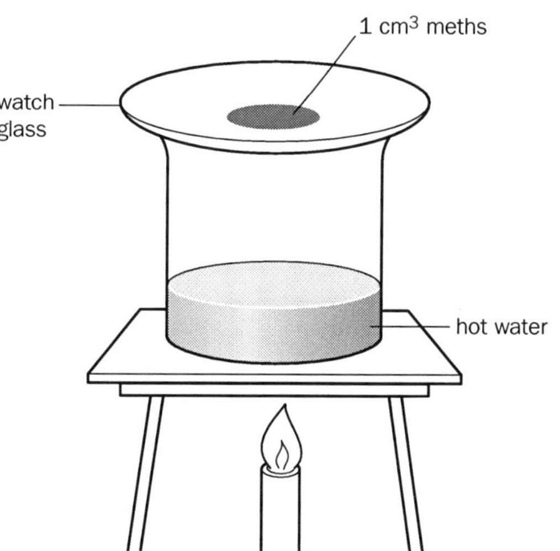

- In which position did the methylated spirits evaporate quickest?
- Why do you think washing dries better when it is hung on a washing line on a warm, windy day?

SAFETY WARNING

Meths is flammable.

Keep away from naked flames.

17.1
Practical notes

Evaporation

This is a simple activity in which pupils investigate the effect of temperature, moving air, and surface area on the rate of evaporation. Methylated spirits is used because it evaporates faster than water. Even when put in a warm place, 1 cm^3 of water will take a long time to evaporate completely. Teachers will need to decide whether or not to put a time limit on this part of the activity. Remind pupils to take care when handling hot water. Make sure that there are no naked flames in the room during this activity, as methylated spirits vapour is highly flammable. Ensure adequate ventilation throughout the activity.

17.1
Technician's notes

Number of pupils:

Number of groups:

Visual aids: ....

ICT resources: ....

Equipment/apparatus needed: ....

Evaporation

Each group will need:

- a copy of worksheet 17.1
- watch glass
- beaker
- test tube
- stop clock
- small syringe or alternative for measuring 1 cm^3
- methylated spirits (meths)
- access hot water
- cold tile or alternative cold surface, e.g. glass
- oven gloves
- safety goggles
- access to fridge.

Safety notes

- **See HAZCARDS for methylated spirits.**
- **Ensure no naked flames in the room during the activity and that there is adequate ventilation.**

CLEAPSS/SSERC SAFETY REFERENCE:

17.2a Pure water from tap water

W/S

Name: Date: Group:

What you need:

Conical flask, bung with delivery tube, anti-bumping granules, beaker, test tube, Bunsen burner, heatproof mat, tripod, gauze mat, safety goggles.

What to do:

1 Put on safety goggles.

2 Put some tap water into the conical flask and add some anti-bumping granules. The granules will help the water to boil smoothly.

3 Fit the bung and delivery tube. Put cold water in the beaker and stand a clean dry test tube in the water. Put the end of the delivery tube in the test tube.

4 Heat the flask until the water boils then adjust the gas tap to keep the water simmering.

5 Turn off the Bunsen burner before all of the water in the flask boils away. Let your apparatus cool down before touching it.

- What can you see inside the test tube?
- What can you see on the walls of the flask?
- What is the job of the cold water in the beaker?
- Is tap water pure?
  Explain your answer.

SAFETY WARNING

The apparatus will get hot during this activity.

17.2a
Practical notes

Pure water from tap water

Distillation is a good example of a physical change (two changes in fact). Ink could be added to the tap water to demonstrate distillation as a separation technique. Water from hard water areas will give very good results, pupils will see plenty of scale around the inside of the flask. Teachers working in soft water areas may need to consider adding some 'impurities' to the water before giving it to pupils. Make sure pupils do not let the apparatus boil dry. Remind pupils that the apparatus will get very hot during this activity and not to touch anything until it has cooled down.

17.2a
Technician's notes

Pure water from tap water

Each pupil will need:

- a copy of worksheet 17.2a
- conical flask
- bung with delivery tube
- anti-bumping granules (pieces of broken porcelain will do)
- 250 cm^3 beaker
- test tube
- Bunsen burner
- heatproof mat
- tripod
- gauze mat
- safety goggles
- access to oven gloves.

Number of pupils: ....Number of groups: ....

Visual aids:

....

....

ICT resources:

....

....

Equipment/apparatus needed:

....

....

Safety notes

- **Make sure pupils do not let the apparatus boil dry.**
- **Pupils should be reminded to take care when handling hot apparatus.**
- **Oven gloves should be provided when clearing up.**

CLEAPSS/SSERC SAFETY REFERENCE:

17.2b Rock from rock salt W/S

Name: Date: Group:

What you need:

Rock salt, pestle and mortar, beaker, Bunsen burner, heatproof mat, tripod, gauze, stirrer, filter funnel, filter paper, evaporating basin, safety goggles.

What to do:

1 Put on safety goggles.

2 Grind up some rock salt using a pestle and mortar.

3 Half fill a beaker with water and add the ground up rock salt. Warm the mixture gently over a low, blue Bunsen flame stirring gently all the time.

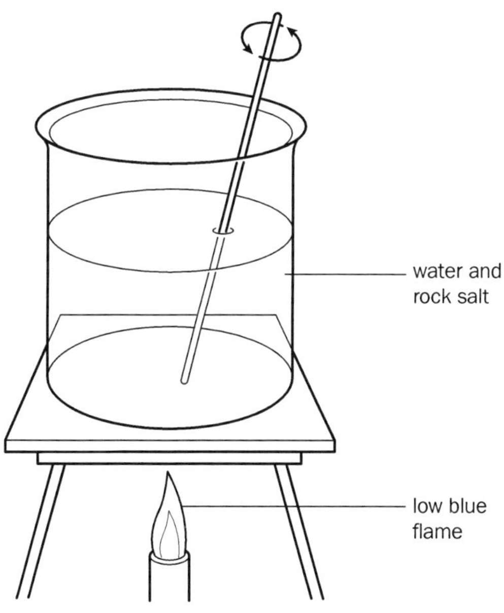

4 Let the mixture cool, then filter it into an evaporating basin.

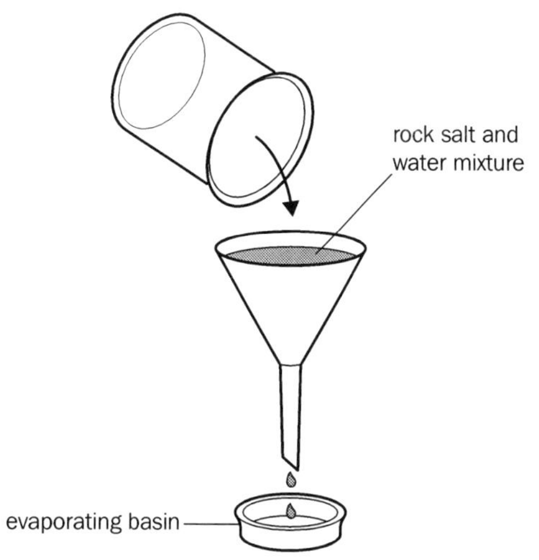

5 Stand the evaporating basin on a tripod and carefully boil the water away.

Stop when the salt starts to 'spit'.

SAFETY WARNING

This equipment will get very hot after only a short time.

Beware of hot, spitting salt.

Wear safety goggles.

17.2b
Practical notes

Rock from rock salt

This activity requires careful application of practical skills which some pupils may not yet have fully acquired. Teachers will need to consider whether or not a demonstration of the techniques involved would be appropriate beforehand. Certainly it is important to stress the importance of not overheating the mixture in the evaporating basin. Ready ground rock salt could be provided to save on time or equipment.

17.2b
Technician's notes

Rock from rock salt

Each group will need:

- a copy of worksheet 17.2b
- rock salt (2 – 3 g should do)
- pestle and mortar
- 250 cm^3 beaker
- Bunsen burner
- heatproof mat
- tripod
- gauze
- stirrer
- filter funnel
- filter paper
- evaporating basin
- safety goggles
- access to oven gloves.

Number of pupils:

Number of groups:

Visual aids: ....

ICT resources: ....

Equipment/apparatus needed: ....

Safety notes

- **This equipment will get very hot after only a short time.**
- **Beware of hot, spitting salt.**
- **Wear safety goggles at all times.**

CLEAPSS/SSERC SAFETY REFERENCE:

17.2c Temperature and solubility W/S

Name: Date: Group:

What you need:

Test tube, bung, beaker, thermometer, Bunsen burner, heatproof mat, tripod, gauze mat, sugar, spatula, safety goggles.

What to do:

1 Copy this table into your book. Put your results in the table.

Temperature in °C	Number of spatulas of sugar
20	
40	
60	

2 Put on safety goggles.

3 Half fill the test tube with cold water. Take the temperature of the water and write it in the table. Add one spatula full of sugar and put the bung in the test tube. Shake the tube until the sugar dissolves.

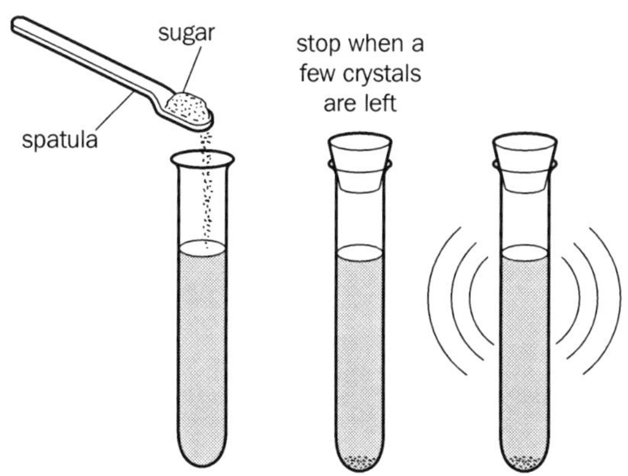

4 Add another spatula full of sugar and shake again. Keep on doing this until no more sugar will dissolve. Write the number of spatulas full of sugar used in the table.

SAFETY WARNING

Remove the bung before heating the test tube.

5 Take the bung off the test tube and stand it in a beaker of cold water. Heat the water gently until the water in the test tube is about 40 °C. Add another spatula full of sugar to the test tube, replace the bung, and shake the tube until the sugar dissolves. Keep on doing this until no more sugar will dissolve. Write the number of spatulas full of sugar used in the table.

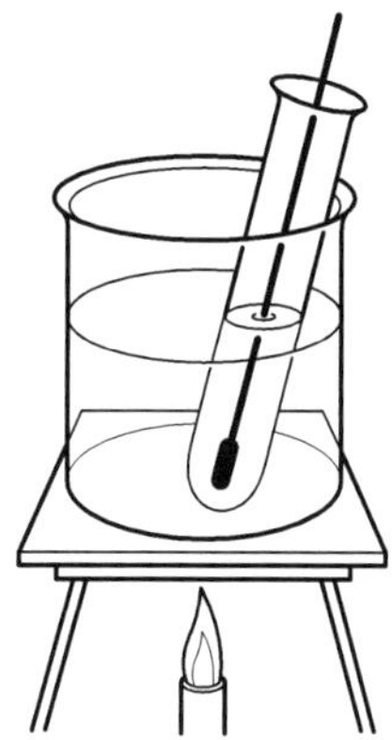

6 Do the same at 60 °C and put your results in the table.

7 Draw a line graph of your results. Label your axes like this:

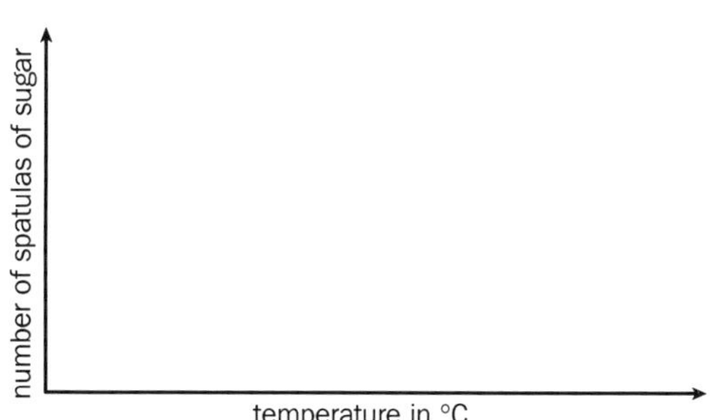

How does temperature affect the way sugar dissolves?

17.2c
Practical notes

Temperature and solubility

This is a straightforward activity designed to enable pupils to see the link between temperature and solubility of sugar. Sugar lumps could be used as an alternative, it makes counting easier and cuts down on the risk of spillages. Artificial sweeteners have been tried and found unsatisfactory for this activity. They are very soluble and costs may be prohibitive. Remind pupils to remove bungs from test tubes before heating them in water baths. An interesting extension could be to use sodium chloride instead of sugar. Pupils will then see that solubility does not increase with temperature in all cases.

17.2c
Technician's notes

Temperature and solubility

Each pupil will need:

- a copy of worksheet 17.2c
- test tube
- bung to fit test tube
- beaker
- thermometer
- Bunsen burner
- heatproof mat
- tripod
- gauze mat
- access to supply of sugar
- spatula
- safety goggles
- access to oven gloves.

Number of pupils:

Number of groups:

Visual aids: ....

....

....

ICT resources: ....

....

....

Equipment/apparatus needed: ....

....

....

Safety notes

- **Bungs must be removed from test tubes before heating them in a water bath.**
- **Pupils should be reminded to take care when handling hot apparatus.**
- **Oven gloves should be provided when clearing up.**

CLEAPSS/SSERC SAFETY REFERENCE:

17.2d Growing crystals

W/S

Name: **Date:** **Group:**

What you need:

Two beakers, spatula, alum (potassium aluminium sulphate), Bunsen burner, heatproof mat, tripod, gauze mat, filter funnel, filter paper, foil, thin thread, glass rod, safety goggles, oven gloves.

What to do:

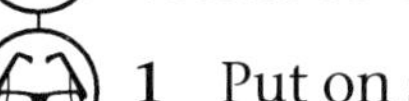

1 Put on safety goggles.

2 Put 150 cm^3 of water into the beaker. Warm the water but do not boil it.

3 Add a few crystals of alum and stir. When no more alum will dissolve, turn off the Bunsen burner and let the solution cool down. You now have a saturated solution of alum.

4 Filter the solution into a clean beaker and cover it with foil.

5 Carefully tie a crystal of alum to the glass rod so that it hangs in the middle of the alum solution. Replace the foil cover, label your beaker and put it in a safe place for a few days.

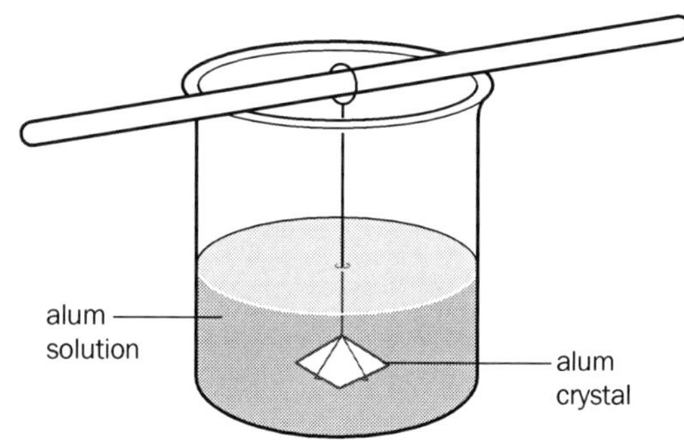

6 Check your beaker every day.

If the level of alum solution gets low, top it up with more.

If your crystal starts to dissolve it means your solution is not saturated.

17.2d
Practical notes

Growing crystals

This is an enjoyable way of showing a physical change; crystallisation. So long as the starter solutions are saturated, the procedure is foolproof. Pupils usually like to keep their crystals or at least preserve them for display etc. This can be done by painting with nail varnish after drying. Copper sulphate is a colourful alternative to alum. It is useful to have a stock of saturated solution available as back up. Remind pupils to cover their beakers at all times as specks of dust will provide centres for crystallisation.

17.2d
Technician's notes

Growing crystals

Each pupil will need:

Number of pupils: ....Number of groups: ....

Visual aids:

ICT resources:

Equipment/apparatus needed:

- a copy of worksheet 17.2d
- two 250 cm^3 beakers
- spatula
- alum (potassium aluminium sulphate)
- filter funnel
- filter paper
- aluminium foil (enough to make lid for beaker)
- thin thread
- glass rod or similar to hang crystal from
- Bunsen burner
- heatproof mat
- tripod
- gauze mat
- safety goggles
- access to oven gloves.

Safety notes

- **Pupils should be reminded to take care when handling hot apparatus.**
- **Oven gloves should be provided when filtering saturated solution.**
- **Safety goggles should be worn at all times.**

CLEAPSS/SSERC SAFETY REFERENCE:

17.3a Making an indicator

W/S

Name: **Date:** **Group:**

What you need:

Large beaker, small beaker, three test tubes, test tube rack, red cabbage, Bunsen burner, heatproof mat, tripod, gauze mat, stirrer, filter funnel, filter paper, hydrochloric acid, sodium hydroxide solution, water, dropper, labels, safety goggles, oven gloves.

What to do:

1 Put on safety goggles.

2 Tear up some red cabbage leaves into small pieces and put them in the large beaker. The beaker should be about half full of leaves.

3 Add enough water to cover the leaves. Heat the water and stir the leaves as the mixture heats up. The water will become purple as the colour comes out of the leaves.

Turn off the Bunsen burner when the leaves have lost most of their colour; they will be almost white.

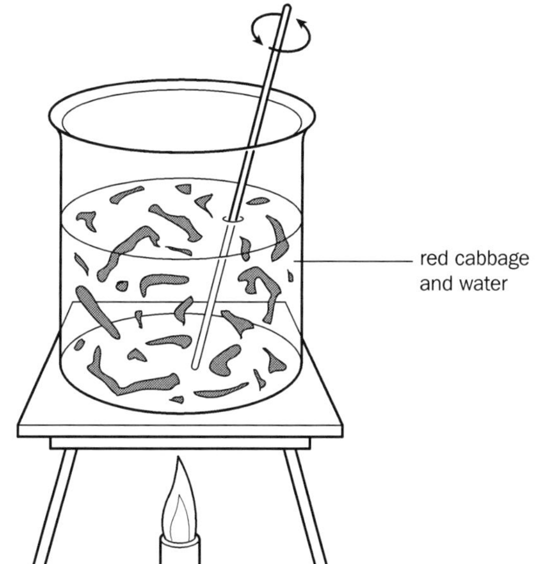

SAFETY WARNING

Acids and alkalis are corrosive.

Safety goggles must be worn at all times.

4 When it has cooled, filter the mixture and collect the purple solution in a small beaker. This solution is your indicator.

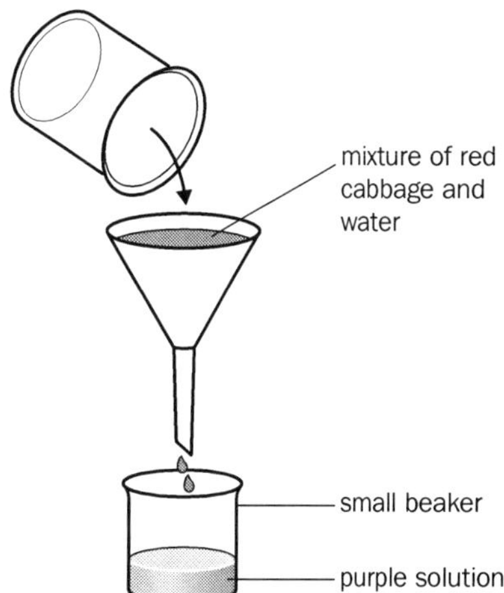

5 Label one test tube 'Acid' and quarter fill it with hydrochloric acid.

Label another test tube 'Alkali' and quarter fill it with sodium hydroxide solution.

Label another test tube 'Neutral' and quarter fill it with water.

6 Use a dropper to add some of your red cabbage indicator to each of the test tubes.

7 Copy this table into your book. Put your results in the table.

Solution	Colour of red cabbage indicator
Acid	
Alkali	
Neutral	

17.3a
Practical notes

Making an indicator

In this activity pupils extract and use the juice from red cabbage as an indicator. Most indicators come from plant material. If there is sufficient time, or a circus arrangement is adopted, pupils could apply the same technique to make indicators from a range of plant material. Beetroot, elderberries, blackcurrents, blackberries and a variety of red, blue or purple flower petals usually give good results. Remind pupils to let their indicator cool down before filtering.

17.3a
Technician's notes

Making an indicator

Each pupil will need:

Number of pupils: ....Number of groups: ....

Visual aids: ....

ICT resources: ....

Equipment/apparatus needed: ....

- a copy of worksheet 17.3a
- 250 cm^3 beaker
- 100 cm^3 beaker or glass jar
- three test tubes
- test tube rack
- red cabbage (enough when torn up to half fill a 250 cm^3 beaker)
- Bunsen burner
- heatproof mat
- tripod
- gauze mat
- stirrer (glass rod or similar)
- filter funnel
- filter paper (coarse)
- access to dilute hydrochloric acid
- access to dilute sodium hydroxide solution
- access to water/distilled water (this must be neutral)
- dropper
- labels
- safety goggles
- access to oven gloves.

Safety notes

- **See HAZCARDS for hydrochloric acid and sodium hydroxide solution.**
- **Acids and alkalis are corrosive.**
- **Safety goggles must be worn at all times.**
- **Pupils should be reminded to take care when handling hot apparatus.**
- **Oven gloves should be provided when clearing up.**

CLEAPSS/SSERC SAFETY REFERENCE:

17.3b Neutralisation

W/S

Name: Date: Group:

What you need:

Universal indicator paper, scissors, white tile, universal indicator chart, beaker, dropper, hydrochloric acid, sodium hydroxide solution, stirrer, Bunsen burner, heatproof mat, tripod, gauze mat, evaporating basin, safety goggles.

What to do:

1 Put on safety goggles.

2 Cut up the universal indicator paper into small pieces and arrange them in a row on the tile.

3 Put about 25 cm^3 of sodium hydroxide into the beaker. Add a few drops of hydrochloric acid and stir the mixture. Use the stirrer to put one drop of the mixture on to a piece of indicator paper. The indicator will go blue because there is more alkali than acid.

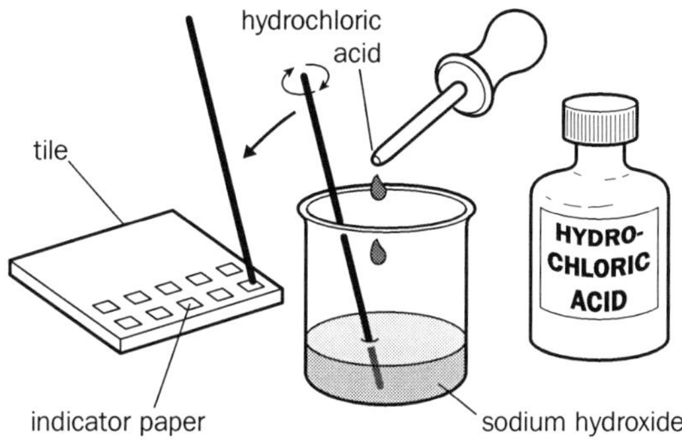

4 Add more acid, stir, and test the pH of the solution again. Keep on doing this until you get a neutral solution.

?

How will you know when the solution is neutral?

5 Pour your neutral solution into an evaporating basin and carefully bring it to the boil. Adjust the gas tap so the solution is just simmering.

Turn off the Bunsen burner when the solution starts to spit.

6 Leave your equipment to cool down.

7 Check with your teacher to see if you can taste the white solid left in the evaporating basin.

SAFETY WARNING

The apparatus will get hot during this activity.

17.3b
Practical notes

Neutralisation

This simple technique gives pupils the opportunity of investigating neutralisation if burettes, conical flasks etc. are either not available or in short supply. Remind pupils not to put pieces of indictor into the solution, it will colour the liquid and spoil the rest of the activity. Teachers will need to make their own risk assessment before allowing pupils to taste their salt product. Certainly apparatus must be thoroughly cleaned before use if this is to be the case. A 'pre-prepared' sample might be a useful alternative and cuts out any potential hazards. Remind pupils not to boil the solution and to extinguish Bunsen burners as soon as the salt starts to spit.

17.3b
Technician's notes

Neutralisation

Each pupil will need:

Number of pupils: ....Number of groups: ....

Visual aids: ....

ICT resources: ....

Equipment/apparatus needed: ....

- a copy of worksheet 17.3b
- universal indicator paper (two or three strips)
- scissors
- white tile
- universal indicator chart
- 100 cm^3 beaker
- dropper
- dilute hydrochloric acid
- dilute sodium hydroxide solution
- stirrer (glass rod)
- Bunsen burner
- heatproof mat
- tripod
- gauze mat
- evaporating basin
- safety goggles
- access to oven gloves.

Safety notes

- **Pupils should be reminded to take care when handling hot apparatus.**
- **Oven gloves should be provided when clearing up.**

CLEAPSS/SSERC SAFETY REFERENCE:

17.2a Soluble or insoluble? H/W

Name: Date: Group:

What you need to know...

Something that dissolves is described as soluble. Something that will not dissolve is called insoluble. A substance that dissolves is called a solute. The liquid doing the dissolving is called a solvent. Solute and solvent together make a solution.

What to do:

1 Investigate what happens when salt, sugar, and flour are put into cold water.

Put some cold water in a glass beaker and add some salt, about half a teaspoonful will do. Stir the mixture and watch what happens.

Do the same for sugar and flour.

a For each experiment, name the
  i solvent
  ii solute used.

b Which of the foods are insoluble?

c Which of the foods make solutions?

2 Investigate what happens when tea and coffee are put into warm water.

Put some warm water (warm tap water will do) in a glass beaker and add a tea bag. Stir the mixture and watch what happens.

Do the same with a teaspoonful of instant coffee.

a Describe what happens when you put the
  i tea bag
  ii instant coffee in the warm water.

Use these words in your answer:

- solute
- solvent
- solution
- insoluble
- soluble

b Predict what would have happened if you had used cold water instead of warm water in this experiment.

HANDY HINTS

Watch carefully as you add the items to the water and as they are stirred.
Allow the mixtures to settle before answering the questions.

17.2b How does temperature affect dissolving? H/W

Name: Date: Group:

What you need to know...

For some solutes such as common salt (sodium chloride) temperature has very little effect on the amount that will dissolve in a certain volume of water. However, other solutes such as sugar dissolve better in hot water than in cold water. For gases, the opposite is true. Less gas can dissolve in warm water than cold water.

Solubility is the amount of solute that dissolves in a certain volume of solvent at a particular temperature to form a saturated solution. A graph of solubility against temperature is called a solubility curve.

What to do:

1 The graph shows the solubility curve of three substances.

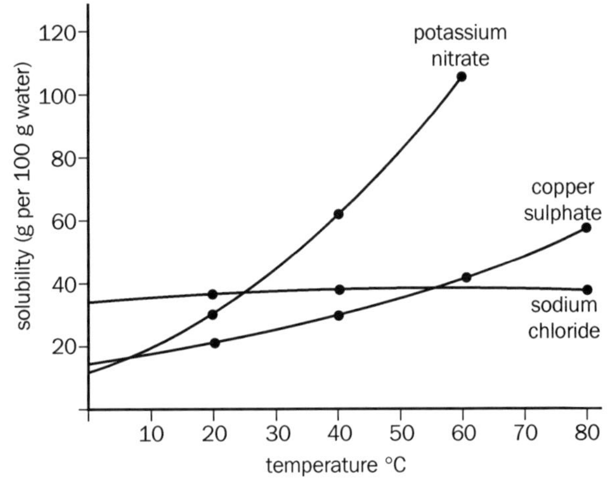

a Which substance dissolves best at
  i 0 °C
  ii 60 °C?

b Which substance has a solubility that is not really affected by temperature?

c What is the solubility of copper sulphate at 50 °C?

d At what temperature do potassium nitrate and sodium chloride have the same solubility?

2 The table shows the solubility of three gases in water:

Solubility (cm^3 per litre of water)	0 °C	20 °C	40 °C	60 °C
oxygen	50	30	18	12
carbon dioxide	1400	848	620	300
nitrogen	25	15	10	8

a Draw a solubility curve for nitrogen.

b How does the solubility of the gases change with temperature?

c What volume of oxygen dissolves in 1 litre of water at 20 °C?

d Which of the three gases is
  i most
  ii least soluble?

e Explain why bubbles appear when you heat water.

f Fish use their gills to remove dissolved oxygen from water. Explain why fish come to the surface in hot weather.

HANDY HINTS

Look at all of the information carefully before answering the questions.

17.3 Investigating pH

H/W

Name: Date: Group:

What you need to know...

You can find out if a substance is an acid or an alkali by using universal indicator paper and a universal indicator chart. The pH scale is used to measure how acidic or alkaline a substance is.

What to do:

1 The ends of the sentences below have been mixed up.

Match the beginning with the correct end of each sentence and write them in your book:

The pH of a liquid...	...a high pH (13 or 14).
A very acid solution has...	...the pH goes down.
A very alkaline solution has...	...a low pH (1 or 2).
When water is added to an acid...	...tells how acidic or alkaline it is.
When water is added to an alkali...	...the pH goes up.

2 Ask your teacher for some universal indicator paper and a universal indicator chart.

Use small pieces of universal indicator paper to find out the pH of some common household chemicals.

For safety, wear some rubber gloves and take care not to splash any chemicals on your skin or in your eyes.

Put your results in a table using these headings:

Name of chemical	pH

HANDY HINTS

Look for things like soft drinks, liquids used in cooking, and chemicals used for cleaning

17.4 Physical and chemical changes H/W

Name: **Date:** **Group:**

What you need to know...

In a physical change, the appearance of a substance changes but it is still the same substance. However in a chemical change, a new substance is formed.

What to do:

1 Sort out these changes. Write two lists, one headed 'Physical changes' and one headed 'Chemical changes':

- water boiling in a kettle
- sugar dissolving in tea
- metal going rusty
- gas burning on a cooker
- water evaporating from clothes on a washing line
- a piece of chalk dissolving in acid
- toast burning under a grill
- crystals forming from a saturated solution of salt.

2 The diagram shows some apparatus that can be used to make distilled water.

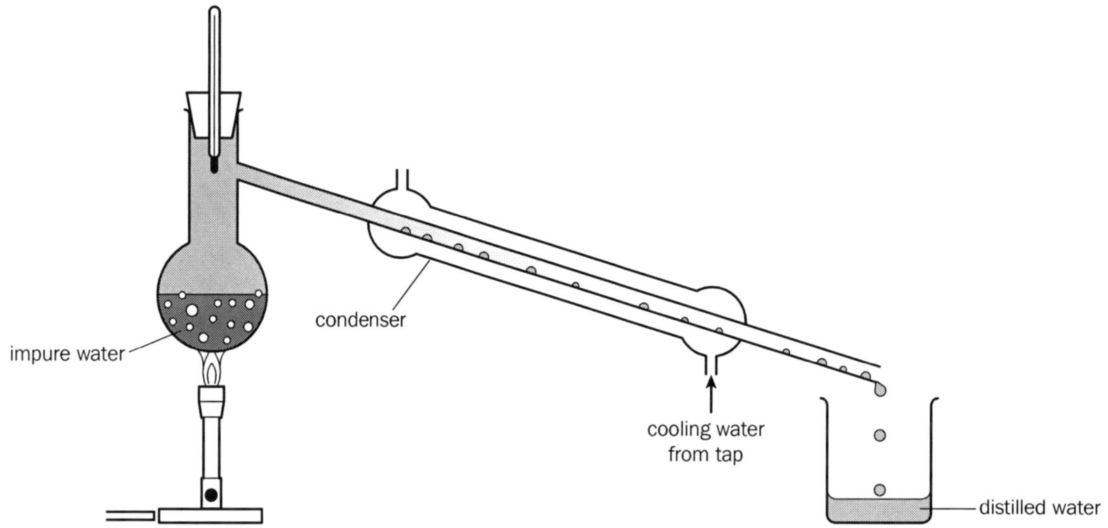

a Describe how this apparatus works.

b Suggest why the tube with a water jacket round it is called a condenser.

c Give one place where a
  i chemical change
  ii physical change is taking place.

d Explain why this apparatus is like the water cycle in nature.

HANDY HINTS

Look where and why water is evaporating and condensing in the apparatus. Where and why does this happen in nature?

Chapter 17 ▶ Investigation 17F Speeding up reactions

	Most chemical reactions take place at a rate which can be speeded up or slowed down.
In this investigation:	you are going to find out if there is a link between the size of pieces of marble and the rate of reaction.

Preparation: Predict

Finish the sentences in the box.

What I think will happen is…

I think this because…

Preparation: Plan

Write a short plan of your investigation.

Think about:

- the apparatus you are going to use
- how one variable depends on another variable
- what you are going to measure and how you are going to measure it
- how many readings you are going to take
- how you are going to record your results
- how you are going to make your investigation fair
- how you are going to make your investigation safe.

Show your plan to your teacher before going on.

Carry out

Carry out your investigation and record your results.

Present your results in an appropriate way.

Report

Write a report on your investigation.

Here are some things you should include:

- a diagram of your apparatus
- what you did
- what happened
- explain your results
- if your prediction was correct or not
- how reliable your results were
- what you could have done if you had more time.

Chapter 17 ► Investigation 17E Speeding up reactions

	Most chemical reactions take place at a rate which can be speeded up or slowed down.
In this investigation:	you are going to find out if there is a link between the size of pieces of marble and the rate of reaction.

Preparation: Predict

Finish the sentences in the box.

What I think will happen is...

I think this because...

Preparation: Plan

Write a short plan of your investigation.

Think about:

- the apparatus you are going to use
- what you are going to measure and how you are going to measure it
- how many readings you are going to take
- how you are going to record your results
- how you are going to make your investigation fair
- how you are going to make your investigation safe.

Show your plan to your teacher before going on.

Carry out

Carry out your investigation and record your results in a table.

Draw a bar graph of your results.

Report

Write a report on your investigation.

Here are some things you should include:

- a diagram of your apparatus
- what you did
- what happened
- explain your results
- if your prediction was correct or not
- how reliable your results were
- what you could have done if you had more time.

Chapter 17 ► Investigation 17D Speeding up reactions

	Most chemical reactions take place at a rate which can be speeded up or slowed down.
In this investigation:	you are going to find out if there is a link between the size of pieces of marble and the rate of reaction.

Preparation: Predict

Finish the sentence in the box.

I think that there (is/is not) a link between the size of pieces of marble and the rate of reaction because…

You are going to use this equipment to find out if there is a link between the size of pieces of marble and the rate of reaction:

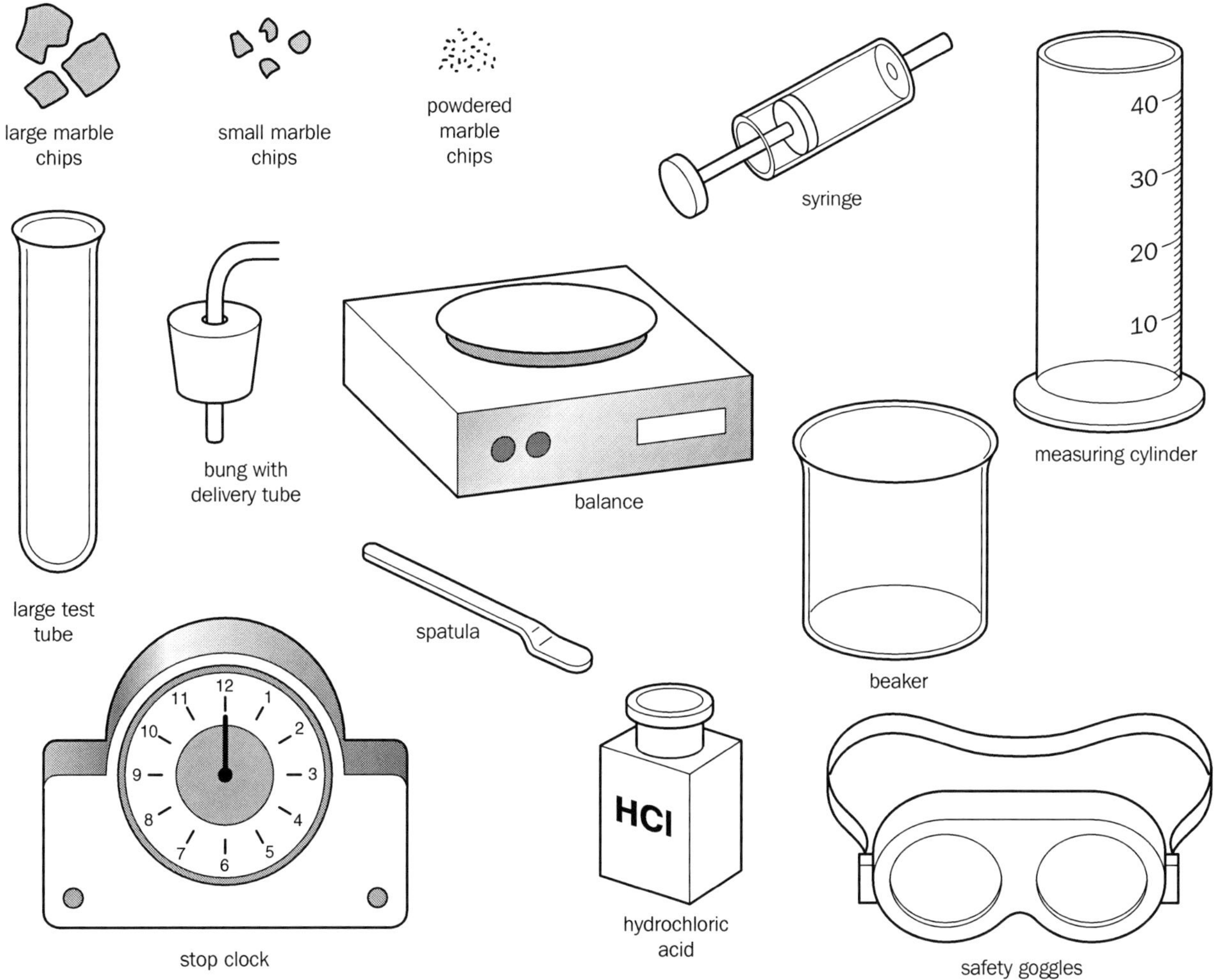

Chapter 17 ▶ Investigation 17D Speeding up reactions

Preparation: Plan

Finish the sentences in the box.

I will measure...

Things I will keep the same are...

My investigation will be fair because...

My investigation will be safe because...

Carry out

Set up your apparatus like this. See if there is a difference in the time it takes to produce 2 cm^3 of gas by each size of marble chip.

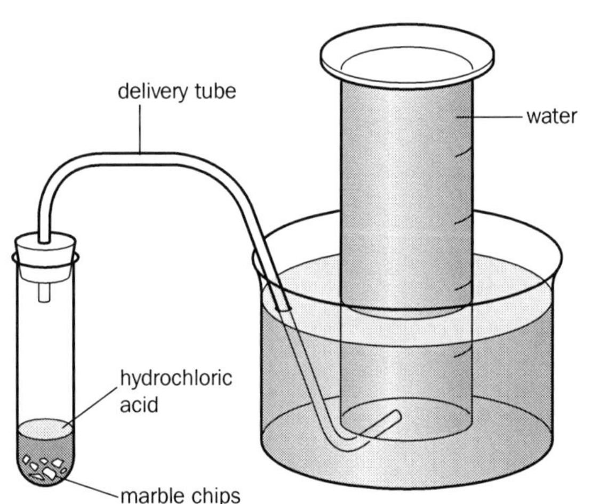

Put your results in a table like this:

Size of pieces of marble chips	Time taken to collect 2 cm^3 of gas (in s)
Large	
Small	
Powdered	

Draw a bar graph of your results on a piece of graph paper.

Label the axes like this:

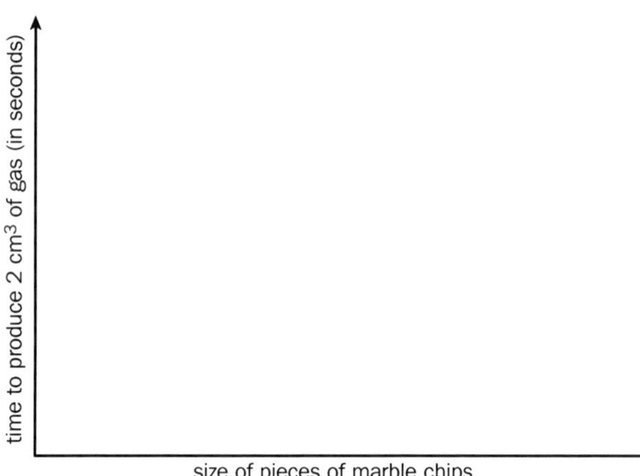

Report

Write a report on your investigation.

Here are some things you should include:

- a diagram of your apparatus
- what you did
- what happened
- explain your results
- if your prediction was correct or not
- what you could do to improve the investigation
- what you could have done if you had more time.

Chapter 17 ► Investigation 17C Speeding up reactions

	Most chemical reactions take place at a rate which can be speeded up or slowed down.
In this investigation:	you are going to find out if there is a link between the size of pieces of marble and the rate of reaction.

Preparation: Predict

Finish the sentence in the box.

I think that there (is/is not) a link between the size of pieces of marble and the rate of reaction because…

You are going to use this equipment to find out if there is a link between the size of pieces of marble and the rate of reaction:

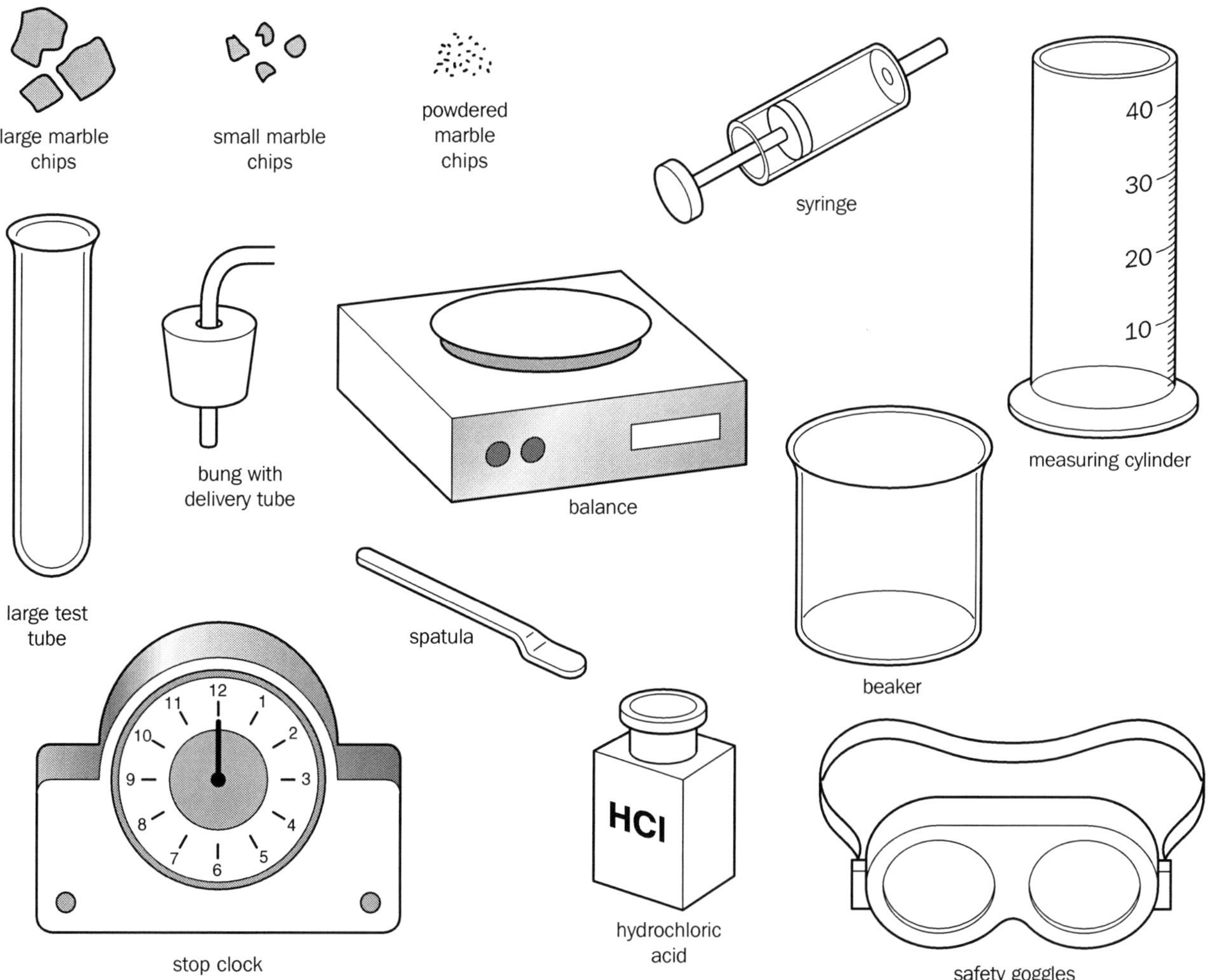

Chapter 17 ► Investigation 17C Speeding up reactions

Preparation: Plan

Finish the sentences in the box.

I will measure...

Things I will keep the same are...

My investigation will be fair because...

My investigation will be safe because...

Carry out

- Put on safety goggles.
- Weigh out 1 g of large marble chips, 1 g of small marble chips, and 1 g of powdered marble chips and put them on separate pieces of paper.
- Set up your apparatus like this:

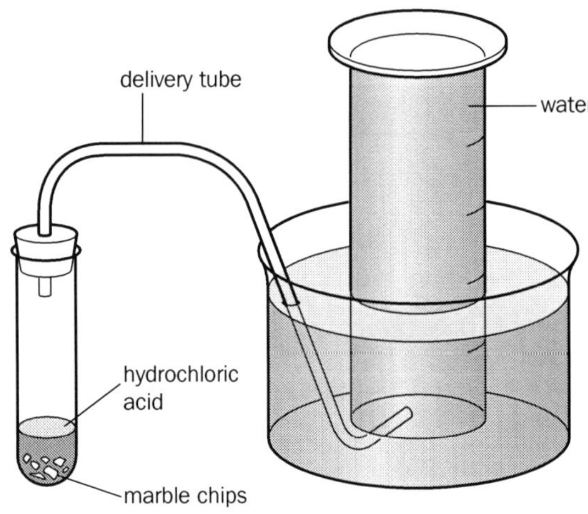

- Quickly but carefully use the syringe to add 10 cm^3 of hydrochloric acid to the large marble chips in the test tube.
- Fit the bung and start timing.
- See how long it takes to collect 2 cm^3 of gas in the measuring cylinder.
- Do the same with the small marble chips and the powdered marble chips. Use fresh acid each time.

  Put your results in a table like this:

Size of pieces of marble chips	Time taken to collect 2 cm^3 of gas (in s)
Large	
Small	
Powdered	

Chapter 17 ► Investigation 17C Speeding up reactions

Draw a bar graph of your results on this grid.
Use different colours for each bar.

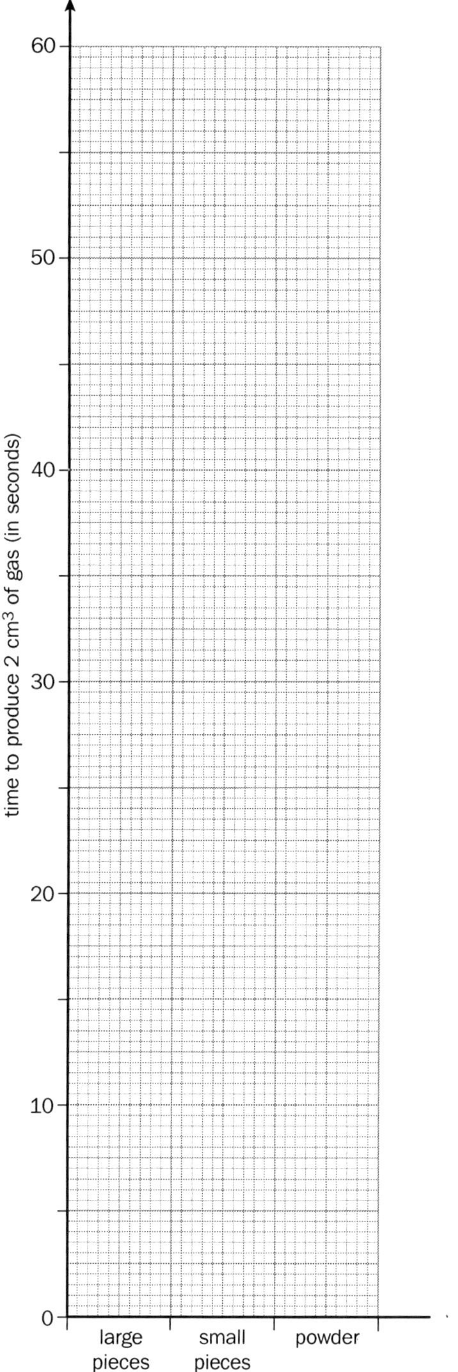

Report

Finish the sentences in the box.

What I did was…

What happened was…

From my results I found out that there (is/is not) a link between the size of pieces of marble and the rate of reaction. I know this because…

My prediction (was/wasn't) correct. If I could do the investigation again I would…

Investigation 17
Practical notes

Speeding up reactions

Investigation 17
Technician's notes

Speeding up reactions

Each group will need:

Number of apparatus sets:

Number of pupils:

Number of groups:

Visual aids:

ICT resources:

Equipment/apparatus needed:

- large marble chips
- small marble chips
- powdered marble chips
- large test tube
- hydrochloric acid
- bung with piece of rubber tubing attached
- 250 cm^3 beaker
- 10 cm^3 measuring cylinder
- syringe (10 cm^3)
- stop clock
- balance
- spatula
- access to a large container for collection of waste acid and marble chips
- safety goggles.

Safety notes

- **See HAZCARDS for dilute hydrochloric acid.**
- **Pupils must wear safety goggles.**
- **Watch for silly behaviour with syringes.**

CLEAPSS/SSERC SAFETY REFERENCE:

Chapter 17 ▶ Test

White

Changing materials

1 The diagram shows the arrangement of water molecules in a liquid state.

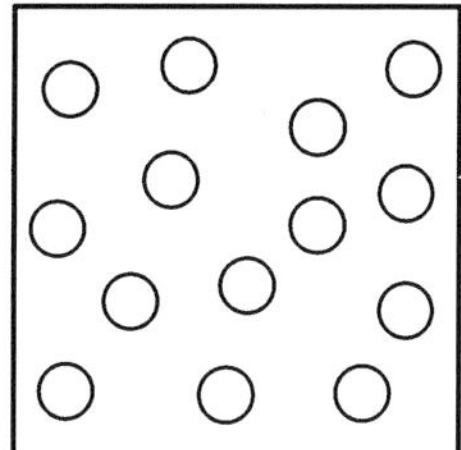

a Draw similar diagrams showing how water molecules are arranged in
  i ice (solid)
  ii water vapour (gas).

b i What happens to water molecules if the temperature rises?
  ii Explain why.

c The change from liquid water to gas is called evaporation. Give two ways that evaporation can be speeded up.

6 marks

2 Which of these words match the gaps in the passage that follows?

insoluble soluble solute
solution solvent

When sugar dissolves completely in water a ___(a)_________ is formed. The solid sugar is called the _____(b)________ and the water is called the _____(c)________. Because it dissolves, sugar is described as ____(d)_______. Sand will not dissolve in water, it is described as ____(e)_______ .

5 marks

3 The diagram shows the water cycle.

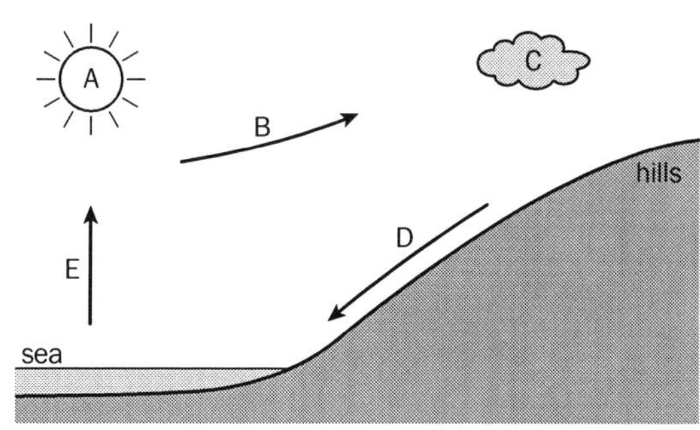

a Which letter on the diagram shows where there is
  i condensation
  ii evaporation
  iii cooling of air
  iv runoff
  v a source of heat energy?

b Explain why it tends to rain more in hilly areas.

c The water in clouds usually falls as rain. Give two other forms in which water can fall to the ground.

9 marks

4 Evaporating, filtering, and distilling are three methods of separating mixtures.

Which method would you use to:

a get sugar from sugar solution?

b get water from muddy water?

c separate sand and gravel?

d get ink powder from ink?

e get water from ink?

5 marks

5 You are given a bucket of sandy sea water. Your job is to get as much clean salt as possible from the sandy liquid.

a Arrange these sentences in the correct order so they best describe how you would do the experiment.

Put filtrate into an evaporating basin.

Collect salt from evaporating basin.

Filter the sandy mixture.

Heat evaporating basin until all water has gone.

Put filter paper in filter funnel.

b Give one safety precaution that you should take when doing your experiment.

5 marks

Chapter 17 ▶ Test
Changing materials

1 The table contains some acids and alkalis found in the home.

Household item	pH	Colour of universal indicator paper	Acid, alkali, or neutral
Oven cleaner	a	b	c
Tap water	d	e	f
Vinegar	g	h	i
Lemon juice	j	k	l
Baking powder	m	n	o

Which of these words match the gaps in the table?

blue	green	orange
purple	red	1
3	7	9
13	neutral	strong acid
strong alkali	weak acid	weak alkali

10 marks

2 Bee stings contain acid. People often put bicarbonate of soda (sodium hydrogen carbonate) on bee stings to neutralise them.

a What kind of substance is bicarbonate of soda?

b What does neutralise mean?

c Name the two products of a neutralisation reaction.

d Give one other example of a neutralisation reaction.

5 marks

3 Marble chips dissolve in hydrochloric acid at a steady rate.

a Explain how
  i breaking the chips into smaller pieces
  ii using more concentrated acid
  iii warming up the materials, will speed up the chemical reaction between the marble chips and the acid.

b In the chemical industry catalysts are often used in chemical reactions.
  i What is a catalyst?
  ii What is the advantage of using catalysts in industrial chemical reactions?

8 marks

4 A small piece of magnesium ribbon is put into a test tube with some sulphuric acid. Bubbles of hydrogen gas are produced. Magnesium sulphate solution remains in the test tube when the reaction has finished.

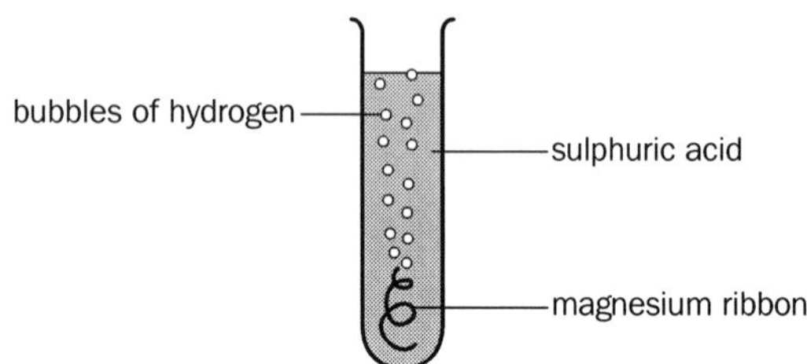

a The test tube gets hot. What does this tell you about the reaction between magnesium and sulphuric acid?

b Explain why it is impossible to get the magnesium back after the reaction has finished.

c The word equation for this reaction is:

magnesium + sulphuric acid ⟶ magnesium sulphate + hydrogen

What are the missing parts of this symbol equation for the same reaction?

$Mg + H_2SO_4 \longrightarrow$ ___(i)___ + ___(ii)___

d If the hot magnesium sulphate solution is left to cool after the experiment, crystals of magnesium sulphate will appear.
  i What is this process called?
  ii Explain why this is called a physical change.

7 marks

Chapter 17 ► Mark scheme Changing materials

White

Question	Answer	Marks	Level
1 a i	diagram with molecules close together	1	
ii	diagram with molecules far apart	1	C
b i	move faster	1	
ii	more energy	1	C
c	increase temperature, moving air, increase surface area (any two)	2	C
		6	
2 a	solution	1	C
b	solute	1	C
c	solvent	1	C
d	soluble	1	C
e	insoluble	1	C
		5	
3 a i	C	1	
ii	E	1	
iii	B	1	
iv	D	1	
v	A	1	C
b	air forced up/colder air (1) condensation (1)	2	C
c	snow, hail, sleet (any two)	2	C
		9	

Question	Answer	Marks	Level
4 a	evaporating	1	D
b	filtering	1	D
c	filtering	1	D
d	evaporating	1	D
e	distilling	1	D
		5	
5 a	Put filter paper in filter funnel Filter the sandy mixture Put filtrate into an evaporating basin Heat evaporating basin until all water has gone Collect salt from evaporating basin. (all correct = 4 marks, one wrong = 3 marks, two wrong = 2 marks, three wrong = 1 mark)	4	D
b	any correct safety precaution, e.g. safety goggles, handle hot apparatus with oven gloves etc.	1	D
		5	
	TOTAL 30 marks		

Suggested grade/level boundaries

C = 18/30

D = 25/30

Chapter 17 ► Mark scheme Changing materials

Blue

Question	Answer	Marks	Level
1 a	13		
b	purple		
c	strong alkali		
d	7		
e	green		
f	neutral		
g	3		
h	orange		
i	weak acid		
j	1		
k	red		
l	strong acid		
m	9		
n	blue		
o	weak alkali	10	E
		10	
2 a	alkali	1	E
b	cancel each other out	1	E
c	salt (1) and water (1)	2	E
d	correct example, e.g. indigestion tablets	1	E
		5	
3 a i	increased surface area (1) more particles exposed to react (1)	2	
ii	more particles to react	1	
iii	increased temperature (1) particles collide more (1)	2	F
b i	chemical which speeds up a reaction (1) without being used up (1)	2	F
ii	faster reaction without added cost, etc.	1	F
		8	

Question	Answer	Marks	Level
4 a	heat is produced/ exothermic	1	F
b	it is a chemical change (1) new substance(s) formed (1)	2	F
c i	$MgSO_4$	1	
ii	H_2	1	F
d i	crystallisation	1	F
ii	change of form/ not a new chemical	1	F
		7	
	TOTAL 30 marks		

Suggested grade/level boundaries

E= 13/30

F = 23/30

18 Electromagnetism/electronics

Pupils have already been introduced to electricity in Book 1 and should now be familiar with simple circuits, current, voltage, resistance, the heating effect of electricity, and the basic safety rules when using electricity. This chapter continues this work by looking at electromagnetism and its uses. The main part of the chapter deals with modern electronic systems, their component parts, their construction, and how they have advanced the quality of our lives.

Assessment opportunities

Formative assessment opportunities are provided by worksheets, homework sheets, and an investigation.

The **worksheets** cover material at level F only for attainment targets for knowledge and understanding. This is because Chapter 18 only covers material at this level. Teachers may wish to use these worksheets not only as part of practical activities but also to provide evidence of pupil achievement.

Worksheet	Level
18.1a	F
18.1b	F
18.3a	F
18.3b	F
18.3c	F
18.3d	F

The **homework sheets** cover material at level F only for attainment targets for knowledge and understanding. These homework sheets can be used individually as a follow-up to work done in class or assembled into a homework booklet allied closely to schemes of work.

Homework sheet	Level
18.1a	F
18.1b	F
18.3a	F
18.3b	F

The **investigation** covers all three skill areas at levels C, D, E, and F. It is written in a way that allows for pupils to be assessed in all three skill areas at one level. Alternatively, customised assessments can be constructed enabling pupils to be assessed at different levels in all three skills. The latter approach is more time consuming, but it does provide the opportunity for pupils to show evidence of achievement at different levels in different skills in the same investigation. Teachers will need to use their professional judgement when deciding which level is appropriate to individual pupils. It is envisaged that pupils will show progression through the levels as they work through their science course.

A single **summative test** is provided at level F only. This is because Chapter 18 only covers material at this level. The test has a total of 30 marks and will take about 30 minutes for pupils to complete, although this can be varied depending on pupil ability.
A mark scheme is provided together with suggested grade/level boundaries. It is envisaged that this test will be given to pupils on completion of the material covered in Chapter 18.

ICT opportunities

The use of data loggers/remote sensors can extend the range, speed, and sensitivity of measurements in many of the worksheets for this chapter. Once downloaded onto a PC, data-handling programs can be used to analyse information gathered, data can be manipulated, and appropriate graphs etc. presented. The Internet provides pupils with access to a huge range of scientific information. A list of suitable websites is included in this Teacher's Guide.

Students' book chapter 18 contents and guide levels

18.1	Electromagnetism	*Starting off*	F
	Electromagnets at work	*Going further*	F
	Design one for yourself	*For the enthusiast*	F
18.2	Electronic systems	*Starting off*	F
	The input	*Going further 1*	F
	The output	*Going further 2*	F
	Analogue processors	*For the enthusiast*	F
18.3	Digital processors (1) – counters	*Starting off*	F
	Digital processors (2) – logic gates	*Going further*	F
	More logic gates	*For the enthusiast*	F

18.1a The magnetic field of a current W/S

Name: Date: Group:

What you need:

Wire and connectors, power supply (6 V), card, scissors, tripod, clamp stand and clamp, plotting compass, iron filings.

What to do:

1 Cut a slit in the card so a wire can pass through the middle.

2 Use the plotting compass to draw six circles around the centre of the card like this:

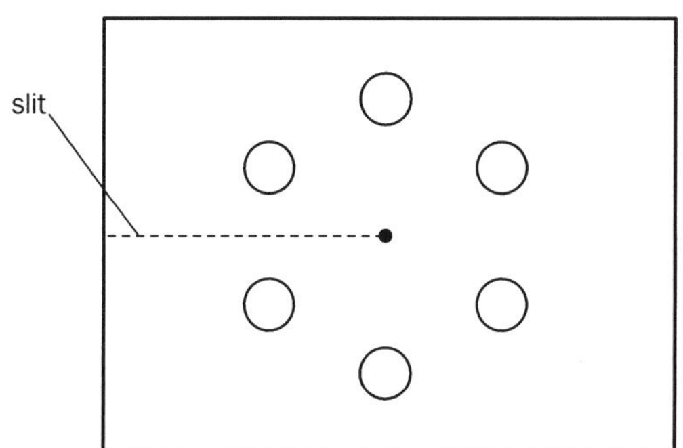

3 Pass the wire through the card and put the card on the tripod. Hold the wire in a clamp so that it falls vertically down through the middle of the card.

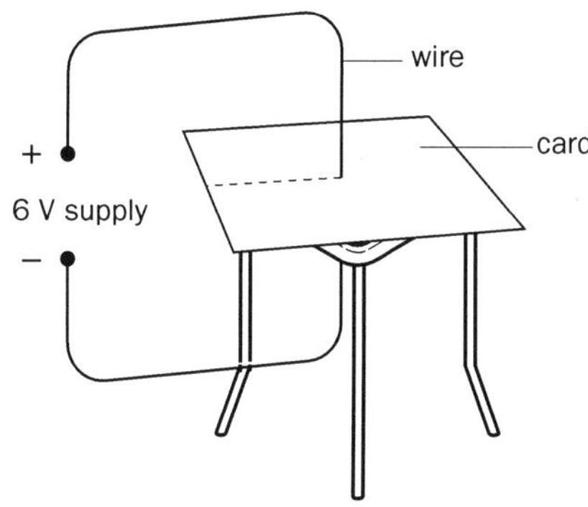

4 Connect the ends of the wire to the power supply so the current goes down (positive to negative) through the card. Switch on the power supply.

5 Put the plotting compass in each of the circles in turn. Draw the positions of the pointers on the card.

- What happens to the compass pointers?
- Is there a pattern to the magnetic field?
- What is the pattern?

6 Switch off the power supply and check the direction of the compass pointers again.

What happens to the compass pointers now?

7 Change the direction of the current by reversing the connections on the power supply. Switch on the power supply.

8 Check the direction of the compass pointers again and draw them in the circles (use a different colour pencil).

- What happens to the compass pointers now?
- Is there a pattern to the magnetic field?
- What is the pattern?

9 Leave the power supply switched on and remove the compass. Sprinkle some iron filings onto the card around the wire. Gently tap the card and watch the iron filings.

- What happens to the iron filings?
- Why do they do this?

SAFETY WARNING

Check electrical equipment before use.

Report any faulty equipment.

18.1a
Practical notes

The magnetic field of a current

The magnetic field produced in this activity is weak so pupils will need to be told to look carefully at the compass needle. Tell pupils to make sure that there is plenty of vertical wire passing through the card so the magnetic field is not interfered with. Remind pupils to use iron filings sparingly to avoid unnecessary mess.

18.1a
Technician's notes

The magnetic field of a current

Each group will need:

- a copy of worksheet 18.1a
- insulated wire and connectors (wire must be at least 50 cm long to allow for vertical drop)
- power supply (6 V)
- piece of thick white card approximately 15 cm × 15 cm
- scissors
- tripod
- clamp stand and clamp
- plotting compass
- iron filings in sprinkler.

Safety notes

- **Check electrical equipment before use.**
- **Report any faulty equipment.**

CLEAPSS/SSERC SAFETY REFERENCE:

Number of apparatus sets:

Number of pupils:

Number of groups:

Visual aids:

ICT resources:

Equipment/apparatus needed:

18.1b Making an electromagnet W/S

Name: Date: Group:

What you need:

Wire, connectors, power pack, iron nail, piece of plain paper, plotting compass, iron filings, sticky tape.

What to do:

1 Wrap wire around the nail about 20 to 30 times to make a coil. Leave enough wire at each end to connect to the power supply. Wrap some sticky tape around the ends of the coil to hold it in place.

2 Put your electromagnet on the piece of paper and connect the ends to the power supply.

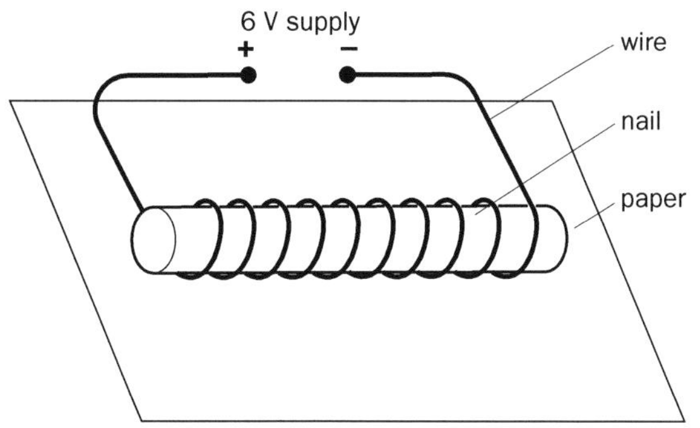

3 Switch on the power supply.

4 Put the plotting compass at each end of the electromagnet.

- Which way do the compass pointers point?
- Which end of the electromagnet is a north-seeking pole and which end is a south-seeking pole?

5 Leave the power supply switched on and remove the plotting compass. Sprinkle some iron filings on to the paper around your electromagnet. Gently tap the paper and watch the iron filings.

- What happens to the iron filings?
- Draw the shape of the magnetic field in your book.

6 Switch off the power supply and disconnect your electromagnet. Pour the iron filings carefully from the paper back into the shaker. Remove the nail from your electromagnet leaving just the coil of wire.

7 Put the coil of wire on the paper and connect the ends to the power supply.

8 Switch on the power supply and sprinkle some iron filings on to the paper around the coil. Tap the paper gently and watch the iron filings.

- Do the iron filings make a clear pattern now?
- What can you say about the strength of the magnetic field produced by the coil with and without a nail inside it?

SAFETY WARNING

Check electrical equipment before use.

Report any faulty equipment.

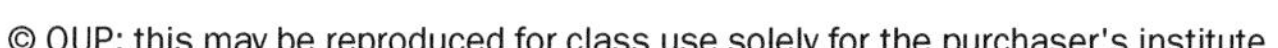

18.1b
Practical notes

Making an electromagnet

Some trial work with voltage and thickness and length of wire is recommended if unnecessary overheating of the coil is to be avoided. In any case pupils should be told to keep power supplies switched on for the shortest possible time during the stages of this activity. Links can usefully be made with magnetic fields produced by permanent magnets. Warn pupils that the coil may get hot during the activity.

18.1b
Technician's notes

Making an electromagnet

Each group will need:

- a copy of worksheet 18.1b
- insulated wire (approximately 1 m)
- connectors
- power pack
- iron nail (approximately 10 cm long)
- piece of plain paper
- plotting compass
- iron filings in shaker
- access to sticky tape.

Number of apparatus sets: ....

Number of pupils:

Number of groups:

Visual aids: ....

....

ICT resources: ....

....

Equipment/apparatus needed: ....

....

Safety notes

- **Check electrical equipment before use.**
- **Report any faulty equipment.**
- **Warn pupils that the coil may get hot during the activity.**

CLEAPSS/SSERC SAFETY REFERENCE:

18.3a Electronic systems 1 W/S

Name: Date: Group:

What you need:

Microelectronics board, power supply.

What to do:

1 Connect the power supply to the microelectronics board. Check with your teacher if you are not sure how to do this.

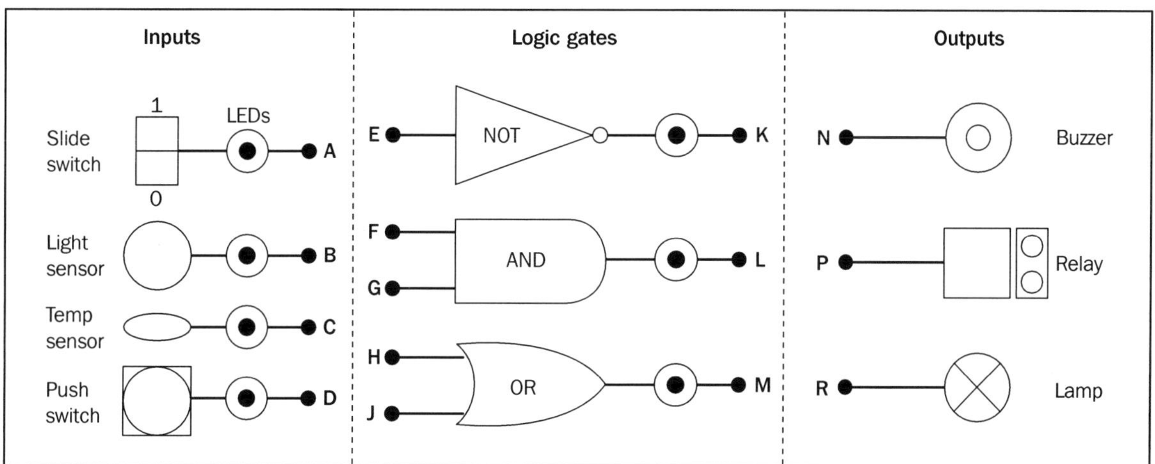

2 Move the slide switch to position 1 and watch the LED (light emitting diode).

What happens when the slide switch moves between positions 0 and 1?

3 Cover the light sensor with your finger tip and watch the LED.

What happens when the light sensor is covered/in the dark?

4 Breathe warm air gently onto the temperature sensor and watch the LED.

What does a temperature sensor do when the temperature rises?

5 Press the push switch and watch the LED.

- What happens when the push switch is pressed?
- What is the difference between the way in which the push switch and slide switch work?

SAFETY WARNING

Check electrical equipment before use.

Report any faulty equipment.

18.3a
Practical notes

Electronic systems 1

This activity sheet is based on the popular Microelectronics for all (MFA) equipment. Other boards are available in which the components may be positioned and/or labelled differently. In this activity pupils examine the functions of four common input components (a slide switch, a light sensor, a temperature sensor, and a push switch). Some pupils may need help connecting the power supply. It is advisable to have each board checked for correct operation before starting the activity.

18.3a
Technician's notes

Electronic systems 1

Each group will need:

- a copy of worksheet 18.3a
- microelectronics board
- power supply.

Number of apparatus sets:

Number of pupils:

Number of groups:

Visual aids:

ICT resources:

Equipment/apparatus needed:

Safety notes

- **Check electrical equipment before use.**
- **Report any faulty equipment.**

CLEAPSS/SSERC SAFETY REFERENCE:

18.4b Electronic systems 2 W/S

Name: **Date:** **Group:**

What you need:

Microelectronics board, power supply, motor unit, connecting wire.

What to do:

1 Connect the power supply to the microelectronics board. Check with your teacher if you are not sure how to do this.

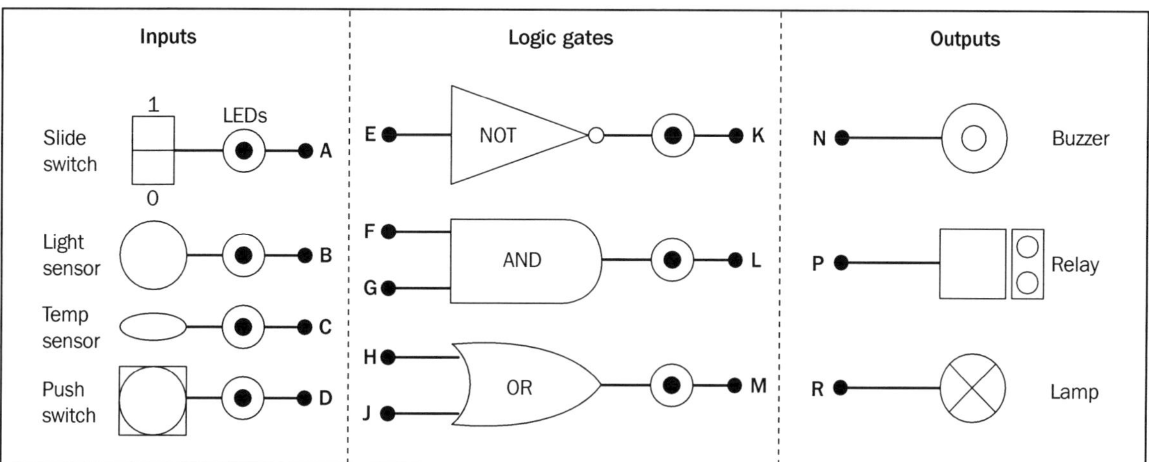

2 Connect a wire from the slide switch (connector A) to the buzzer (connector N).

3 Move the slide switch to position 1 then back to position 0.

What happens when you move the slide switch?

4 Change the connections so that the slide switch (connector A) is connected to the lamp (connector R).

5 Move the slide switch to position 1 then back to position 0.

What happens when you move the slide switch?

6 Change the connections so that the slide switch (connector A) is connected to the relay (connector P).

7 Move the slide switch to position 1 then back to position 0.

What happens when you move the slide switch?

8 Connect a motor to the relay. Ask your teacher if you are not sure how to do this.

9 Move the slide switch to position 1 then back to position 0.

What happens when you move the slide switch?

SAFETY WARNING

Check electrical equipment before use.

Report any faulty equipment.

18.3b
Practical notes

Electronic systems 2

This activity sheet is based on the popular Microelectronics for all (MFA) equipment. Other boards are available in which the components may be positioned and/or labelled differently. In this activity pupils examine the functions of three common output components (a buzzer, a lamp, and a relay). Each component is controlled by a push switch, although other inputs could be used as well. Other output devices are also available, including numerical display units. Some pupils may need help connecting the power supply and in particular, attaching the motor to the relay. It is advisable to have each board checked for correct operation before starting the activity.

18.3b
Technician's notes

Electronic systems 2

Each group will need:

- a copy of worksheet 18.3b
- microelectronics board
- power supply
- motor unit
- connecting wire.

Number of apparatus sets:

Number of pupils:

Number of groups:

Visual aids:

ICT resources:

Equipment/apparatus needed:

Safety notes

- **Check electrical equipment before use.**
- **Report any faulty equipment.**

CLEAPSS/SSERC SAFETY REFERENCE:

18.3c Electronic systems 3

W/S

Name: **Date:** **Group:**

What you need:

Microelectronics board, power supply, connecting wire.

What to do:

1 Copy this table into your book:

Binary code	Electric current	voltage	LED	Slide switch	Light sensor	Temp. sensor	Push switch	buzzer	relay	lamp
0	off	Low/zero	unlit							
1	on	high	lit							

2 Connect the power supply to the microelectronics board. Check with your teacher if you are not sure how to do this.

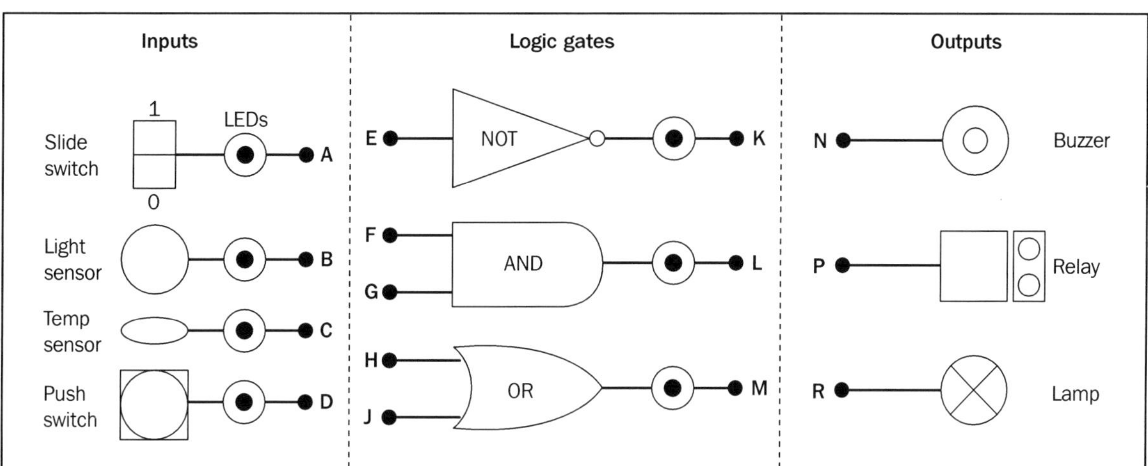

3 Check each input in turn to find out what conditions make it binary code 1 or binary code 0. Put your results in the table.

4 Connect a wire from the slide switch (connector A) to each output in turn. Find out what each one does when it is in binary code 1 and binary code 0. Put your results in the table.

SAFETY WARNING

Check electrical equipment before use.

Report any faulty equipment.

18.3c
Practical notes

Electronic systems 3

This activity sheet is based on the popular Microelectronics for all (MFA) equipment. Other boards are available in which the components may be positioned and/or labelled differently. In this activity pupils examine three common output components (a buzzer, a lamp, and a relay) and assign binary codes to their function. Each component is controlled by a push switch, although other inputs could be used as well. The activity develops the ideas covered in activity 18.3d and could be combined with it depending upon pupil ability. Some pupils may need help connecting the power supply. It is advisable to have each board checked for correct operation before starting the activity.

18.3c
Technician's notes

Electronic systems 3

Each group will need:

- a copy of worksheet 18.3c
- microelectronics board
- power supply
- connecting wire.

Number of apparatus sets:

Number of pupils:

Number of groups:

Visual aids:

ICT resources:

Equipment/apparatus needed:

Safety notes

- **Check electrical equipment before use.**
- **Report any faulty equipment.**

CLEAPSS/SSERC SAFETY REFERENCE:

18.3d Electronic systems 4

W/S

Name: **Date:** **Group:**

What you need:

Microelectronics board, power supply, two connecting wires.

What to do:

1 Copy this table into your book:

input	LED (lit or unlit)	Binary code (0 or 1)	Logic gate	LED (lit or unlit)	Binary code (0 or 1)
Slide switch			NOT		
Slide switch			NOT		
Slide switch			AND		
Push switch			AND		
Slide switch			OR		
Push switch			OR		

2 Look carefully at the logic gates on the microelectronics board. At the end of each gate is an LED which lights up when its output is binary code 1 (on).

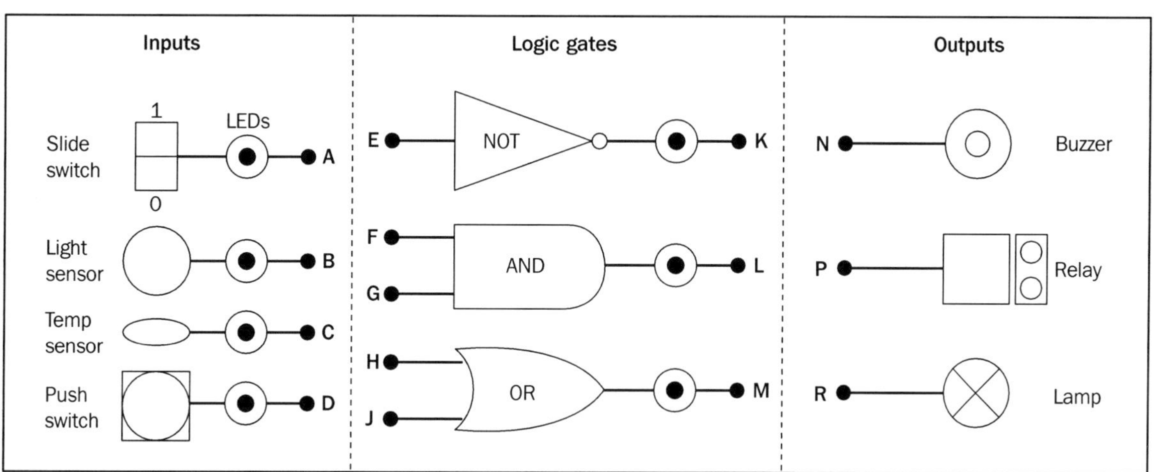

3 Connect the power supply to the microelectronics board. Check with your teacher if you are not sure how to do this.

4 Connect a wire from the slide switch (connector A) to the NOT gate (connector E).

5 Move the slide switch to position 1 and back to position 0. Watch what happens to the LED at the end of the NOT gate. Write your observations in the table.

6 The AND gate needs two inputs. Connect the slide switch (connector A) to the AND gate (connector F). Connect the push switch (connector D) to the AND gate (connector G).

7 Use the two switches to find out what positions make the LED at the end of the AND gate light up. Put your results in the table.

8 The OR gate also needs two inputs. Connect the slide switch (connector A) to the OR gate (connector H). Connect the push switch (connector D) to the OR gate (connector J).

9 Use the two switches to find out what positions make the LED at the end of the OR gate light up. Put your results in the table.

18.3d
Practical notes

Electronic systems 4

This activity sheet is based on the popular Microelectronics for all (MFA) equipment. Other boards are available in which the components may be positioned and/or labelled differently. In this activity pupils use the two switch inputs to examine the function of the NOT, AND, and OR gates. Some pupils may need help connecting the power supply. It is advisable to have each board checked for correct operation before starting the activity.

18.3d
Technician's notes

Electronic systems 4

Each group will need:

- a copy of worksheet 18.3d
- microelectronics board
- power supply
- connecting wires.

Number of apparatus sets:

Number of pupils:

Number of groups:

Visual aids:

ICT resources:

Equipment/apparatus needed:

Safety notes

- **Check electrical equipment before use.**
- **Report any faulty equipment.**

CLEAPSS/SSERC SAFETY REFERENCE:

18.1a Electromagnets

H/W

Name: **Date:** **Group:**

What you need to know ...

When an electric current flows through a wire, the wire has a magnetic field round it. This magnetic field is weak, but can be made stronger by wrapping wire around an iron core.

What to do:

1 Copy this diagram into your book.

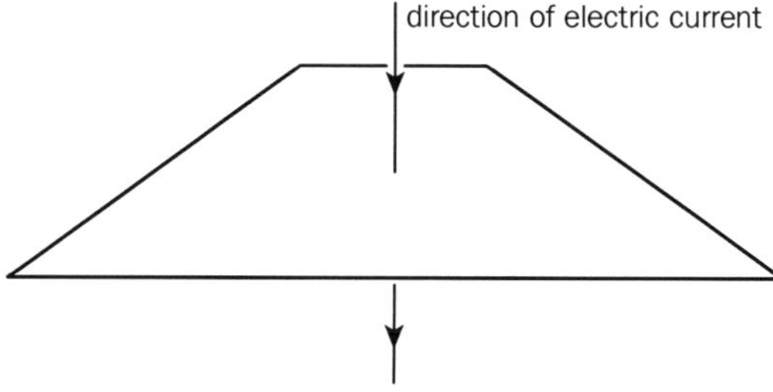

Draw lines on the diagram to show the shape of the magnetic field produced when a current passes through the wire.

2 Copy this diagram of a simple electromagnet.

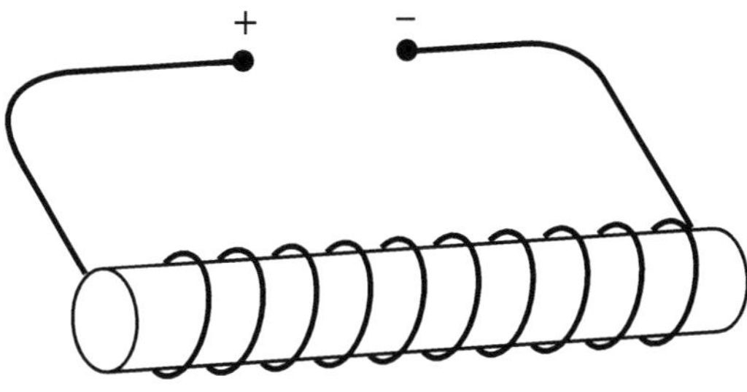

Draw lines on the diagram to show the shape of the magnetic field produced when a current passes through the wire coil.

a What happens to the poles of the electromagnet if you change the direction of current flow?

b What changes could you make to the electromagnet shown in the diagram to make it stronger?

3 Electromagnets are useful in scrapyards. They are hung on the ends of cranes to pick up scrap iron objects, separating them from other metals.

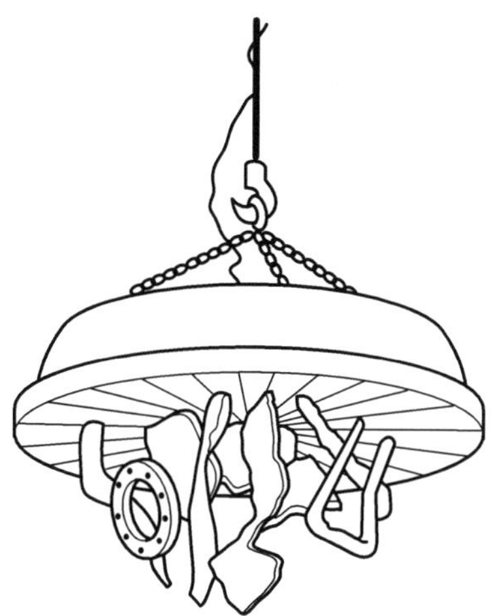

a What is the main difference between an electromagnet and a permanent magnet that makes it good for its job in a scrapyard?

b Why is it useful to be able to sort out scrap metal?

4 Electrical appliances in the home make use of the heating and magnetic effects of an electric current.

Make a list of three things in your home that use the heating effect and a list of three things that use the magnetic effect of an electric current.

HANDY HINTS

Electromagnets are used in electric motors.

18.1b Using electromagnets

H/W

Name: **Date:** **Group:**

What you need to know …

A relay is a switch operated by an electromagnet. With a relay it is possible to use a tiny switch with thin wires to turn on a circuit with a much higher current or voltage. An electric bell uses a sort of 'make and break' circuit that repeatedly switches an electromagnet on and off.

What to do:

1 The diagram shows a relay.

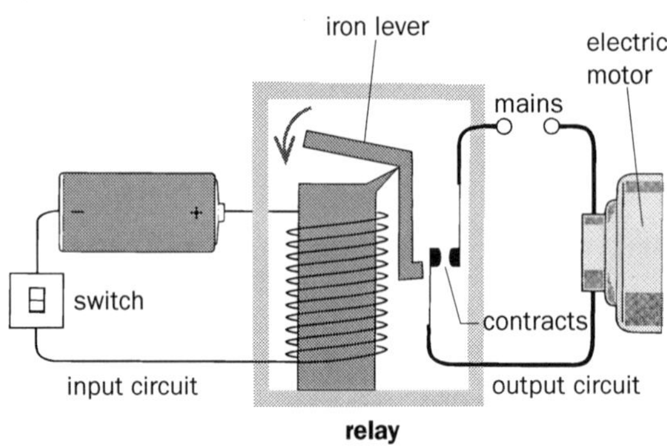

a Copy the passage below filling in the gaps using these words:

- attracts
- current
- electric motor
- electromagnet
- input
- output
- touch

The switch in the input circuit is closed which causes a _______________ to flow in the _______________ circuit. This makes the coil become an _______________ . The coil _______________ the iron lever which pivots causing the contacts to _______________ each other. This allows a current to flow in the _______________ circuit making the _______________ work.

b Explain why a relay is a safety device.

2 The diagram shows an electric bell.

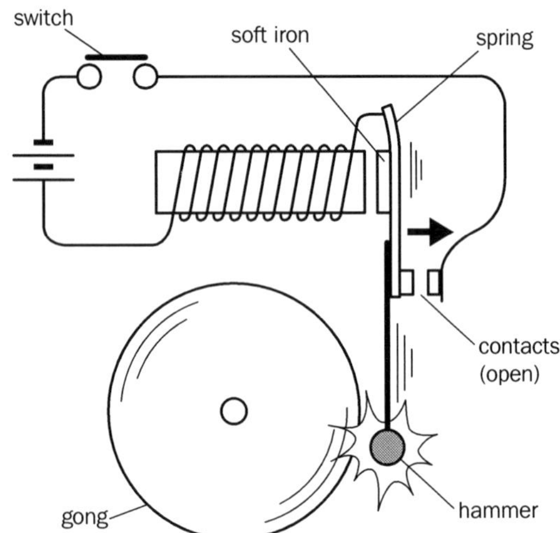

a Describe the series of events that make the bell ring. Start from the moment the switch is closed.

b Why has the hammer arm got soft iron on it?

c Explain why it is important for the contacts to be kept clean?

HANDY HINTS

In Question 2 use the diagram of the bell to help you.
Write a separate sentence for each thing that happens.

18.3a Using logic gates

H/W

Name: Date: Group:

What you need to know ...

Logic gates are the decision makers of electronic systems. Different types of logic gate give different outputs for the same input, and this can be used to design useful circuits. When the input of a NOT gate is 1 its output is 0. The output of an AND gate is 1 when both inputs are 1. The output of an OR gate is 1 when either one of two inputs is 1.

What to do:

Design the following three circuits.

Circuit 1

Design a circuit that will make a lamp come on automatically when it goes dark. Show the connections between the components clearly on the diagram below. List the inputs, logic gates, and outputs you have used in your circuit.

Circuit 2

Design a circuit for a cooling fan that will come on automatically when it gets warm. You must be able to turn the fan off when you want to. The fan will be driven by a motor connected to the relay. List the inputs, logic gates, and outputs you have used in your circuit.

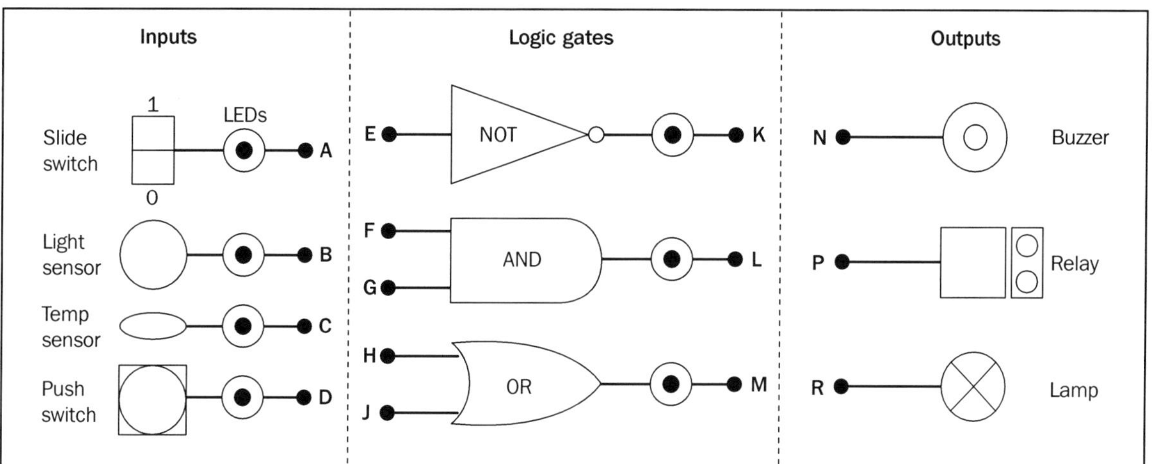

Circuit 3

Design a circuit to control a motorised machine. The machine is operated by a switch so that it can be switched on or off for long periods. Sometimes mechanics need to make tiny adjustments and need to be able to move the machine a bit at a time using a push switch.

List the inputs, logic gates, and outputs you have used in your circuit.

18.3b Logic gates and truth tables

Name: Date: Group:

What you need to know ...

Logic gates are the decision makers of electronic systems. Different types of logic gate give different outputs for the same input, and this can be used to design useful circuits. When the input of a NOT gate is 1 its output is 0. The output of an AND gate is 1 when both inputs are 1. The output of an OR gate is 1 when either one of two inputs is 1. Truth tables are a shorthand way of showing the possible outputs of logic gates for all possible inputs.

What to do:

1 Copy and complete this truth table for a NOT gate:

Input	Output
0	
1	

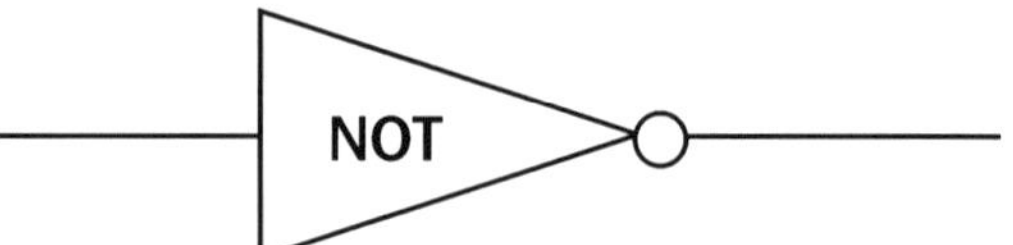

2 Copy and complete this truth table for an OR gate:

Input 1	Input 2	Output
0	0	
0	1	
1	0	
1	1	

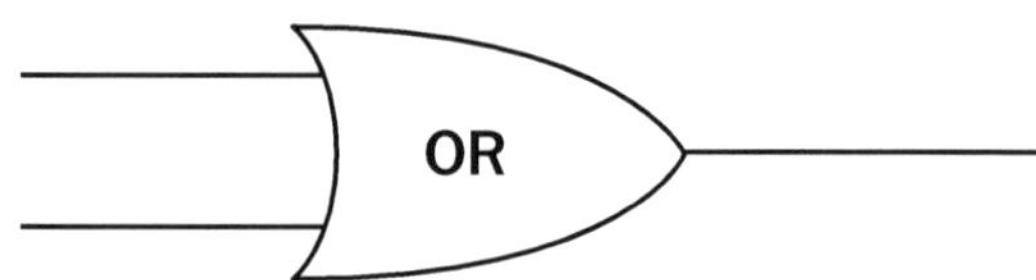

3 Copy and complete this truth table for an AND gate:

Input 1	Input 2	Output
0	0	
0	1	
1	0	
1	1	

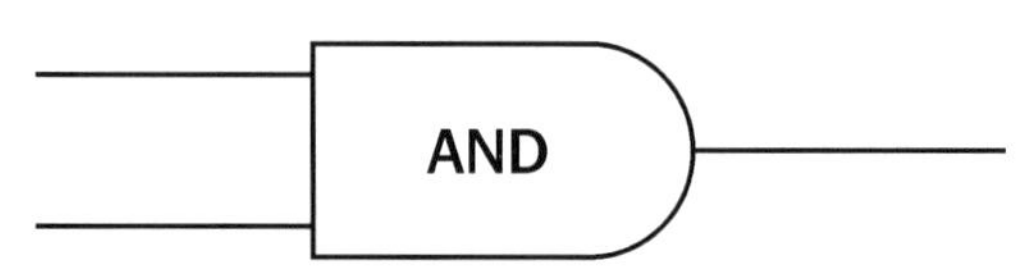

4 Design a circuit for a fire alarm that detects smoke or heat. Use a NOT gate and an OR gate.

Write a truth table for your circuit.

Chapter 18 ► Investigation 18F Electromagnets

An electromagnet is made by wrapping a coil of wire around a soft iron core.
A magnetic field is produced when a current passes through the wire.

In this investigation: you are going to find out how the strength of an electromagnet is affected by:

- the number of turns of wire on the coil
- the size of the current passing through the wire.

Preparation: Predict

Finish the sentences in the box.

What I think will happen is…

I think this because…

Preparation: Plan

Write a short plan of your investigation.

Think about:

- the apparatus you are going to use
- how one variable depends upon another variable
- what you are going to measure and how you are going to measure it
- how many readings you are going to take
- how you are going to record your results
- how you are going to make your investigation fair
- how you are going to make your investigation safe.

Show your plan to your teacher before going on.

Carry out

Carry out your investigation and record your results.

Present your results in an appropriate way.

Report

Write a report on your investigation.

Here are some things you should include:

- what you did
- what happened
- explain your results
- if your prediction was correct or not
- how reliable your results were
- what you could have done if you had more time.

Chapter 18 ► Investigation 18E Electromagnets

	An electromagnet is made by wrapping a coil of wire around a soft iron core. A magnetic field is produced when a current passes through the wire.
In this investigation:	you are going to find out how the strength of an electromagnet is affected by: • the number of turns of wire on the coil • the size of the current passing through the wire.

Preparation: Predict

Finish the sentences in the box.

What I think will happen is…

I think this because…

Preparation: Plan

Write a short plan of your investigation.

Think about:

- the apparatus you are going to use
- what you are going to measure and how you are going to measure it
- how many readings you are going to take
- how you are going to record your results
- how you are going to make your investigation fair
- how you are going to make your investigation safe.

Show your plan to your teacher before going on.

Carry out

Carry out your investigation and record your results in a table.

Draw a bar graph of your results.

Report

Write a report on your investigation.

Here are some things you should include:

- what you did
- what happened
- explain your results
- if your prediction was correct or not
- how reliable your results were
- what you could have done if you had more time.

Chapter 18 ► Investigation 18D Electromagnets

	An electromagnet is made by wrapping a coil of wire around a soft iron core. A magnetic field is produced when a current passes through the wire.
In this investigation:	you are going to find out how the strength of an electromagnet is affected by: • the number of turns of wire on the coil • the size of the current passing through the wire.

Preparation: Predict

Finish the sentence in the box.

I think that the strength of an electromagnet (is/is not) affected by:
- *the number of turns of wire on the coil*
- *the size of the current passing through the wire because…*

You are going to use this equipment to find out how the strength of an electromagnet is affected by:

- the number of turns of wire on the coil
- the size of the current passing through the wire.

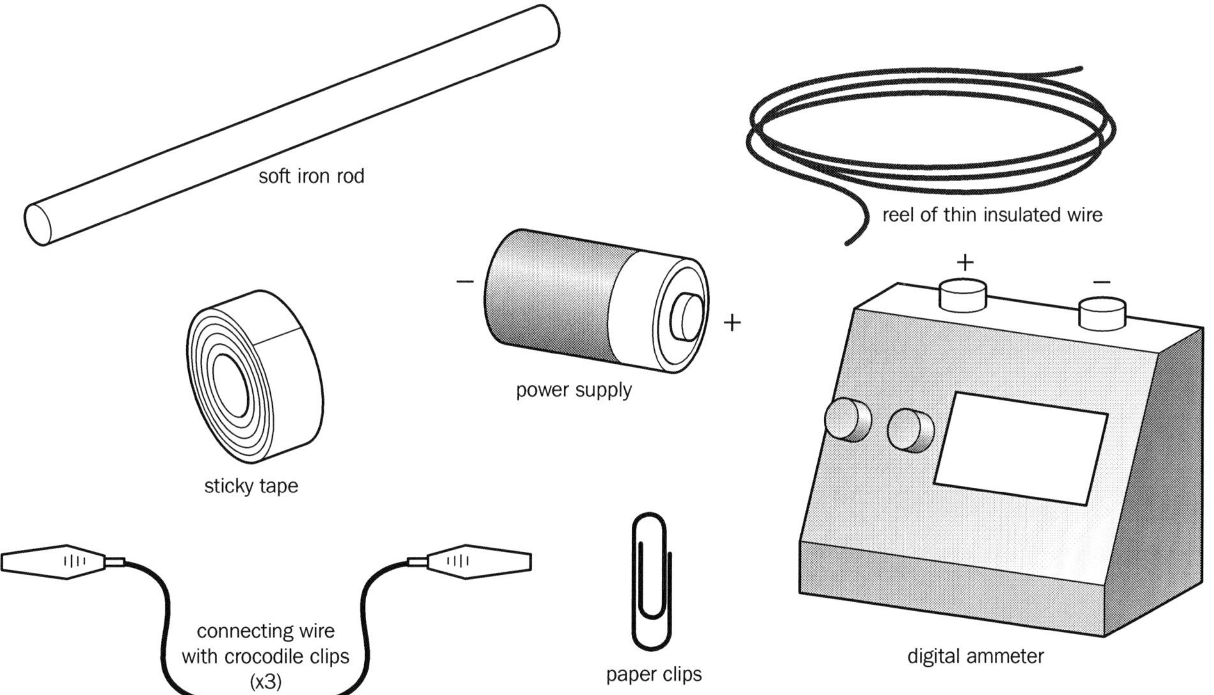

Chapter 18 ▶ Investigation 18D Electromagnets

Preparation: Plan

Finish the sentences in the box.

I will measure…

Things I will keep the same are…

My investigation will be fair because…

My investigation will be safe because…

Carry out

Do your experiment first with a few turns of wire around the soft iron rod then double the number of turns. Use a low current then a larger current for each of your two coils. The stronger your electromagnet, the more paper clips it will pick up.

Put your results in a table like this:

Current	Number of turns in coil	Number of paper clips held
	10	
	10	
	20	
	20	

Draw a bar graph of your results on a piece of graph paper.

Use a key with different colours for each result. Label the axes like this:

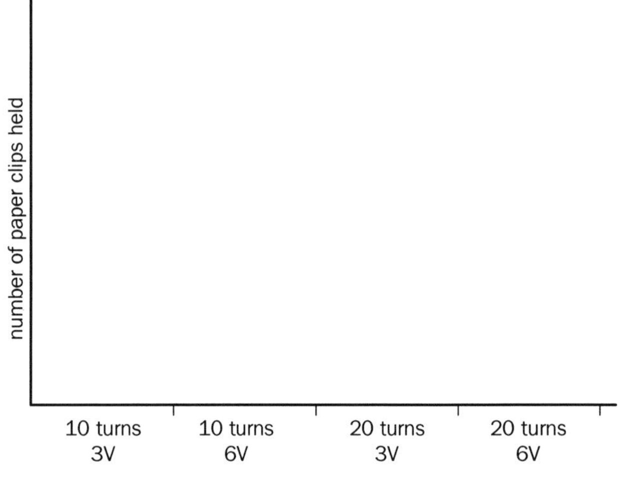

Report

Write a report on your investigation.

Here are some things you should include:

- what you did
- what happened
- explain your results
- if your prediction was correct or not
- what you could do to improve the investigation
- what you could have done if you had more time.

Chapter 18 ▶ Investigation 18C Electromagnets

	An electromagnet is made by wrapping a coil of wire around a soft iron core. A magnetic field is produced when a current passes through the wire.
In this investigation:	you are going to find out how the strength of an electromagnet is affected by: • the number of turns of wire on the coil • the size of the current passing through the wire.

Preparation: Predict

Finish the sentence in the box.

I think that the strength of an electromagnet (is/is not) affected by:
- *the number of turns of wire on the coil*
- *the size of the current passing through the wire because…*

You are going to use this equipment to find out how the strength of an electromagnet is affected by:

- the number of turns of wire on the coil
- the size of the current passing through the wire.

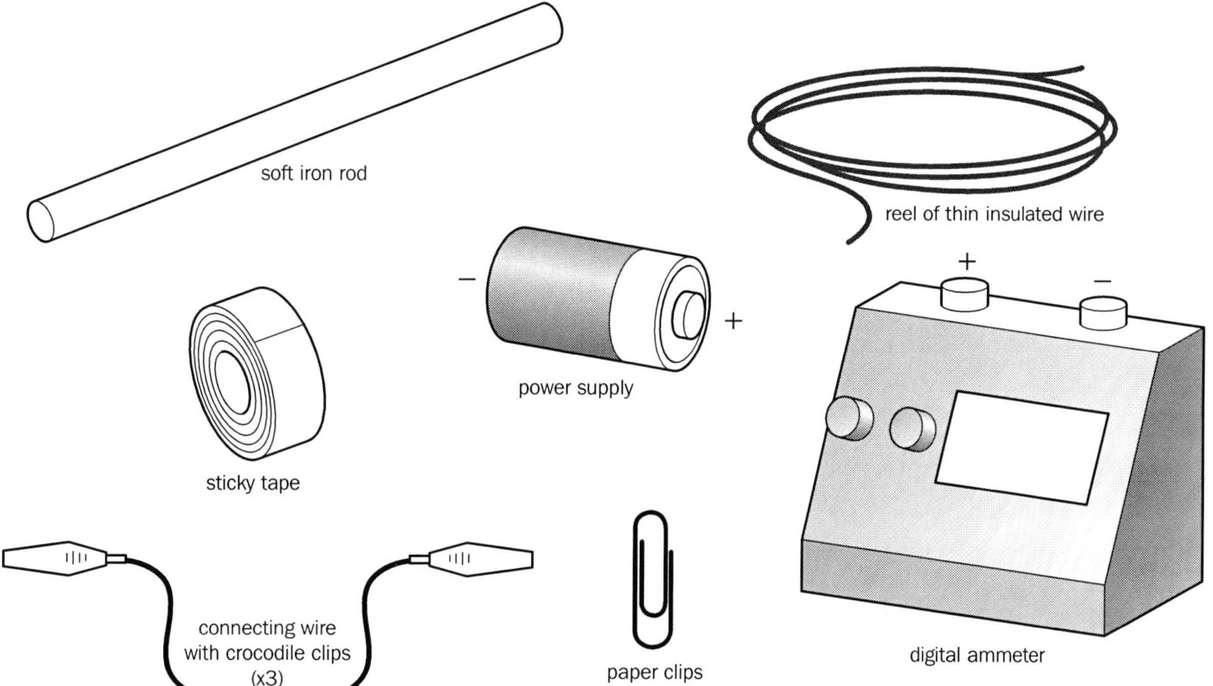

Chapter 18 ► Investigation 18C Electromagnets

Preparation: Plan

Finish the sentences in the box.

I will measure…

Things I will keep the same are…

My investigation will be fair because…

My investigation will be safe because…

Carry out

- Wind ten turns of wire around the soft iron rod.
- Make this circuit:

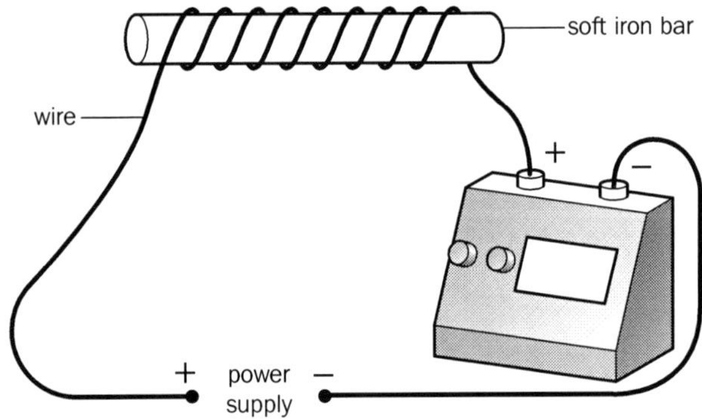

- Set the power supply to 3 volts and switch it on.
- See how many paper clips your electromagnet will hold.
- Increase the voltage to 6 volts and see how many paper clips your electromagnet will hold this time.
- Switch off the power supply and disconnect your electromagnet.
- Wind 20 turns of wire around the soft iron rod.
- See how many paper clips your electromagnet will hold at 3 volts and 6 volts.

Put your results in a table like this:

Voltage	Current	Number of turns in coil	Number of paper clips held
3 volts		10	
6 volts		10	
3 volts		20	
6 volts		20	

Chapter 18 ► Investigation 18C Electromagnets

Draw a bar graph of your results on this grid.

Use different colours for each bar.

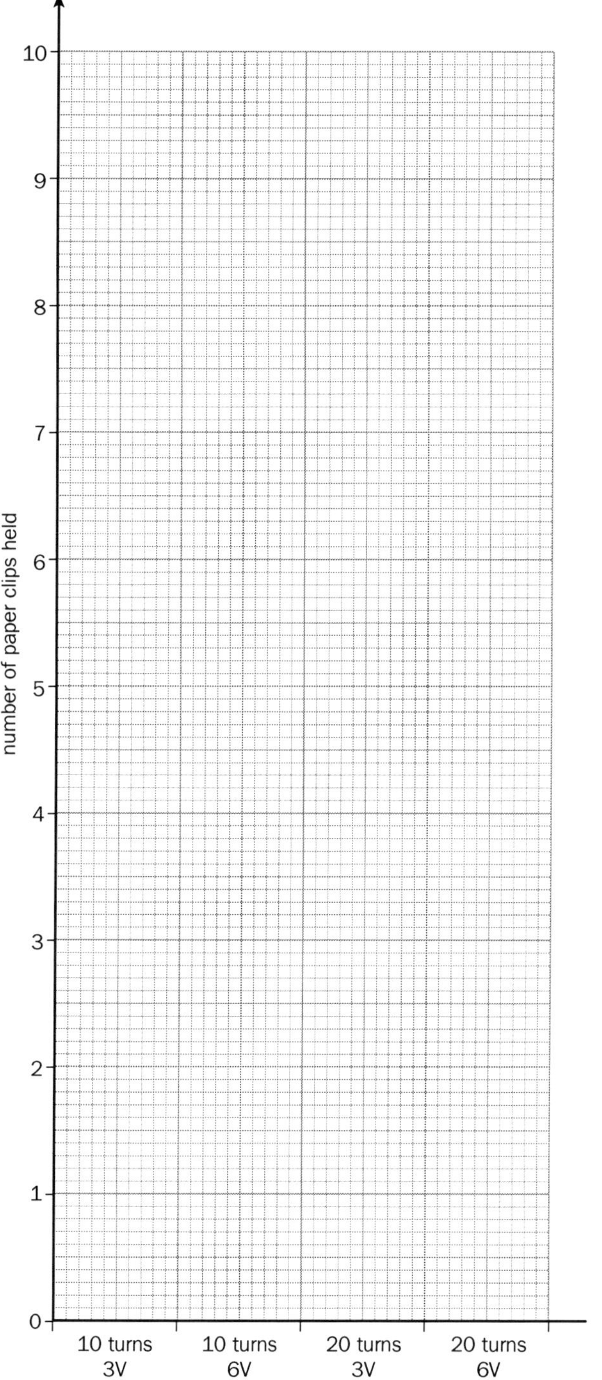

Report

Finish the sentences in the box.

What I did was…

What happened was…

From my results I found out that the strength of an electromagnet (is/is not) affected by:

- *the number of turns of wire on the coil*
- *the size of the current passing through the wire. I know this because…*

My prediction (was/wasn't) correct. If I could do the investigation again I would…

Investigation 18
Practical notes

Electromagnets

Investigation 18
Technician's notes

Electromagnets

Each group will need:

- soft iron rod
- about 1.5 m of thin insulated wire
- power supply to supply 3 V and 6 V
- three connecting wires with crocodile clips
- digital ammeter
- access to lots of paper clips
- sticky tape.

Number of apparatus sets:

Number of pupils:

Number of groups:

Visual aids:

ICT resources:

Equipment/apparatus needed:

Safety notes

Pupils should be warned that the wire may get hot during this activity.

CLEAPSS/SSERC SAFETY REFERENCE:

Chapter 18 ▶ Test
Electromagnetism/electronics

1 The diagram shows a wire passing through a piece of card. The wire is carrying an electric current.

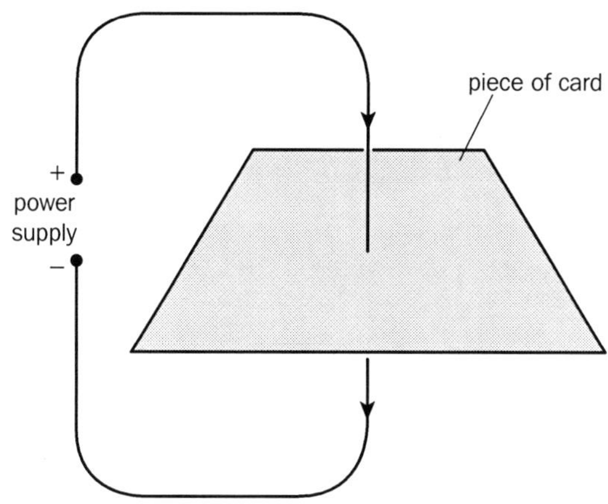

a How could you show that a magnetic field is produced by the current flowing in the wire?

b Draw the shape of the magnetic field produced by the current flowing in the wire.

c What would happen to the
   i direction
   ii strength of the magnetic field if the current in the wire was reversed?

4 marks

2 Diagram A shows a coil of wire. Diagram B shows a coil of wire wrapped round an iron core.

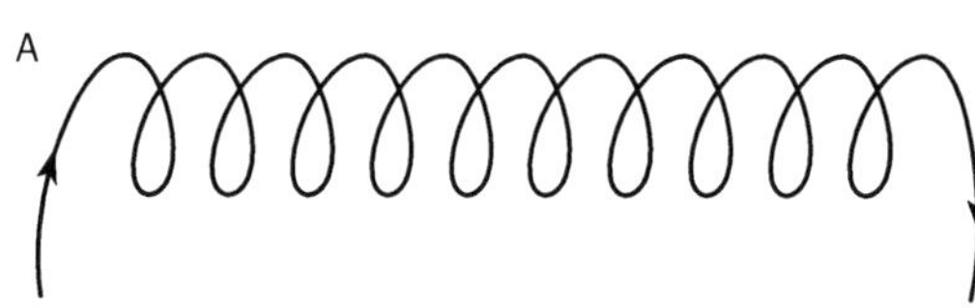

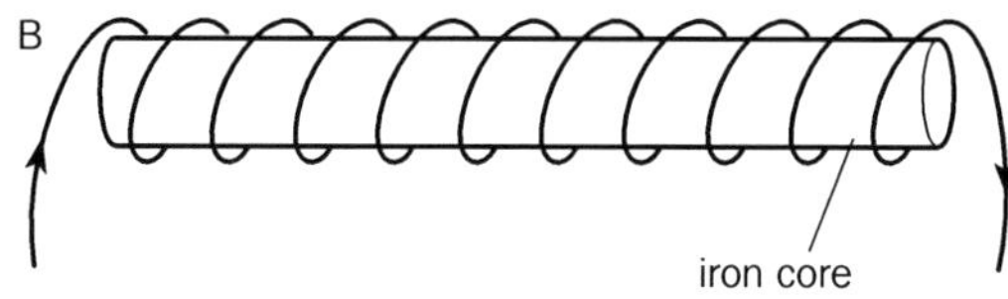

a Explain why the magnetic field around B is stronger than the magnetic field around A.

b Give one reason why the core is not made of steel

c Give two other ways the magnetic field around B can be made stronger.

4 marks

3 The diagram shows an electric bell.

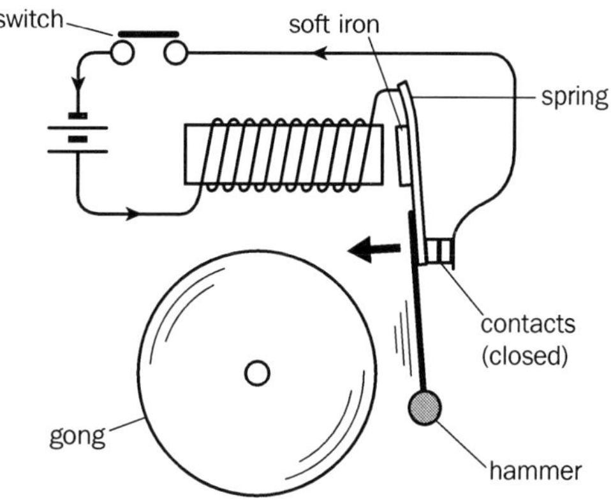

The following sentences describe how the electric bell works. They are in the wrong order. Write the letter for each sentence in the correct order starting with letter C

A The spring pulls the hammer arm back into the starting position.

B The electromagnet attracts the iron and the hammer hits the gong.

C When the switch is pushed a current flows in the coil.

D The electromagnet no longer attracts the iron.

E Because of the movement of the hammer, the contacts separate and no current flows.

F The contacts close and the current starts to flow again.

5 marks

4 Which of these components used in electrical systems are input components?

buzzer
heater
light dependent resistor (LDR)
light emitting diode (LED)
microphone
motor
reed switch
relay
thermistor
loudspeaker.

4 marks

5 Logic gates are the decision makers in electronic systems.

a Give the outputs for each of these logic gates.

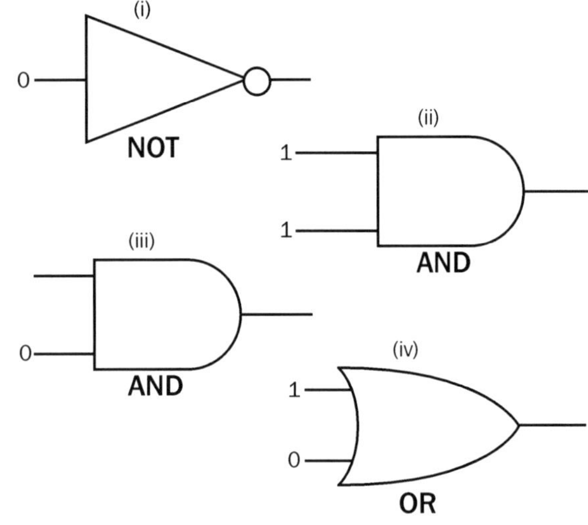

b What sort of logic gates would you use for these electronic systems:

i a cooling fan comes on either when it is switched on or when it gets warm

ii a sprinkler comes on when it is dark and cool?

7 marks

6 The diagram shows an electronic system found in modern motor cars.

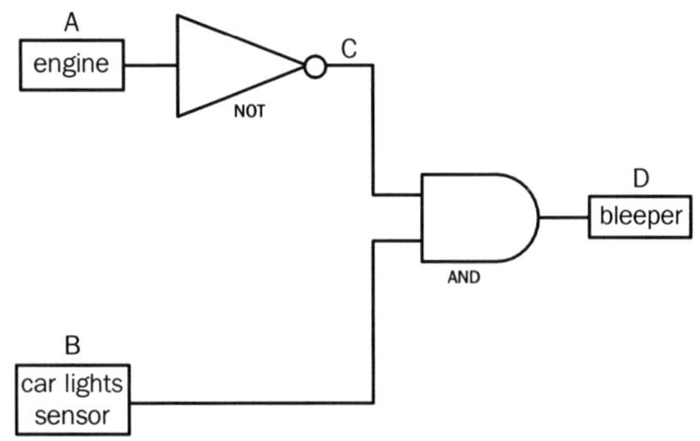

a Describe what the circuit does.

b Copy and complete this truth table for the electronic system.

Input A (engine)	Input B (light sensor)	Input C (from NOT gate)	Output (bleeper)
0			
1			
0			
1			

6 marks

Chapter 18 ► Mark scheme Electromagnetism/electronics

White/Blue

Question	Answer	Marks	Level
1 a	plotting compass/ iron filings	1	F
b	circular	1	F
c i	changes to opposite direction	1	
ii	stays same	1	F
		4	
2 a	iron has magnetic field/ is magnetised	1	F
b	steel stays magnetised when current is off	1	F
c	increased current (1) more turns on coil (1)	2	F
		4	
3	C B E D A F (all in order = 5 marks, 4 in order = 4 marks etc.)	5	F
		5	
4	LDR (1) microphone (1) reed switch (1) thermistor (1)	4	F
		4	

Question	Answer	Marks	Level
5 a i	1 (on)	1	
ii	1 (on)	1	
iii	0 (off)	1	
iv	1 (on)	1	F
b i	NOT gate	1	
ii	NOT gate(s) (1) and AND gate (1)	2	F
		7	
6 a	bleep if lights on (1) and engine off (1)	2	F
b	(see table below) (1 mark for each correct row)	4	F
		6	
	TOTAL 30 marks		

Input A (engine)	Input B (light sensor)	Input C (from NOT gate)	Output (bleeper)
0	0	1	0
1	0	0	0
0	1	1	1
1	1	0	0

Suggested grade/level boundaries

E = 13/30*

F = 25/30

* although this is only a 'Level F' chapter as much of the material depends on previous knowledge at 'Level E'.

(This page is intentionally blank.)

19 Our environment

This chapter builds on work begun in Book 1. It begins by looking at the way human activities influence our environment. The impact of population growth and pollution is considered along with ways of reducing their impact. The chapter goes on to look at interactions between animal and plant species and how they have evolved to adapt or become extinct in an ever changing world.

Assessment opportunities

Formative assessment opportunities are provided by worksheets, homework sheets, and an investigation.

The **worksheets** cover material at levels C and D for attainment targets for knowledge and understanding. Teachers may wish to use these worksheets not only as part of practical activities but also to provide evidence of pupil achievement.

Worksheet	Level
19.1a	C
19.1b	D
19.1c	D
19.2a	D
19.2b	D

The **homework** sheets cover material at levels D and E for attainment targets for knowledge and understanding. These homework sheets can be used individually as a follow-up to work done in class or assembled into a homework booklet allied closely to schemes of work.

Homework sheet	Level
19.1a	D
19.1b	D
19.1c	E
19.2	D

The **investigation** covers all three skill areas at levels C, D, E, and F. It is written in a way that allows for pupils to be assessed in all three skill areas at one level. Alternatively, customised assessments can be constructed enabling pupils to be assessed at different levels in all three skills. The latter approach is more time consuming, but it does provide the opportunity for pupils to show evidence of achievement at different levels in different skills in the same investigation. Teachers will need to use their professional judgement when deciding which level is appropriate to individual pupils. It is envisaged that pupils will show progression through the levels as they work through their science course.

Summative tests are provided at two levels, white and blue. The white test contains questions covering levels C and D. The blue test contains questions covering attainment target levels D and F but can be used for assessing at levels D, E, and F. Each test has a total of 30 marks and will take about 30 minutes for pupils to complete, although this can be varied depending on pupil ability. Mark schemes are provided together with suggested grade/level boundaries. It is envisaged that these tests will be given to pupils on completion of the material covered in Chapter 19.

ICT opportunities

The use of data loggers/remote sensors can extend the range, speed, and sensitivity of measurements in many of the worksheets for this chapter. Once downloaded onto a PC, data-handling programs can be used to analyse information gathered, data can be manipulated, and appropriate graphs etc. presented. The Internet provides pupils with access to a huge range of scientific information. A list of suitable websites is included in this Teacher's Guide.

Students' book chapter 19 contents and guide levels

19.1	The environment	*Starting off 1*	C/D
	Protecting the environment	*Starting off 2*	C/D
	Humans and the environment	*Going further*	C/D
	Rare and extinct	*For the enthusiast*	C
19.2	Adapted for life (1)	*Starting off 1*	D
	Adapted for life (2)	*Starting off 2*	D
	Surviving change	*Going further*	D
19.3	Evolution (1)	*Starting off*	F
	Evolution (2): the theory	*Going further*	F
	Evolution (3): mutations	*For the enthusiast*	F

19.1a How long does it take to rot? W/S

Name: Date: Group:

What you need:

Leaves, grass, paper, cardboard, scissors, plastic mesh, plastic bag ties, trowel, plastic gloves.

What to do:

1 Cut up some grass into small pieces, about 2 cm long.

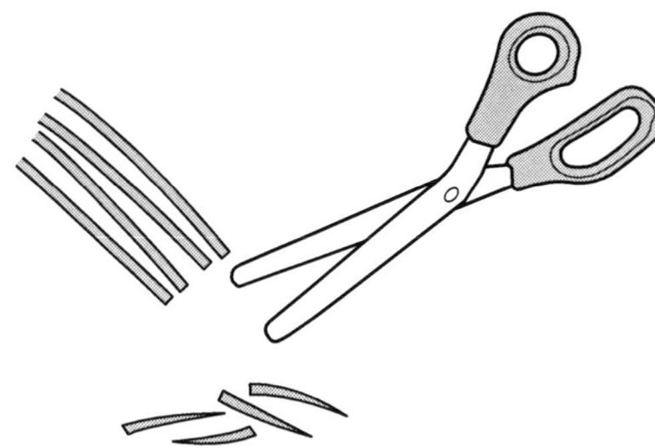

2 Put the pieces of grass into some plastic mesh and tie this into a parcel with a plastic tie.

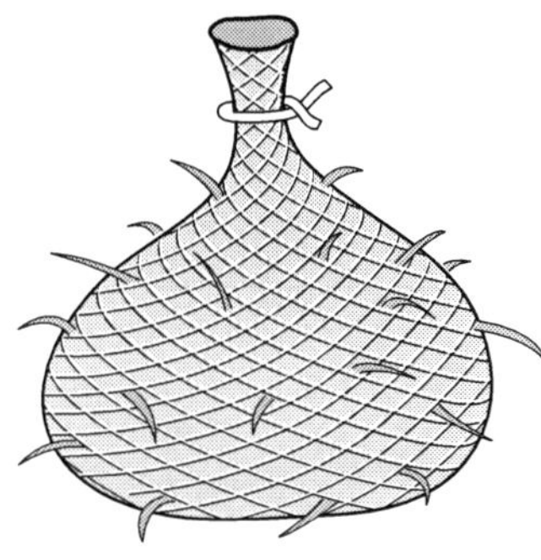

3 Do the same with pieces of leaf, paper, and cardboard, cut them into roughly 2 cm squares.

4 Find a piece of undisturbed ground outside. Dig four holes close together in the soil and bury your parcels. Write your name on a marker peg and stick it into the ground where you buried your parcels.

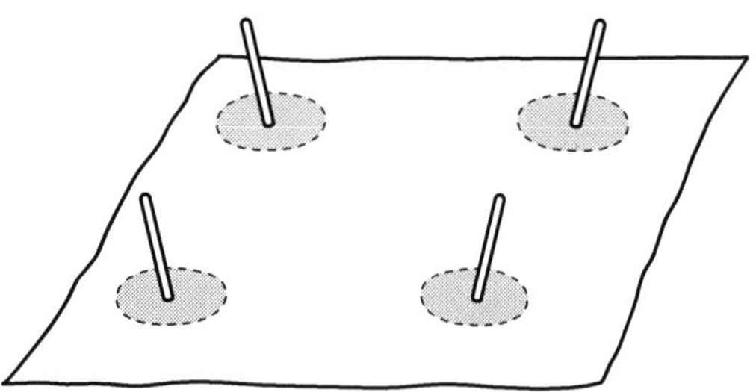

5 After two weeks, dig up your parcels and look carefully at the contents of each one.

Describe what the contents of each parcel look like. Have they got smaller? If so, suggest how.

6 Put the parcels back in the ground and leave them for two more weeks. Don't forget the markers.

- Which material has rotted fastest?
- Why hasn't the plastic mesh rotted?

19.1a Practical notes

How long does it take to rot?

This activity links work on decay and composting with aspects of recycling and the future of landfill sites. It can be modified according to class size and ability. 'Division of labour' seems an obvious way of cutting down on time and cost, and avoids digging up the whole of the school grounds! Plastic mesh 'orange' bags make excellent parcel material. Make sure pupils wear gloves when handling rotting materials.

19.1a Technician's notes

How long does it take to rot?

Each group will need:

- a copy of worksheet 19.1a
- grass
- leaves (large and soft, e.g. horse chestnut)
- paper (newspaper will do)
- cardboard (thick and dense)
- scissors
- plastic mesh ('orange' bags from supermarkets are good)
- plastic bag ties
- trowel
- plastic gloves.

Number of apparatus sets:

Number of pupils: Number of groups:

Visual aids:

ICT resources:

Equipment/apparatus needed:

Safety notes

- **Pupils must wear gloves when handling rotting materials.**
- **Arrange for safe disposal of any rotten material at the end of the activity.**

CLEAPSS/SSERC SAFETY REFERENCE:

19.1b Measuring plant growth

W/S

Name: Date: Group:

What you need:

Plant pot, moist compost, wheat seeds, labels, mm ruler.

What to do:

1 Put some compost into the plant pot. Make sure the compost is moist then put some seeds on the compost close to the edge of the pot. Push the seeds slightly below the surface of the compost.

2 Put the pot of seeds in a warm place and leave them until they start to grow. This may take a few days. Remember to keep the compost moist during this time.

3 When ten of the seeds have started to grow throw the other seeds away.

4 Number your seedlings by sticking labels on the side of the pot.

5 Copy this table into your book:

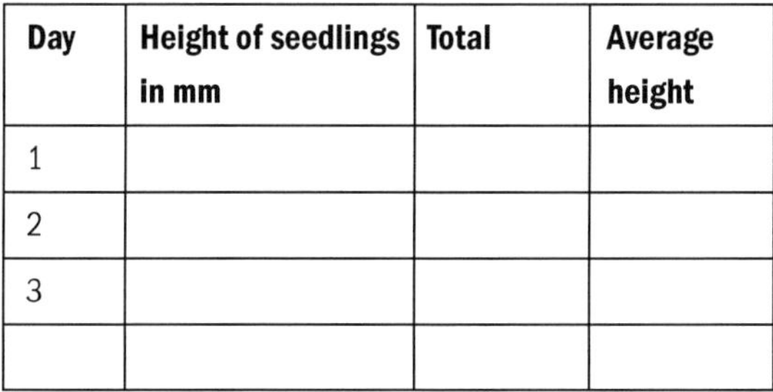

Day	Height of seedlings in mm	Total	Average height
1			
2			
3			

6 Measure the seedlings every day. Put your results in the table.

7 When you have enough results (this will take about two weeks) add all the heights in each row together and put the totals in the table.

Work out the average height using the formula:

$$\text{Average height} = \frac{\text{Total height}}{10}$$

Put these results in the table as well.

8 Draw a line graph of your results. Lay out your graph like this:

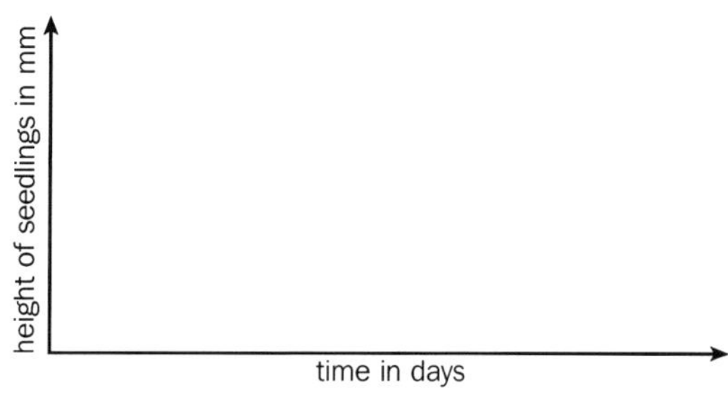

- Did your seedlings grow at a steady rate or did they grow faster on some days than others?
- What sorts of things might affect the speed at which seedlings grow?

19.1b
Practical notes

Measuring plant growth

This activity gives pupils the opportunity to collect data over a two-week period. Make sure that the compost is kept moist, pots could be put in a large tray of water to reduce the risk of spillage when pupils try to water their own. Wheat, oats, or barley seeds are best because the seedlings do not have side shoots. It is advisable to soak the seeds overnight before starting the activity with pupils. The seeds are put around the edge of the pot to make measuring easier. Pupils can measure from the pot edge to the tip of the growing shoot. This activity forms a useful link with Worksheet 19.1c. This gives pupils the opportunity of studying the effect of fertilisers and the reasons for their use/overuse.

19.1b
Technician's notes

Measuring plant growth

Each group will need:

- a copy of worksheet 19.1b
- plant pot or other suitable container with drainage holes
- moist compost
- wheat seeds pre-soaked overnight (barley or oats are also good for this activity)
- labels
- mm ruler.

Number of apparatus sets:

Number of pupils: .... Number of groups: ....

Visual aids: ....

ICT resources: ....

Equipment/apparatus needed: ....

Safety notes

CLEAPSS/SSERC SAFETY REFERENCE:

19.1c Fertilisers and plant growth W/S

Name: Date: Group:

What you need:

Two plant pots, compost moistened with water, compost moistened with fertiliser solution, wheat seeds, labels, mm ruler.

What to do:

1 Label one plant pot 'No fertiliser' and the other plant pot 'Fertiliser'.

2 Put some moist compost into the 'No fertiliser' pot and some compost with fertiliser solution in the 'Fertiliser' pot.

3 Place some seeds (about 15) on the compost in each pot close to the edge. Push the seeds slightly below the surface .

4 Put the two pots of seeds in a warm place and leave them until they start to grow. This may take a few days. Remember to keep the compost moist with water or liquid fertiliser during this time.

5 Number your seedlings by sticking labels on the side of each pot.

6 Measure the seedlings every day. Put your results in the tables.

7 When you have enough results (this will take about two weeks), add all the heights in each row together and put the totals in the tables.

Seeds in ordinary compost			
Day	Height of seedlings in mm	Total	Average height
1			
2			
3			

Seeds in fertiliser compost			
Day	Height of seedlings in mm	Total	Average height
1			
2			
3			

Work out the average height using the formula:

$$\text{Average height} = \frac{\text{Total height}}{10}$$

Put these results in the table as well.

8 Draw line graphs of your results. Lay out your graph like this:

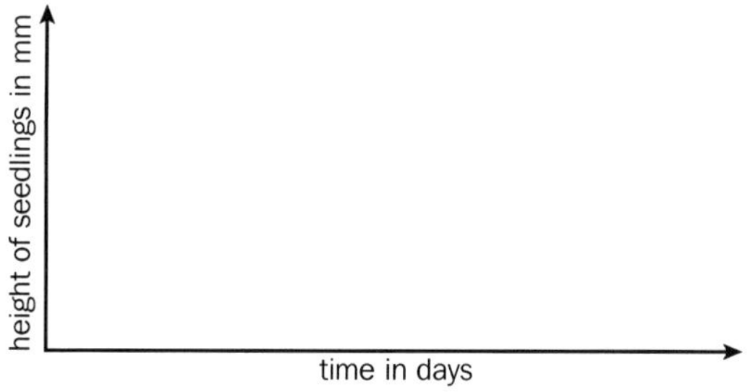

What effect does fertiliser have on plant growth?

19.1c
Practical notes

Fertilisers and plant growth

This activity is an extension of Worksheets 19.1b, once again providing the opportunity of data collection over a period of time. Make sure the compost is kept moist with the correct solution.
Wheat, oats or barley seeds are most suitable as their seedlings do not have side shoots. Soak seeds overnight before starting the activity. Seeds are put around the edges of the pots to make measuring easier.
Pupils will find it easier to measure from the edge of the pot to the growing tip.

19.1c
Technician's notes

Number of apparatus sets:

Number of pupils:

Number of groups:

Visual aids:

ICT resources:

Equipment/apparatus needed:

Fertilisers and plant growth

Each group will need:

- a copy of worksheet 19.1c
- two plant pots or other suitable containers with drainage holes
- compost moistened with water
- compost moistened with fertiliser solution ('Baby Bio' or similar)
- wheat seeds pre-soaked overnight (barley or oats are also good for this activity)
- labels
- mm ruler.

Safety notes

CLEAPSS/SSERC SAFETY REFERENCE:

19.2a Adapted for life W/S

Name: Date: Group:

What you need:

Newspaper, hand lens, leaf litter, dishes or jars with lids, tile or white paper, plastic gloves.

What to do:

1 Draw up a table in your book using these headings:

Name of animal	Tally	Total found	How it is adapted to life in leaf litter

2 Put on the plastic gloves.

3 Put some leaf litter on to the newspaper and carefully sort through it to find some animals. When you find an animal, put it in a dish and put the lid on. Look at the animal and see how it is adapted to life in leaf litter.

4 These drawings will help you find the names of the animals. Fill in the table as you go.

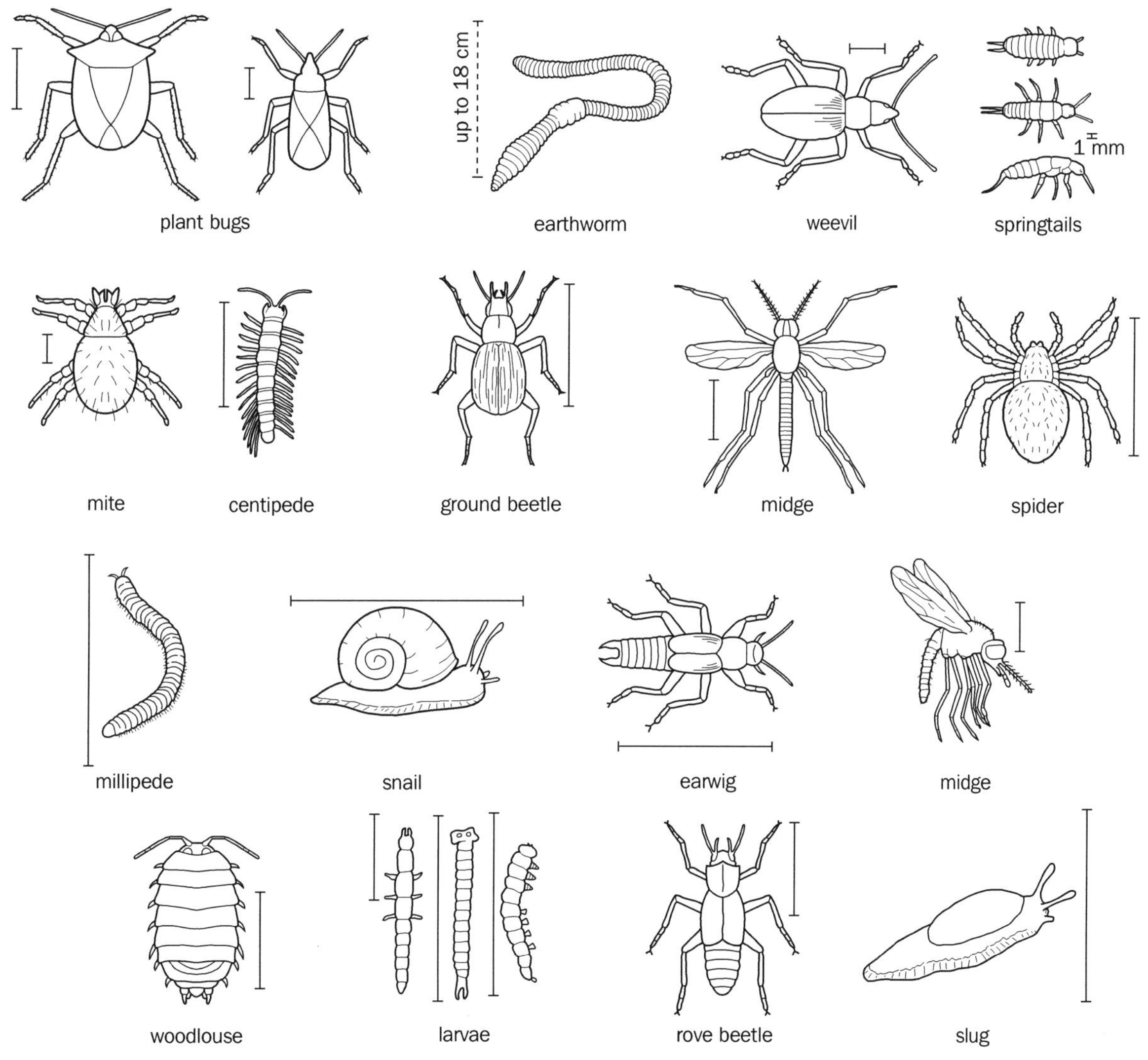

up to 18 cm

1 mm

plant bugs earthworm weevil springtails

mite centipede ground beetle midge spider

millipede snail earwig midge

woodlouse larvae rove beetle slug

19.2a
Practical notes

Adapted for life

Pupils usually enjoy this sort of activity and their enthusiasm may need directing if the aim is to be met. Try to get pupils to look carefully for any adaptations to life in this particular habitat. The drawings show animals most commonly found in leaf litter, but a selection of other resources would be advisable just in case.

This activity could be extended to include a comparison with other habitats, e.g. soil. Watch out for silliness with leaf litter and animals. Remind pupils to put lids on jars and dishes. Make sure all animals are returned to their natural habitat after the activity.

19.2a
Technician's notes

Adapted for life

Each group will need:

- a copy of worksheet 19.2a
- petri dishes and/or jars with lids
- sheets of newspaper to cover tables
- disposable gloves
- tweezers (for pupils who don't like picking up worms etc.)
- tile or piece of white paper
- hand lens
- access to leaf litter (freshly collected leaf litter is best stored in a black plastic bin liner).

Number of apparatus sets:

Number of pupils:

Number of groups:

Visual aids:

ICT resources:

Equipment/apparatus needed:

Safety notes

- **Watch out for silliness with leaf litter/animals.**
- **Ensure all animals are returned to their natural habitat after the activity.**

CLEAPSS/SSERC SAFETY REFERENCE:

19.2b Plants like light

W/S

Name: **Date:** **Group:**

What you need:

Wheat seeds, dish, moist compost, box with hole at one end, aluminium foil.

What to do:

1 Put some moist compost into the dish. Sprinkle some wheat seeds on to the compost.

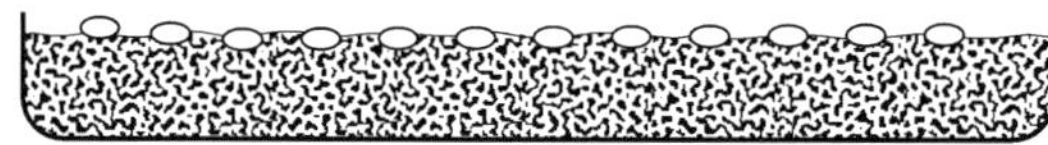

2 Leave the dish for a few days until the seedlings are about 2 – 3 cm tall.

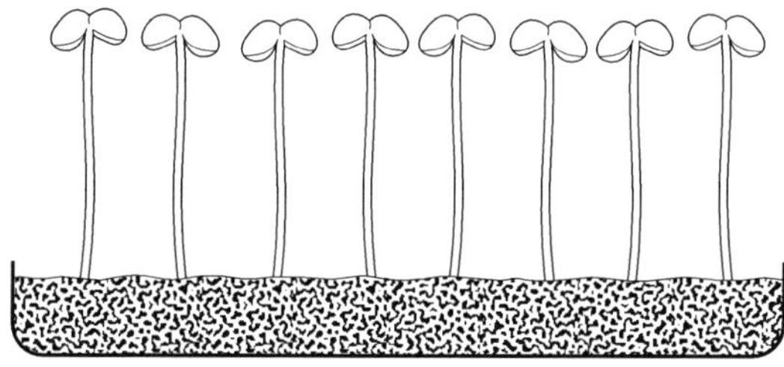

3 Make some 'caps' with foil and put them over the tips of about half the seedlings.

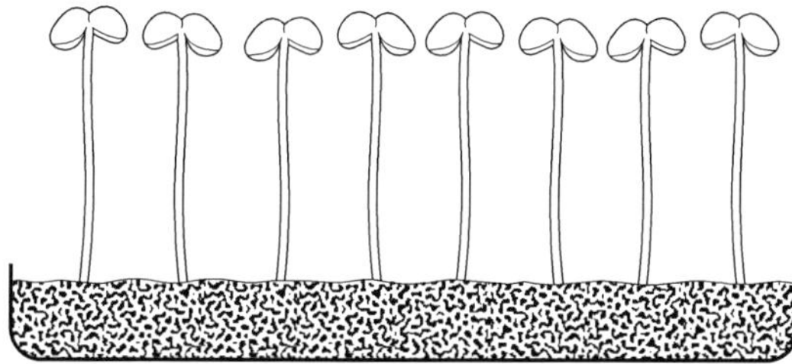

4 Put the dish of seedlings into the box so that they receive light from one direction only.

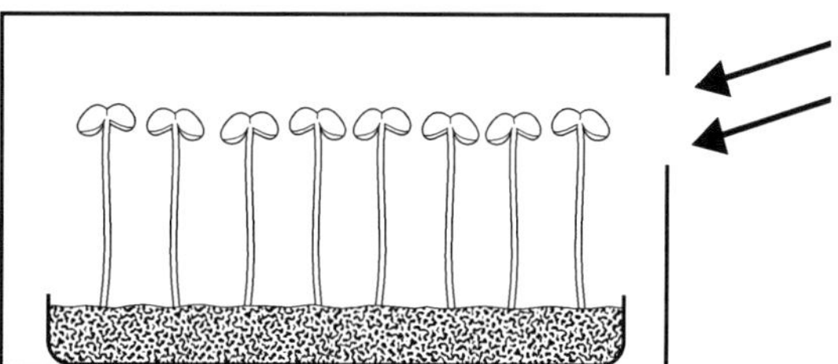

5 After about two days take the dish from the box and look at the seedlings.

- What has happened to the seedlings without caps?
- What has happened to the seedlings with caps?

- Scientists say that plants can detect light at the tips of growing shoots. How do your results support this view?

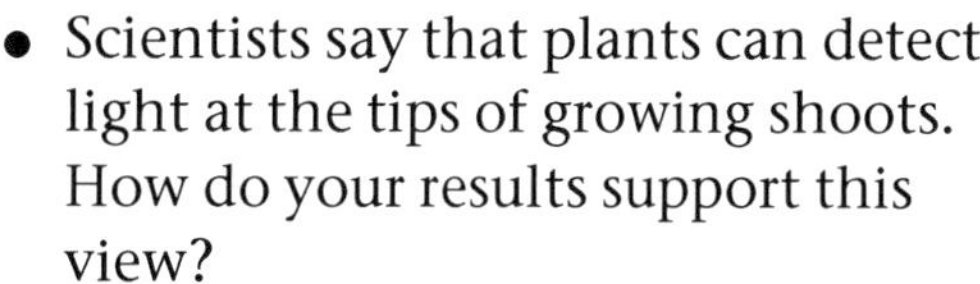

19.2b
Practical notes

Plants like light

This is a simple activity enabling pupils to see how plants respond to changes in their environment; in this case the direction of light. Wheat, oats, or barley seeds are preferable to cress seeds because the aluminium foil 'caps' fit better. Soak seeds overnight before starting the activity. 'Caps' can be made by carefully folding small squares of foil over a spent match stick or similar. More able pupils could go on to research the role of plant hormones in response mechanisms.

19.2b
Technician's notes

Plants like light

Each group will need:

- a copy of worksheet 19.2b
- dish (petri dish or similar)
- moist compost
- 20 wheat seeds pre-soaked overnight (barley or oats are also good for this activity)
- lightproof box with hole cut at one end (an old shoe box is good for this)
- 10 small squares (about 2 cm square) of aluminium foil
- spent match stick or similar for moulding foil 'caps'.

Number of apparatus sets:

Number of pupils:

Number of groups:

Visual aids:

ICT resources:

Equipment/apparatus needed:

Safety notes

CLEAPSS/SSERC SAFETY REFERENCE:

19.1a The number of humans keeps on growing H/W

Name: **Date:** **Group:**

What you need to know …

Humans form part of ecosystems just as other organisms do. However, our numbers keep growing and growing. At the moment the human population is growing by over 1.5 million every week, that is over 150 a minute. As our numbers grow, we are having a greater and greater impact on the environment. The graph shows the growth in the human population over the past 500 000 years.

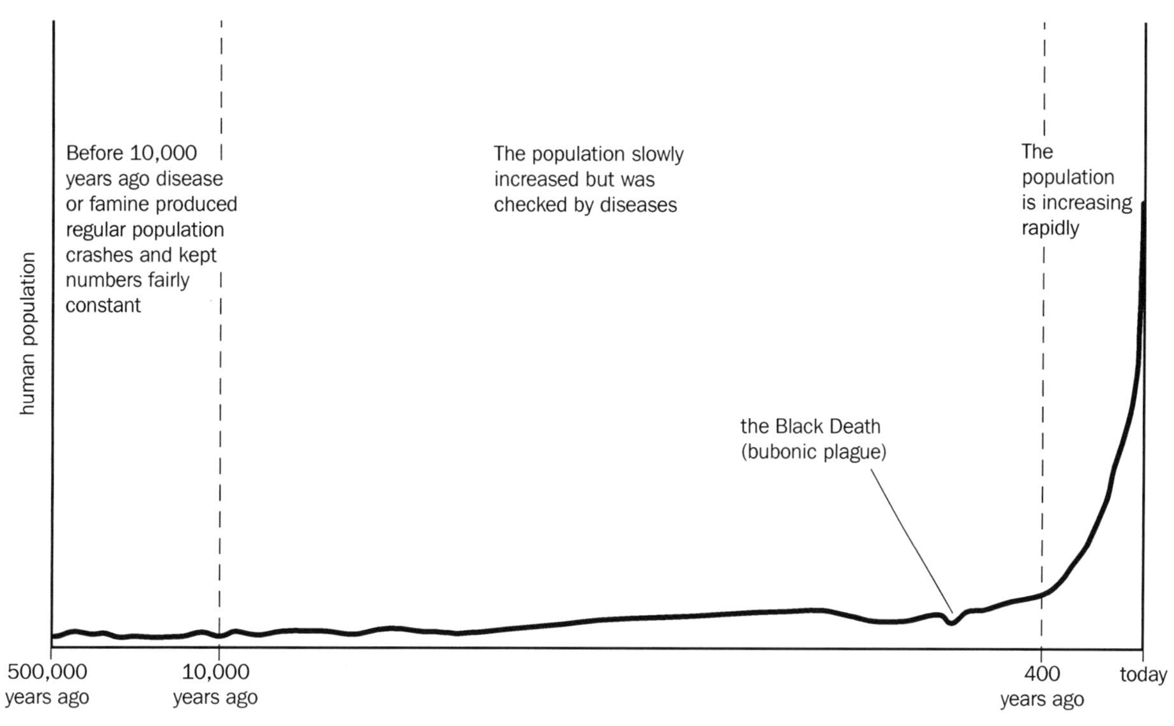

What to do:

1 Answer these questions:

a When was there no real increase in the human population?

b What effect did the Black Death have on the human population?

c Suggest why the population rose only slowly up to about 400 years ago.

d Give two reasons why the human population has got bigger and bigger in the last 400 years.

2 Suggest why the human population is not affected as much as other animal populations by:

a predators

b climate changes, e.g. temperature changes.

3 What do you think should be done to reduce the increase in human population growth?

HANDY HINTS

The information on the graph gives you the answers to Question 1.

19.1b Rubbish survey

H/W

Name: Date: Group:

What you need to know …

About 30 million tonnes of household rubbish are buried in landfill sites each year. There are fewer and fewer places where rubbish can be buried. Recycling as much of our rubbish as we can is the only way to avoid us being buried in our own waste.

What to do:

1 Conduct a survey of the things you throw away at home. Your teacher will tell you how long to do the survey for.

Copy this tally chart into your book. Every time something is thrown in the dustbin, tick the appropriate box in the tally chart.

Type of rubbish	Tally	Total
cans		
cardboard		
cloth		
cooked food		
glass		
metal		
paper		
plant material, e.g. fruit and vegetable peelings		
raw meat		
other items		

2 Make a list of the things in your rubbish that could be recycled.

3 Where is your nearest recycling centre?

SAFETY WARNING

Wear protective gloves when sorting through rubbish and beware of objects that could cut you such as broken glass.

HANDY HINTS

A visit to a recycling centre will give you a clue as to what can be recycled and what cannot.

19.1c Close to extinction H/W

Name: Date: Group:

What you need to know ...

Many of the things that humans do destroy the environment and the animals and plants in it. The number of different animal and plant species is falling. Every year more species become extinct.

What to do:

Read the following article:

11,000 species now facing extinction

The Saiga antelope, the wild Bactrian camel and the Iberian lynx have slipped closer to extinction says the World Conservation Union. A spokesperson says 'These species will disappear unless we can reverse the situation. When they get to these low levels, they reach the point of no return'.

11,167 species are now threatened with extinction, an increase of 121 since 2000. The slender-billed vulture and the Indian vulture have been listed as critically endangered. They have suffered from disease, poisoning and food shortage.

The most alarming move towards extinction is that of the Saiga, an antelope that lives in the deserts of central Asia. It has declined from more than a million in 1993 to 50,000. It is listed as critically endangered and has been hit hard by poaching for meat and horn, which is used in traditional Chinese medicine.

The wild Bactrian camel, listed as endangered since 1996, is now critically endangered. Part of its territory is the Gobi desert, where China carried out nuclear bomb tests. The wild camel competes with domestic animals for water and grazing land. It is also hunted for sport. Habitat is being lost to mining. Cross-breeding with domestic camels has also contributed to its decline.

Also upgraded from endangered to critically endangered is the Iberian lynx, which is close to becoming the first cat species to disappear for 2000 years. The number of lynx has dropped to half the 1200 animals recorded in the early 1990s. Myxomatosis was introduced to Spain to control rabbits, the lynx's main source of food. Destruction of woodland has forced the animals into small, scattered groups in south west Spain.

1 Make a list of the animal species mentioned in the article.

2 How has the number of species threatened with extinction changed since 2000?

3 What has caused the slender-billed vulture and the Indian vulture to become close to extinction?

4 Explain why the Saiga antelope is critically endangered.

5 a What is the habitat of the Bactrian camel?

b Give three reasons for the decline in numbers of Bactrian camels.

6 a What is so worrying about the decline of the Iberian lynx?

b What is the rate of decline of the Iberian lynx?

c Why do you think the destruction of woodlands in Spain has lead to the decline in the Iberian lynx?

7 Explain the difference between 'endangered' and 'critically endangered'.

HANDY HINTS

Read the article a few times before trying to answer the questions.

19.2 Adapted for life

H/W

Name: **Date:** **Group:**

What you need to know …

Living things have developed special features to help them cope with their way of life. They have become adapted to their environment.

What to do:

1 Look at the drawing of a fish. Some features of the fish are labelled.

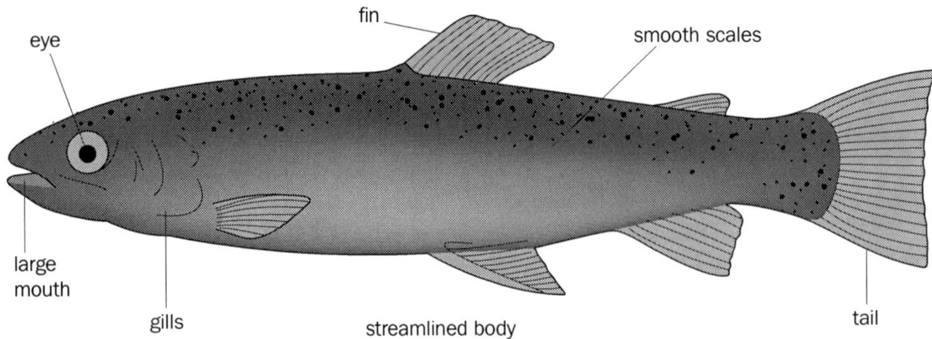

Explain how each feature helps the fish survive in a watery environment. For example, fins help steer the fish in the right direction.

2 Dolphins are mammals just like us. They do all the things that we do except they live in water all the time.

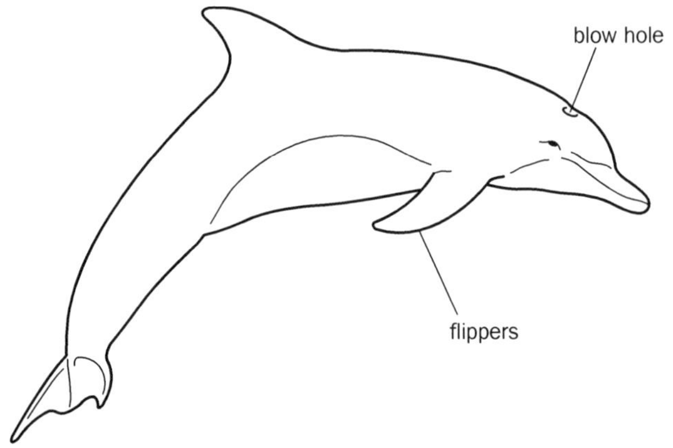

Why do you think the dolphin has:

a its nose (called a 'blowhole') on its back?

b no hair

c flippers instead of arms

d a thick layer of fat (blubber) beneath its skin?

HANDY HINTS

Look at the diagrams carefully before you answer the questions.

Chapter 19 ► Investigation 19F Temperature and plant growth

In this investigation:	you are going to find out if there is a link between temperature and how well plants grow.

Preparation: Predict

Finish the sentences in the box.

What I think will happen is...

I think this because...

Preparation: Plan

Write a short plan of your investigation.

Think about:

- the apparatus you are going to use
- how one variable depends upon another variable
- what you are going to measure and how you are going to measure it
- how many readings you are going to take
- how you are going to record your results
- how you are going to make your investigation fair
- how you are going to make your investigation safe.

Show your plan to your teacher before going on.

Carry out

Carry out your investigation and record your results.

Present your results in an appropriate way.

Report

Write a report on your investigation.

Here are some things you should include:

- what you did
- what happened
- explain your results
- if your prediction was correct or not
- how reliable your results were
- what you could have done if you had more time.

Chapter 19 ► Investigation 19E Temperature and plant growth

In this investigation:	you are going to find out if there is a link between temperature and how well plants grow.

Preparation: Predict

Finish the sentences in the box.

What I think will happen is...

I think this because...

Preparation: Plan

Write a short plan of your investigation.

Think about:

- the apparatus you are going to use
- what you are going to measure and how you are going to measure it
- how many readings you are going to take
- how you are going to record your results
- how you are going to make your investigation fair
- how you are going to make your investigation safe.

Show your plan to your teacher before going on.

Carry out

Carry out your investigation and record your results in a table.

Draw a line graph of your results.

Report

Write a report on your investigation.

Here are some things you should include:

- what you did
- what happened
- explain your results
- if your prediction was correct or not
- how reliable your results were
- what you could have done if you had more time.

Chapter 19 ► Investigation 19D
Temperature and plant growth

In this investigation:	you are going to find out if there is a link between temperature and how well plants grow.

Preparation: Predict

Finish the sentence in the box.

I think there (is/is not) a link between temperature and how well plants grow because…

You are going to use this equipment to find out if there is a link between temperature and how well plants grow:

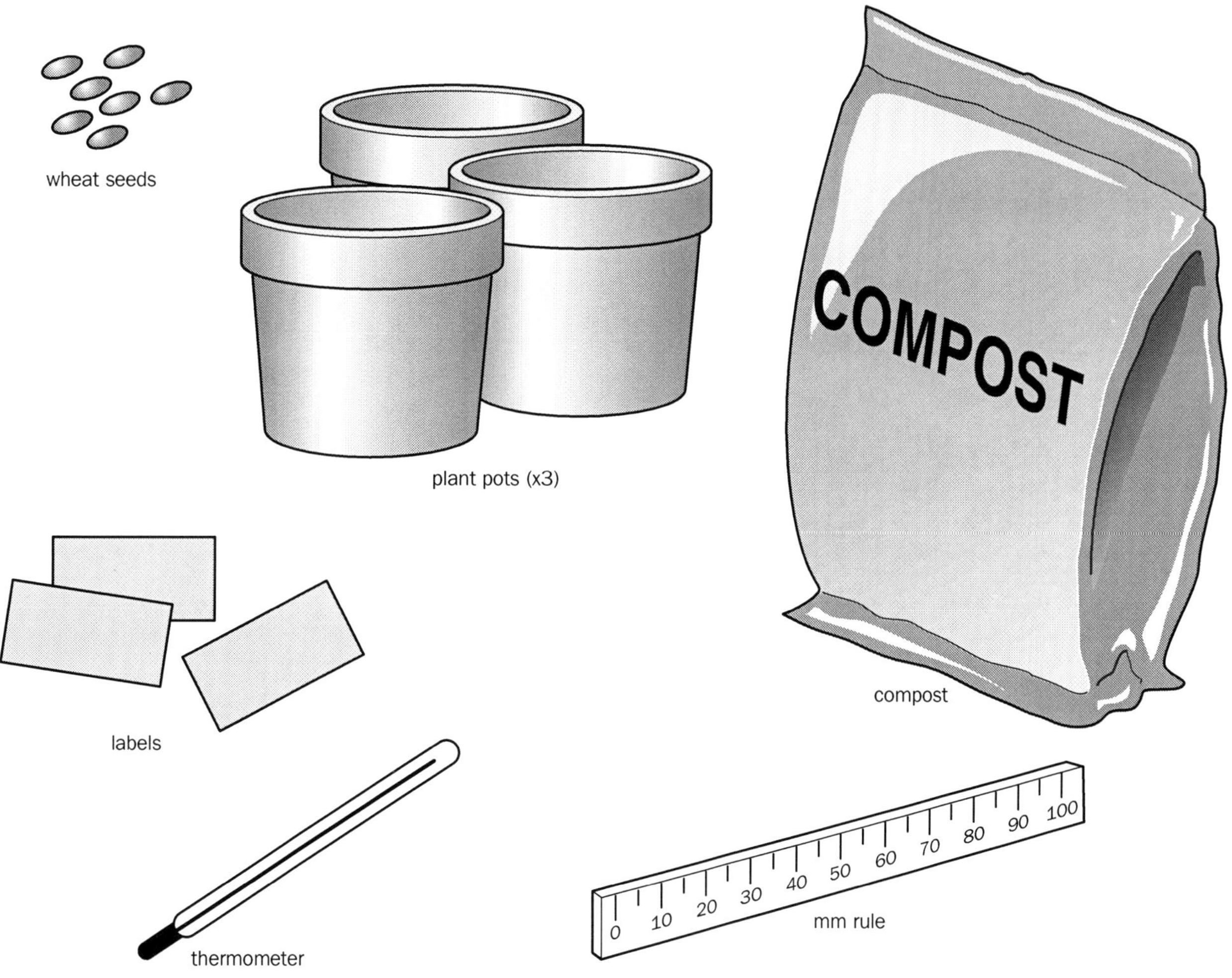

Chapter 19 ► Investigation 19D Temperature and plant growth

Preparation: Plan

Finish the sentences in the box.

I will measure...

Things I will keep the same are...

My investigation will be fair because...

My investigation will be safe because...

Carry out

Plant some seeds in pots. When the seeds have started to grow put the pots in different places and keep records of their growth. You could put the pots in a refrigerator (4 °C), a room (about 15 °C), and a warm place (about 25 °C).

Put your results in a table like this:

Location ______			
Day	**Height of seedlings in mm**	**Total**	**Average (Total 10)**
1			
2			
3			

Draw a line graph of your results on a piece of graph paper.

Use different coloured pencils to draw each line.

Label the lines '4 °C', '15 °C' and '25 °C'.
Label the axes like this:

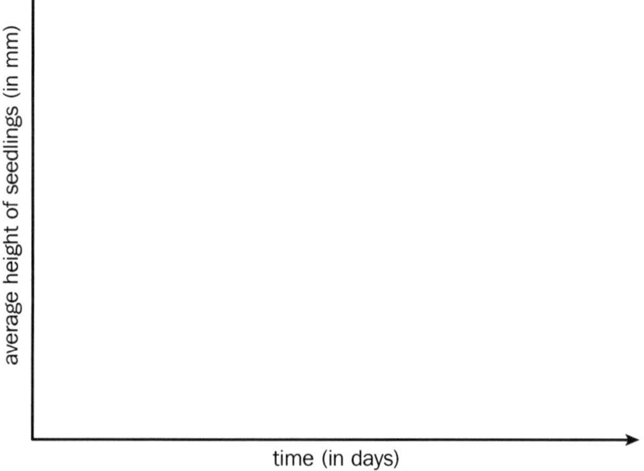

Report

Write a report on your investigation.

Here are some things you should include:

- what you did
- what happened
- explain your results
- if your prediction was correct or not
- what you could do to improve the investigation
- what you could have done if you had more time.

Chapter 19 ► Investigation 19C
Temperature and plant growth

In this investigation:	you are going to find out if there is a link between temperature and how well plants grow.

Preparation: Predict

Finish the sentence in the box.

I think there (is/is not) a link between temperature and how well plants grow because…

You are going to use this equipment to find out if there is a link between temperature and how well plants grow:

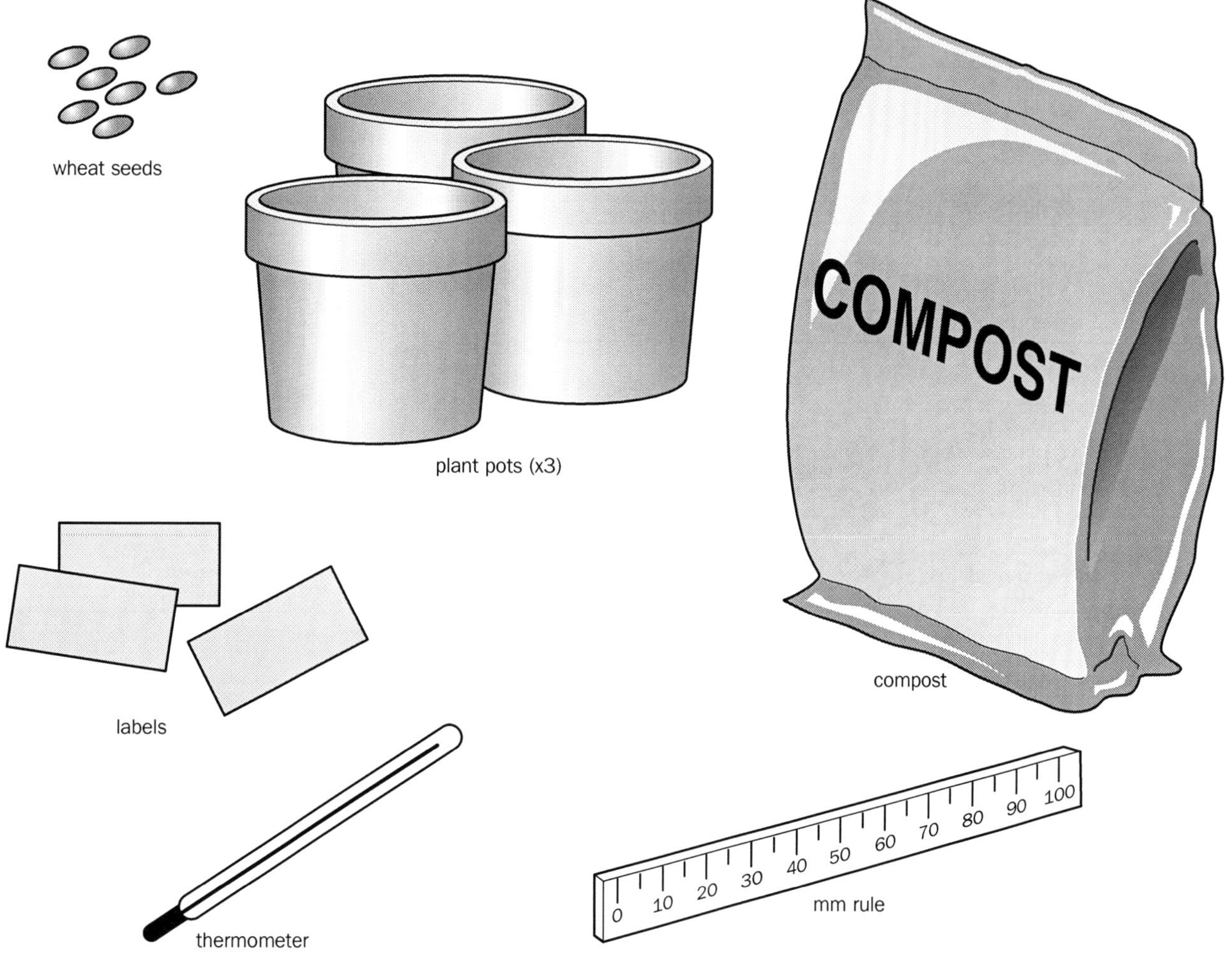

Chapter 19 ► Investigation 19C Temperature and plant growth

Preparation: Plan

Finish the sentences in the box.

I will measure...

Things I will keep the same are...

My investigation will be fair because...

My investigation will be safe because...

Carry out

- Put some compost into the three pots.
- Put about 15 seeds on the compost close to the edge of the pot. Push the seeds slightly below the compost.
- Leave the seeds for a few days until they start to grow.
- When about 10 seeds have started to grow in each pot, number the seedlings by putting labels on the side of the pots.
- Measure the seedlings.
- Put one pot in a refrigerator (4 °C).
- Put one pot in a room (about 15 °C).
- Put one pot in a warm place (about 25 °C).
- Measure the seedlings every day.

  Put your results in a table like this:

Location ________________			
Day	**Height of seedlings in mm**	**Total**	**Average (Total 10)**
1			
2			
3			
4			
5			

Chapter 19 ▶ Investigation 19C Temperature and plant growth

Draw a line graph of your results on this grid.

Use different coloured pencils to draw each line.

Label the lines '4 °C', '15 °C' and '25 °C'.

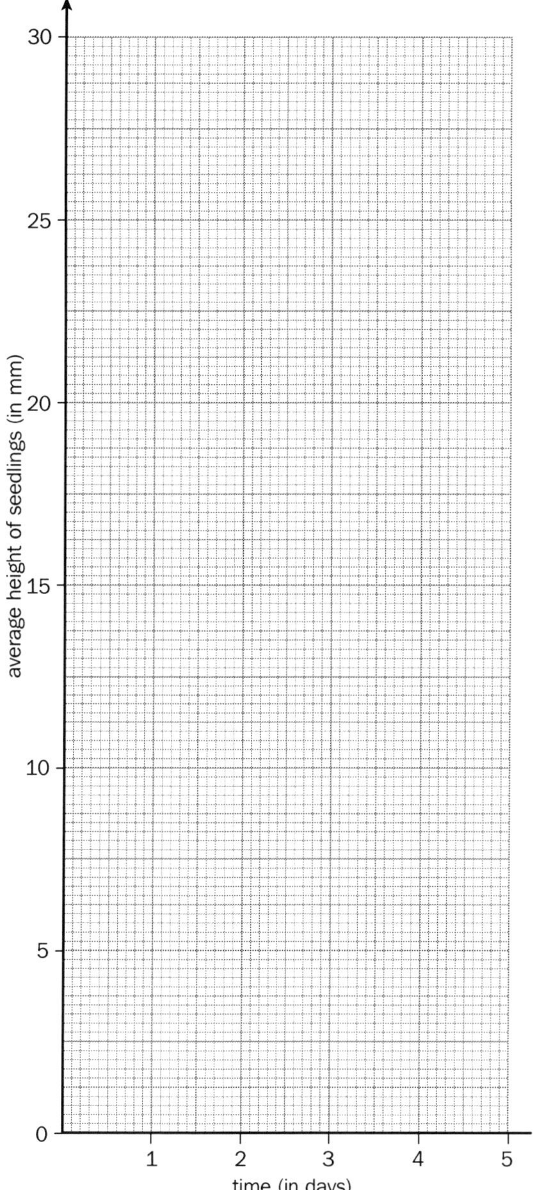

Report

Finish the sentences in the box.

What I did was…

What happened was…

From my results I found out that there (is/is not) a link between temperature and how well plants grow. I know this because…

My prediction (was/wasn't) correct. If I could do the investigation again I would…

Investigation 19 Practical notes

Temperature and plant growth

Investigation 19 Technician's notes

Temperature and plant growth

Number of apparatus sets:

Number of pupils: Number of groups:

Visual aids:

ICT resources:

Equipment/apparatus needed:

Each group will need:

- wheat seeds pre-soaked overnight (barley or oats are also good for this investigation)
- three plant pots or other suitable containers with drainage holes
- moist compost
- thermometer
- labels
- mm rule
- access to refrigerator
- access to warm place (about 25 °C)
- access to room (about 15 °C).

Safety notes

CLEAPSS/SSERC SAFETY REFERENCE:

Chapter 19 ► Test

Our environment

1 The list gives four ways in which humans affect the environment:

cutting down forests quarrying
using fertilisers using pesticides

a Give **one** reason why humans need to do **each** of these.

b Give **one** problem caused by doing **each** of these.

8 marks

2 The wolf is extinct in Scotland.

a What does 'extinct' mean?

b i What was the wolf's habitat?
ii Why did it become extinct?
iii Suggest one thing that could have been done to prevent the wolf from becoming extinct.

c Name one other animal which is extinct in Scotland

d Give one reason why it is important to conserve wild animals and plants

6 marks

3 Which of these words match the gaps in the passage which follows?

colonisers competition grasses
light shrubs water
wind

If an area of land is left alone for hundreds of years, ____ (a) _____ between plants for light, ___(b)__ and minerals will result in it becoming covered with trees.

The first plants are called _____(c)______; they have a short life cycle. These are soon replaced by ____(d)_____ , followed by tall weeds. _____(e)_____ take over by cutting out the ____(f)_____ from the plants beneath them. Finally small trees begin to grow, their seeds will have been carried by the ____(g)____ to land between the shrubs.

7 marks

4 The diagram shows the leaves of a Scots Pine tree.

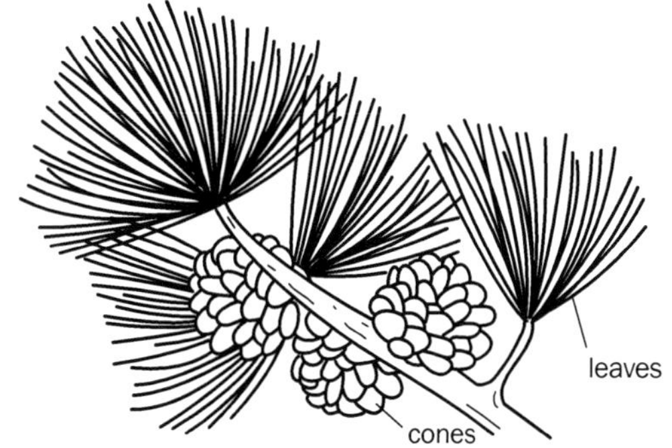

a What is the habitat of the Scots Pine?

b Describe the climate in this habitat.

c Give two ways in which the Scots Pine is adapted to its habitat.

d How do these adaptations help the Scots Pine survive?

5 marks

5 Duncan's mother put her favourite pot plant on a sunny window sill. After a few days, she noticed that the plant was growing towards the window.

a Explain why the plant started to grow towards the window?

b What should Duncan's mother do to get the plant growing straight up.

c What is the best environment for a pot plant?

4 marks

(This page is intentionally blank.)

Chapter 19 ▶ Test

Blue

Our environment

1 The diagram shows the leaves of a Scots Pine tree.

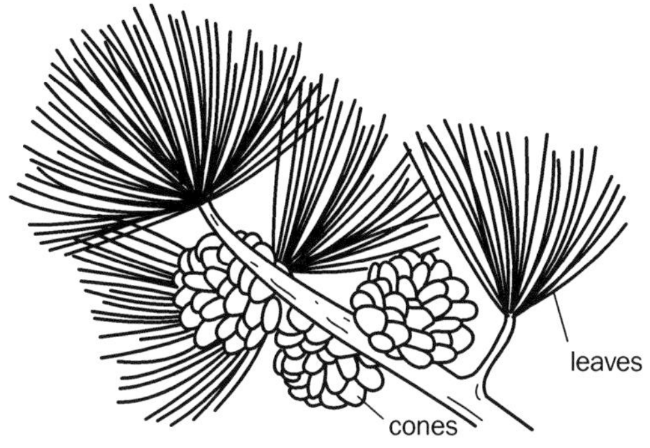

a What is the habitat of the Scots Pine?

b Describe the climate in this habitat?

c Give two ways in which the Scots Pine is adapted to its habitat.

d How do these adaptations help the Scots Pine survive?

5 marks

2 Which of these words match the gaps in the passage which follows?

colonisers	competition	grasses
light	shrubs	water
wind		

If an area if land is left alone for hundreds of years, ____ (a) _____ between plants for light, ___(b)__ and minerals will result in it becoming covered with trees.

The first plants are called _____(c)______; they have a short life cycle. These are soon replaced by ____(d)_____ , followed by tall weeds. _____(e)_____ take over by cutting out the ____(f)_____ from the plants beneath them. Finally small trees begin to grow, their seeds will have been carried by the ____(g)____ to land between the shrubs.

7 marks

3 Duncan's mother put her favourite pot plant on a sunny window sill. After a few days, she noticed that the plant was growing towards the window.

a Explain why the plant started to grow towards the window?

b What should Duncan's mother do to get the plant growing straight up.

c What is the best environment for a pot plant?

4 marks

4 During the day, peppered moths rest on tree trunks. There are two varieties of peppered moth, a light form and a dark form.

In country areas the trees are clean. The light moth is difficult to see and survives to breed. The dark form will be seen by birds and eaten.

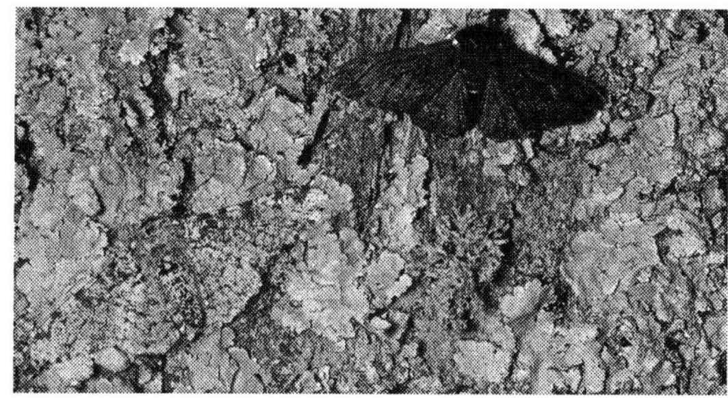

In city areas the trees are blackened by soot. Here the dark moths are better camouflaged than the white moths.

Our environment

a Which form of peppered moth will be more common in

i country areas

ii city areas?

b Explain why this is an example of 'survival of the fittest'.

c How could you show that the light form and the dark form of the peppered moth are the same species?

d Scientists believe the dark form of peppered moth first appeared as a result of a mutation in some light moths.

i What is a mutation?

ii When would this kind of mutation usually happen?

iii Name one thing that can increase the chances of a mutation happening.

9 marks

5 The environment where animals and plants live is affected by many things. Some of these things are living and some are non-living. They keep the environment the same as years go by.

a What is the scientific name for the

i living

ii non-living parts of the environment?

b Name one living part of your environment.

c Name one non-living part of your environment and describe how it can be measured.

5 marks

Chapter 19 ► Mark scheme Our environment

White

Question	Answer	Marks	Level
1 a	space for growing crops/ grazing animals	1	D
	for building materials	1	D
	help crops grow/ greater productivity	1	D
	kill insects and pests/ remove competition	1	D
b	habitat removal/ land erosion	1	D
	damage to landscape/ eyesores	1	D
	run off into rivers/ algal growth, etc.	1	D
	get into food chains/ kill other things as well	1	D
		8	
2 a	dead and gone forever	1	C
b i	forest/woodland	1	C
ii	cut down forests/hunting	1	C
iii	reserves/new forests/ ban hunting etc.	1	C
c	Bear, lynx, etc.	1	C
d	species diversity, source of medicines	1	C
		6	

Question	Answer	Marks	Level
3 a	competition	1	D
b	water	1	D
c	colonisers	1	D
d	grasses	1	D
e	shrubs	1	D
f	light	1	D
g	wind	1	D
		7	
4 a	Scottish Highlands	1	D
b	cold/little available water in winter	1	D
c	reduced leaf area/needles	1	D
	waxy cuticle/coating	1	D
d	reduce water loss	1	D
		5	
5 a	towards light (1) for photosynthesis (1)	2	D
b	turn it regularly	1	D
c	well lit/sufficient water/ correct temperature	1	D
		4	
	TOTAL 30 marks		

Suggested grade/level boundaries

C = 8/30

D = 22/30

Chapter 19 ► Mark scheme Our environment

Blue

Question	Answer	Marks	Level
1 a	Scottish Highlands	1	D
b	cold/little available water in winter	1	D
c	reduced leaf area/needles	1	D
	waxy cuticle/coating	1	D
d	reduce water loss	1	D
		5	
2 a	competition	1	D
b	water	1	D
c	colonisers	1	D
d	grasses	1	D
e	shrubs	1	D
f	light	1	D
g	wind	1	D
		7	
3 a	towards light (1) for photosynthesis (1)	2	D
b	turn it regularly	1	D
c	well lit/sufficient water/correct temperature	1	D
		4	

Question	Answer	Marks	Level
4 a i	light form	1	F
ii	dark form	1	F
b	some better suited to environment (1) so survive/breed (1)	2	F
c	breed them (1) and they will produce fertile offspring (1)	2	F
d i	random change in chromosome/gene	1	F
ii	cell division/gamete formation	1	F
iii	x-rays, UV rays, cigarette smoke, nuclear radiation	1	F
		9	
5 a i	biotic	1	F
ii	abiotic/physical	1	F
b	other people	1	F
c	temperature (1) with a thermometer (1)	2	F
		5	
	TOTAL 30 marks		

Suggested grade/level boundaries

D = 15/30

E = 19/30

F = 23/30

1 The diagram shows the structure of a tooth.

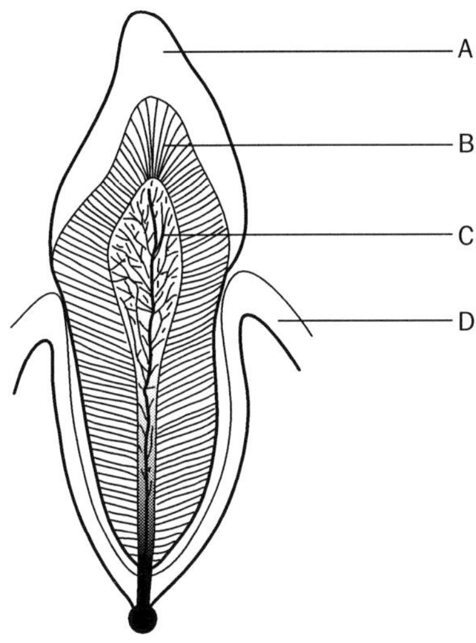

a Which label points to the
  i enamel
  ii dentine
  iii pulp cavity
  iv gum?

b Explain why it is important to clean your teeth regularly.

6 marks

2 Our bodies need a number of different chemical substances to keep them working properly.

Which of these words fill the gaps in the table:

building new cells carbohydrate
energy fat
fish potatoes

Chemical substance	Found in	Used in the body for
a	b	energy
c	butter	d
protein	e	f

6 marks

3 The diagram shows the human breathing system.

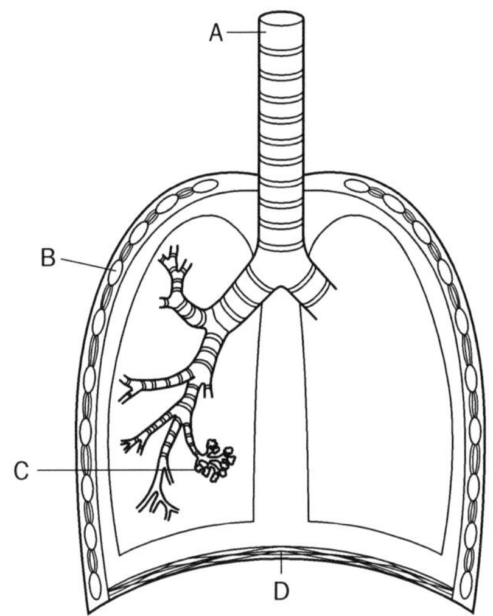

Which letter points to:

a the windpipe

b an alveolus

c a rib

d the diaphragm?

4 marks

4 The diagram shows some living things.

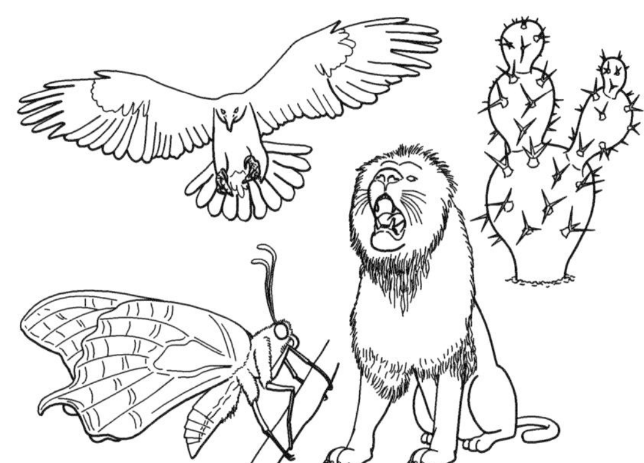

a Describe how
  i the eagle is adapted to catch its food
  ii the cactus is adapted to avoiding losing water in the desert
  iii the butterfly is adapted to suck nectar from flowers
  iv the lion is adapted for tearing flesh and crushing bones.

b Gerbils live in the desert. They do not produce urine. They live in holes during the day and only come out at night.

i Explain how these adaptations help gerbils survive in the desert.

ii Suggest an adaptation that helps gerbils avoid being eaten.

7 marks

5 Read this article from a magazine then answer the questions.

Wildlife threatened by pollution

WIldlife living around the coast of Britain is being threatened by pollution. Sewage, farm waste, and toxic waste from industry are being carried out to sea by rivers in ever increasing amounts.

In a recent report, scientists studying habitats of wading birds have identified high levels of zinc, chromium, and nickel. These have been traced to industrial factories inland.

There are also high levels of phosphates. Undoubtedly these have come from sewage and fertiliser that has run off into rivers. Algae grow rapidly in waters where phosphate levels are high. In a process known as eutrophication, the algae reduce the amount of oxygen available for other life forms in the water.

Research has shown that plankton in particular are threatened. Plankton are at the bottom of every food chain in the sea. Once this supply of food goes, everything else will follow.

Action is needed now. The government must bring in legislation to fight this most serious issue before it is too late.

Source: Target Science Biology 2001

a Name three pollutants mentioned in the article.

b What is eutrophication?

c Explain why the loss of plankton is a serious matter.

d Suggest a law that the government could make that might help the threat to wildlife.

7 marks

6 The graph shows the rate of growth of a potted plant at different temperatures.

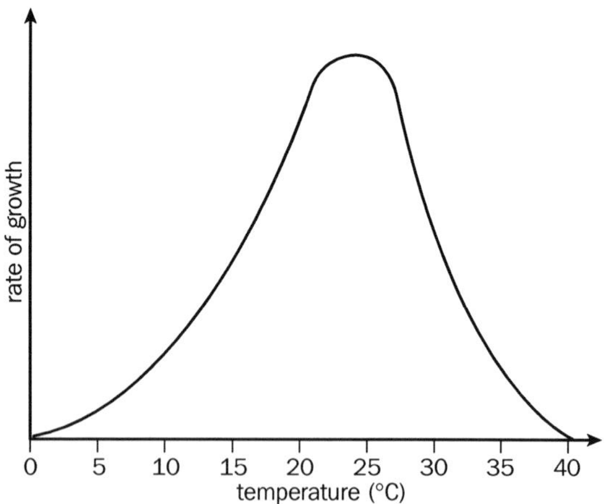

a At what temperature is the plant growing at its fastest rate?

b At what temperatures does the plant not grow at all?

c When it is warmer it is usually brighter. Give one reason for a change in plant growth rate as the temperature changes.

3 marks

7 Put these in order of size starting with the smallest:

Earth galaxy
Moon Sun
Universe

4 marks

8 The table gives some information about four planets in the Solar system.

Planet	A	B	C	D
Time for one orbit around the Sun in Earth years	0.25		12	
Average daytime temperature in °C	350	22	-150	-210

a Suggest how long it takes for planet B to orbit the Sun once.

b i Is the time taken for planet D to orbit the Sun once likely to be more or less than 12 years?
ii Explain your answer.

c i Which planet is nearest the Sun.
ii Explain your answer.

d One of these planets is Earth, which one?

e Give two ways that information about the planets could have been obtained.

8 marks

9 The diagram shows the Sun and two planets, Venus and Earth.

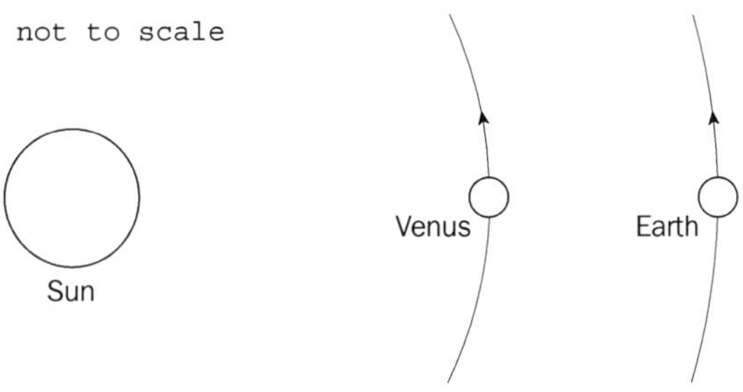

a What force holds the planets in orbit around the Sun?

b i Which of the two planets has the stronger force on it? (assume both planets have the same mass).
ii Explain your answer.

c Explain why Venus can be seen as a bright dot in the night sky even though it is not hot and glowing like a star.

4 marks

10 Describe what happens to the molecules when:

a water changes from a liquid to a gas when it boils

b water changes from a liquid to a solid when it freezes

c water expands when it is heated.

6 marks

11 The diagram shows how we can find out what is produced when a fuel burns.

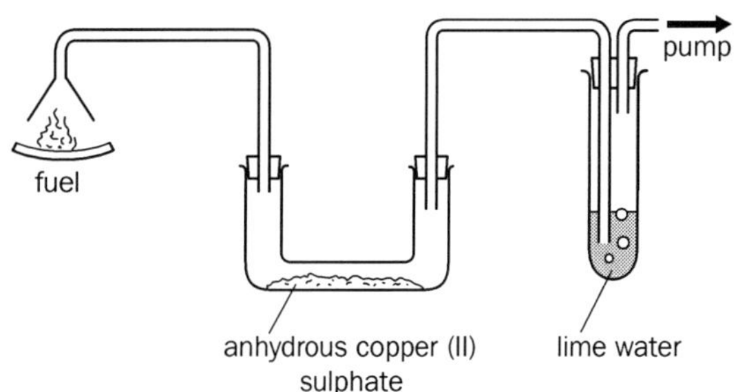

a Which gas from the air is used when a fuel burns?

b The lime water went milky after a short time. What does this show?

c Anhydrous copper sulphate is a powder which is white when dry and blue when wet.
i What colour would you expect the anhydrous copper sulphate to be in the experiment shown in the diagram?
ii Explain your answer.

d Explain why burning is an example of an irreversible chemical change.

e What is the most important product of burning fuel?

6 marks

12 The diagram shows how we can separate water from ink.

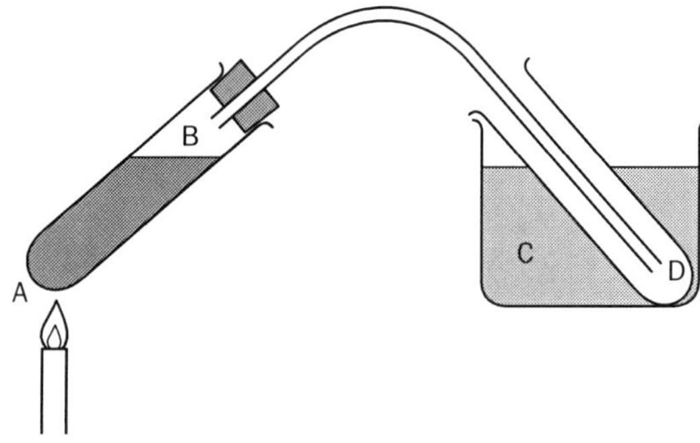

a Ink is a solution of ink powder and water. What does solution mean?

b Where is
i evaporation
ii condensation happening?

c Where is there pure water?

d Give one safety precaution that should be taken when using this apparatus.

5 marks

13 Should friction be high or low in the following cases:

a hands holding bicycle handlebars

b car tyres on the road

c spacecraft returning to Earth

d sliding quickly down a rope

e walking on ice

f pulling a boat up a beach.

6 marks

14 The diagram shows a bobsleigh team ready to start a run.

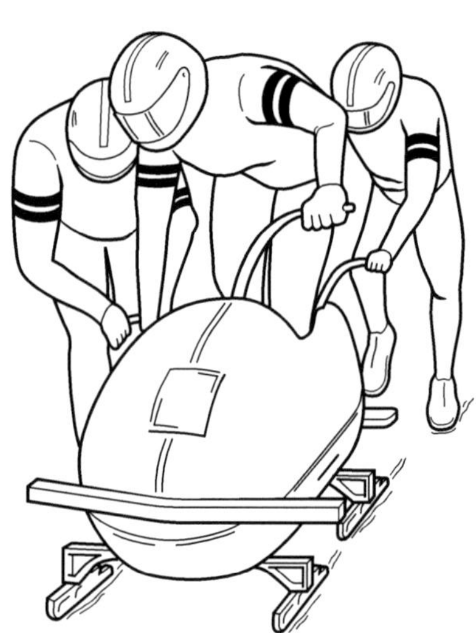

a To get going, the team have to push the bobsleigh. Give one place where the team need
 i as much friction as possible.
 ii as little friction as possible.

b Give two ways that the team can reduce air resistance.

c Explain why the team need
 i a small amount of friction during their run
 ii a large amount of friction at the end of their run?

6 marks

15 An astronaut has a mass of 60 kg. She weighs 600 N on Earth but only 100 N on the Moon.

a What does 1 kg weigh on Earth?

b A piece of Moon rock weighs 20 N on Earth. What is its mass?

c How much bigger is the mass of Earth than the mass of the Moon?

d How much will the astronaut weigh in outer space?

4 marks

16 The list gives some materials:

brick	cling film
cotton wool	glass
Perspex	wood

a Name two transparent materials from the list.

b Name two opaque materials from the list.

c Explain why you cannot see round corners.

3 marks

17 The diagram shows a single ray of light hitting a plane mirror.

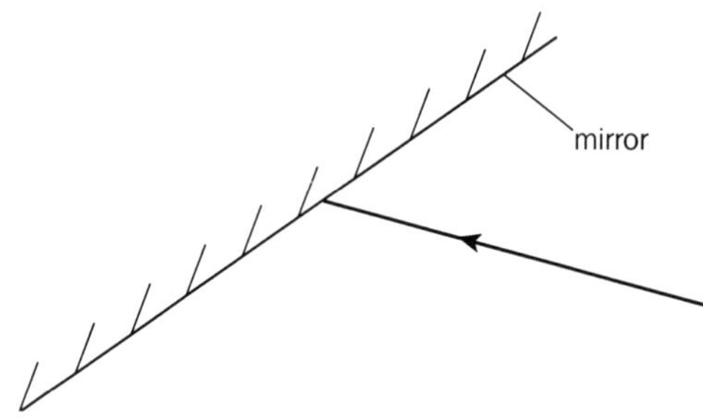

a Copy and complete the diagram by drawing the reflected ray.

b Label the angle of incidence on your diagram.

c What is special about the angle of incidence and the angle of reflection?

d How would the reflected ray differ if the incident ray was shining on a rough surface?

4 marks

18 The diagrams show light rays passing through a convex and a concave lens.

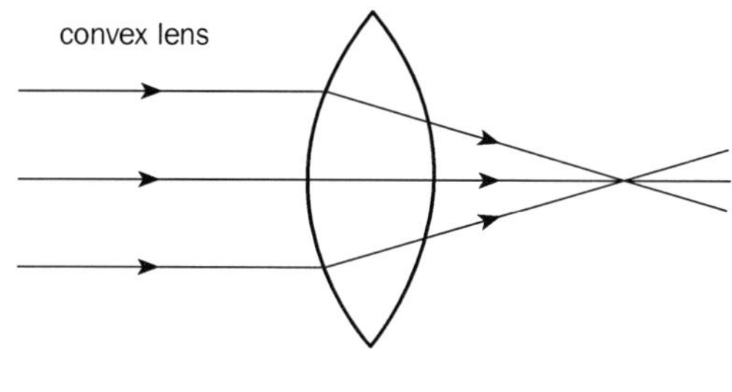

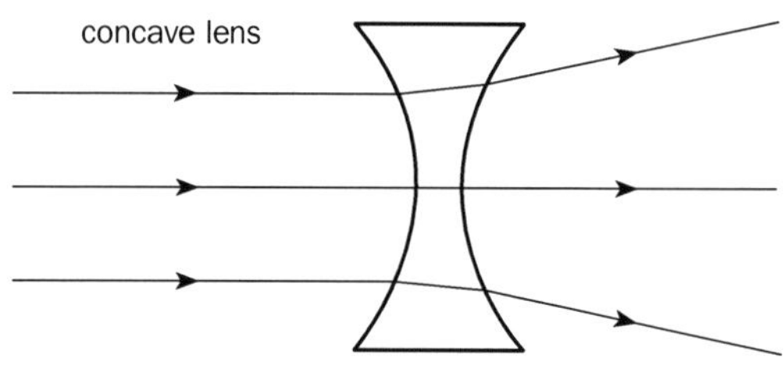

a Give one thing that the two lenses
  i have in common
  ii do not have in common.

b Which of the two lenses can be used as a magnifying lens?

c Name a piece of science equipment that uses lenses.

4 marks

19 The diagram shows a loudspeaker. When music is played through the loudspeaker, the cone vibrates backwards and forwards.

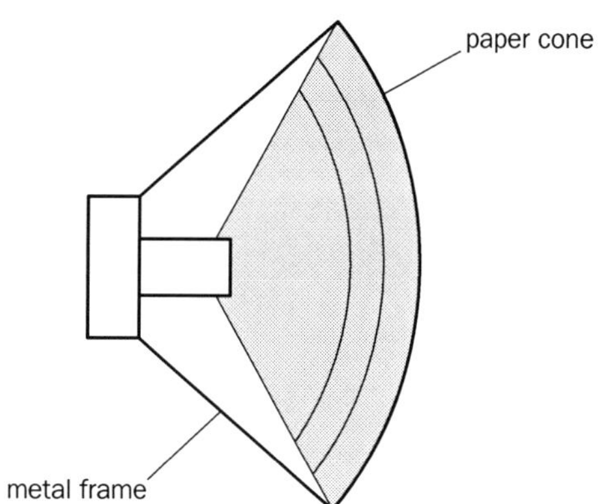

a What happens to the music if the volume is increased?

b What happens to the movements of the cone if the volume is increased?

c Describe the change in movements of the cone when high pitched music changes to music with a low pitch.

d Name one musical instrument that produces sound by
  i vibrating strings
  ii vibrating air.

6 marks

(This page is intentionally blank.)

Book 2 Examination (sheet 1) Blue

1 The diagrams show two kinds of microorganisms; bacteria and a virus.

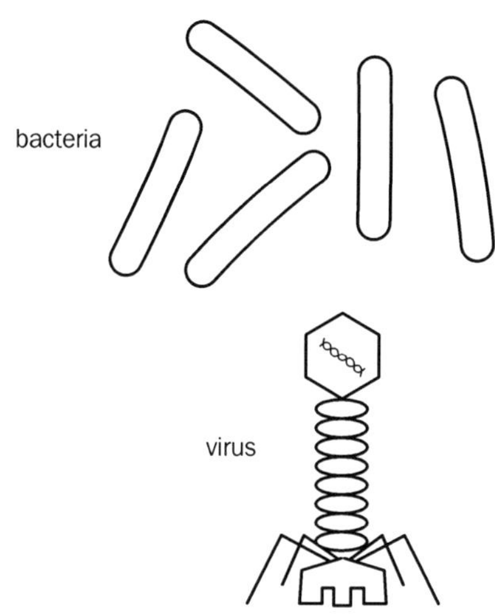

a What are microorganisms?

b Give one way that bacteria and viruses are
  i similar
  ii different.

c Give one way in which microorganisms can be
  i useful
  ii harmful.

5 marks

2 The diagram shows the stages in genetic engineering.

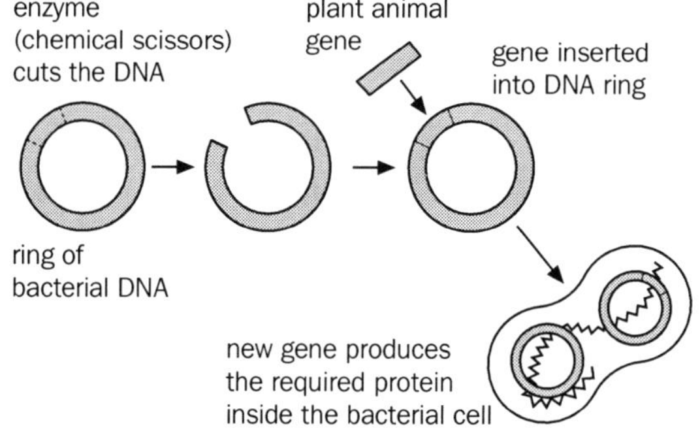

a What is genetic engineering?

b What microorganisms are used in genetic engineering?

c What is the advantage of using these microorganisms?

d Suggest why the enzymes are called 'chemical scissors'.

e Soya plants can be given genes which make them resistant to weed killers.
  i What is a gene?
  ii What is the advantage of making soya plants resistant to weed killers?

6 marks

3 Read the following:

'The common land snail is found in woods, fields, hedges, sand dunes, and rough ground all over Europe. The shells of this snail show a lot of variation. The background colour may be various shades of brown, pink, or yellow and there may be up to five dark bands. Both colour and banding are inherited. Snails which do not blend in with their surroundings are seen and eaten by birds.'

a Which form of snail will be more common on sand dunes.

b This is an example of 'survival of the fittest'. What does this mean?

c How could you show that the different forms of land snail belong to the same species?

d Scientists believe that the various forms of land snail have been formed as a result of mutations.
  i What is a mutation?
  ii Name one thing that can increase the chances of a mutation happening.

7 marks

4 Look around at your environment.

a Name
  i one biotic factor
  ii one abiotic factor in your environment.

b Describe how you would measure the effect of one environmental change on an organism of your choice.

4 marks

5 The diagram shows a piece of apparatus used to compare the amount of carbon dioxide in the air you breathe in and the air you breathe out.

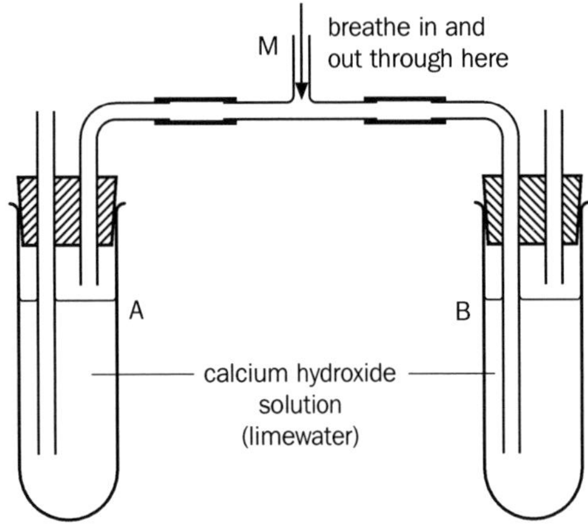

a What happens to lime water when carbon dioxide bubbles through it?

b Which lime water will change first in this experiment?

c What does this show?

d This apparatus must be used carefully. Only gentle, slow breaths are needed to give a good result. Suggest what might happen if someone blows too hard into tube M.

e **i** What cellular reaction produces carbon dioxide as a waste product?
ii How does carbon dioxide get from the cells to the lungs?

6 marks

6 The graph shows the effect of temperature on the way an enzyme works.

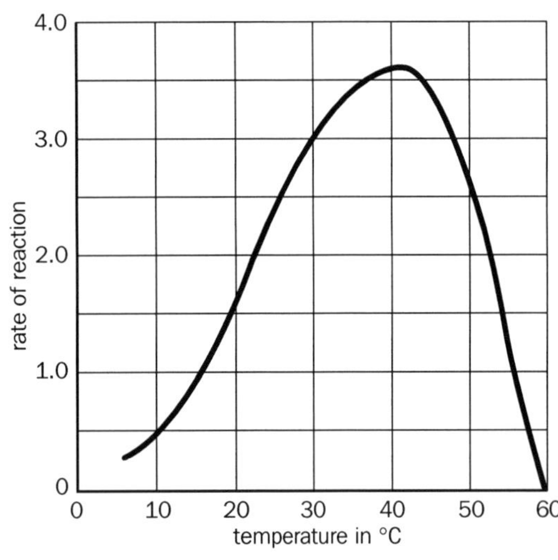

a What is an enzyme?

b At what temperature does this enzyme work fastest?

c Why do you suppose 'cold blooded' animals like lizards need to warm themselves up by sunbathing before they can move quickly?

d Pepsin is an enzyme produced in the stomach.
i What does pepsin do in the stomach?
ii Explain why pepsin won't work as well anywhere else in the digestive system.

5 marks

7 The diagram shows the Earth at four positions in its orbit, three months apart.

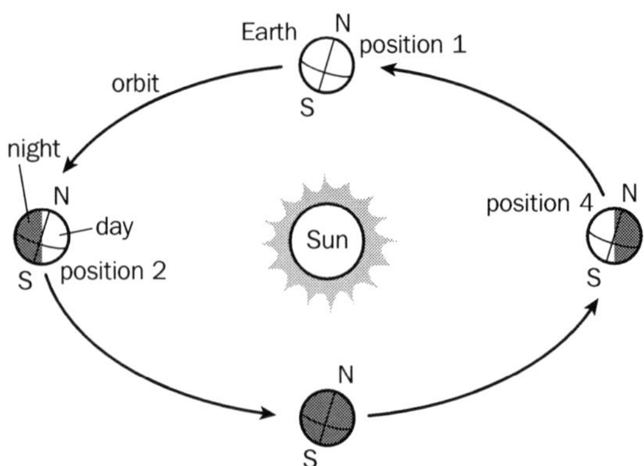

Which position represents:

a June?

b winter in the southern hemisphere?

c summer in the northern hemisphere?

d 24-hour daylight at the South Pole?

4 marks

Book 2 Examination (sheet 3)

Blue

8 The diagram shows the Sun with the Earth and Moon in two different positions in their orbits.

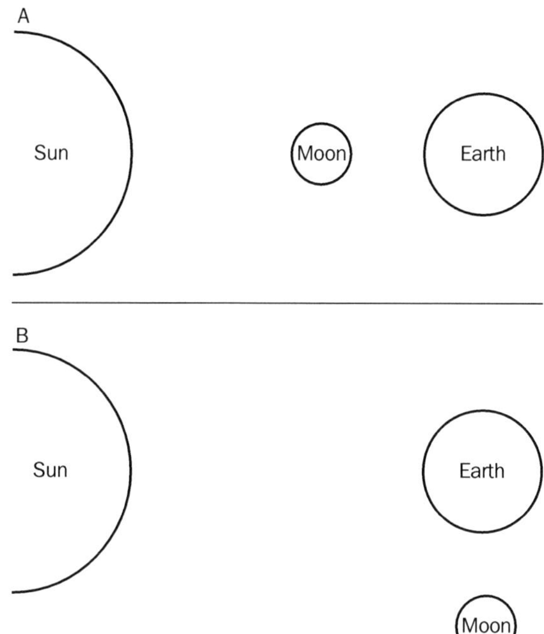

a How long does it take the Earth the orbit the Sun once?

b What do we call the time taken by the Moon to orbit the Earth once?

c Much of the Earth is covered with water which rises and falls as tides twice a day. The height of the tides is affected by the gravitational pull of the Moon and the Sun. Explain why tides travel much further up a beach when the Sun, Earth, and Moon are as shown in diagram A than when they are as shown in diagram B.

4 marks

9 The diagram shows a boy blowing up a balloon. The balloon has black spots painted over it.

Explain how the balloon can be used to model the 'big bang' theory for the evolution of the Universe.

3 marks

10 The diagram shows a lithium atom.

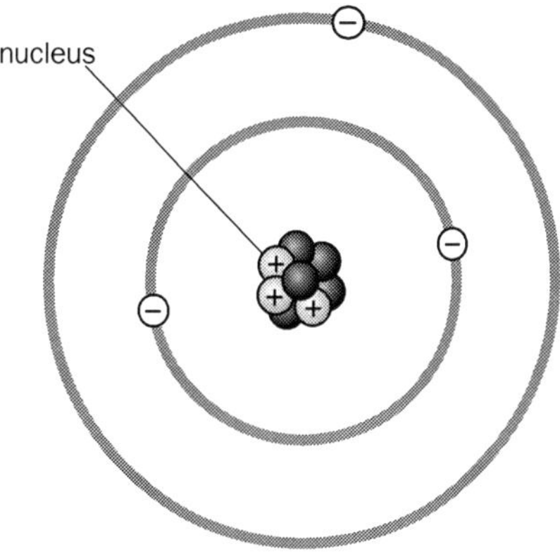

a How many electrons does a lithium atom have?

b What does the nucleus of a lithium atom contain?

c What is the atomic mass number for lithium?

d Name one famous scientist who made an important discovery about atoms.

5 marks

11 The diagram shows a blast furnace that changes iron ore into iron.

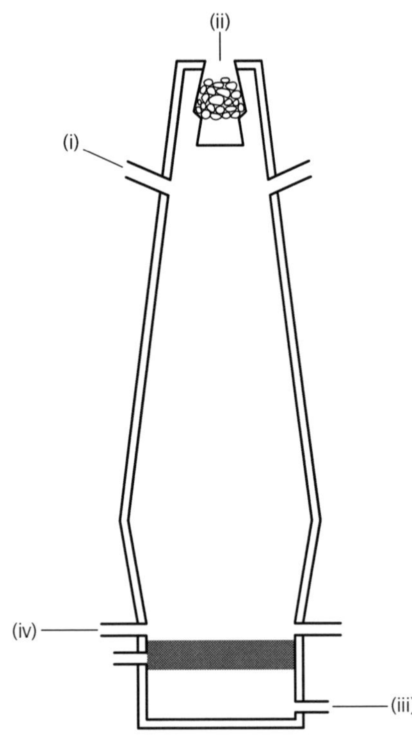

a Use these words to label the diagram:

hot air   iron ore
molten iron   waste gases

b i Explain why aluminium cannot be separated from its ore in a blast furnace.

ii How is aluminium separated from its ore?

6 marks

12 Identical samples of metal are put into solutions of different pH. The table shows how much metal has reacted after 5 minutes.

Solution	A	B	C	D	E	F	G
% of metal reacted	65	60	55	50	30	20	5
pH of solution	1	2	3	4	5	6	7

a i Which of the solutions is most acidic?

ii Explain your answer.

b Explain how you would make a solution less corrosive.

c i What gas is given off when an acid reacts with a metal?

ii Describe how you would test for this gas.

5 marks

13 Eggshells contain calcium carbonate. The shells from three eggs were broken up into small pieces and reacted with hydrochloric acid at 20 °C in a flask. Carbon dioxide gas was collected in a syringe. Calcium chloride was left in the flask at the end of the reaction. The graph shows how much carbon dioxide was given off over a period of time.

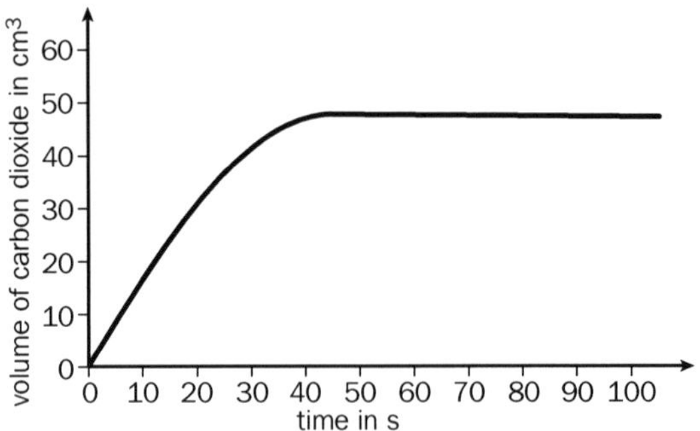

a What volume of gas was collected after 30 seconds?

b How long did the reaction take?

c How would the shape of the line graph change if the reaction took place at a higher temperature.

d Explain why the reaction would have been faster if stronger acid had been used.

e Write a word equation for this reaction.

f Why is this reaction described as a chemical reaction and not a physical reaction?

6 marks

Book 2 Examination (sheet 5)

Blue

14 The diagram shows a model tower crane. The crane has a moveable counterbalance.

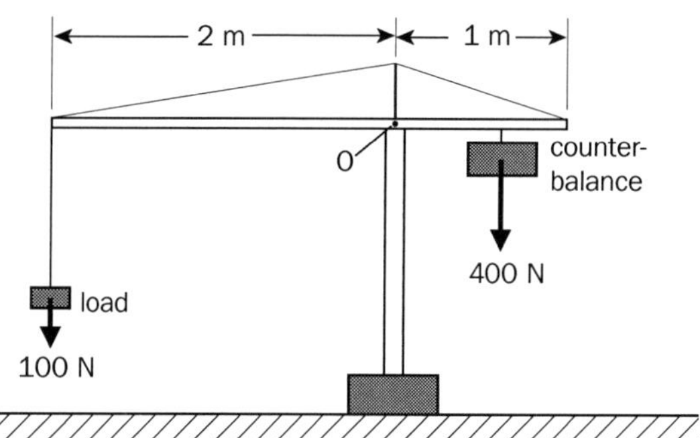

a Why does the crane need a counterbalance?

b What is the turning force of the 100 N load?

c To balance the crane, how far from the pivot should the counterbalance be placed?

d What is the maximum load the crane should lift?

7 marks

15 The diagram shows a hammer being used to knock a nail into a piece of wood.

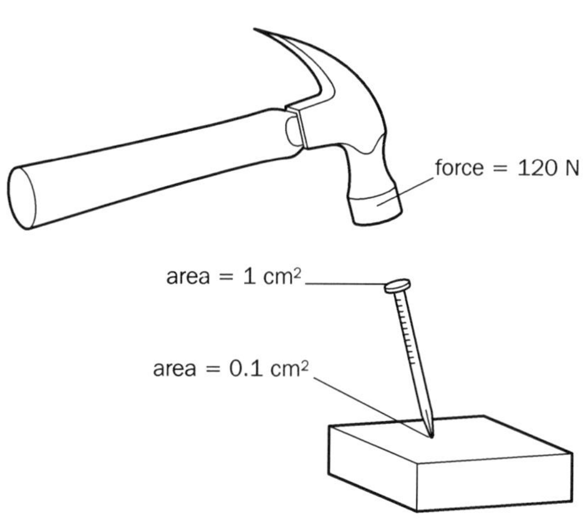

a What is the pressure on the **top** of the nail when the hammer hits it?

b What pressure is being exerted on the piece of wood by the point of the nail when it is hit by the hammer?

4 marks

16 a Give three things that you need to make an electromagnet.

b Give two ways that you could make your magnet stronger.

5 marks

17 A gardener designs a control system that will switch on a greenhouse heater when it gets cold. To save energy, the heater will only come on at night. The diagram shows an outline of the control system, but the names of the sensors and logic gates are missing.

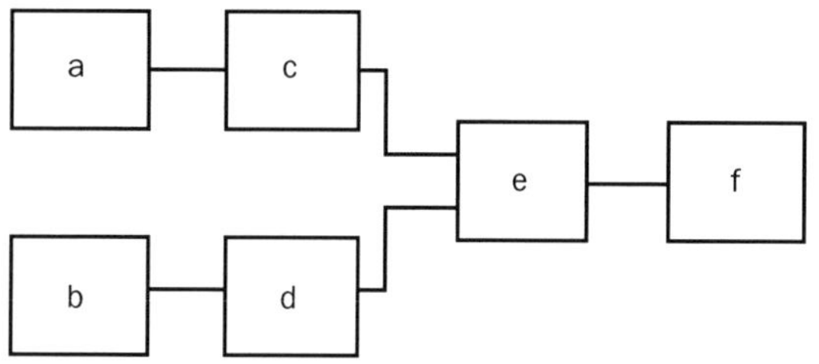

Use words from this list to fill the gaps in the control system. Some words may be used more than once or not at all.

AND gate
LDR
OR gate
heater
NOT gate
thermistor

6 marks

18 The diagram shows a ray of light hitting a glass block.

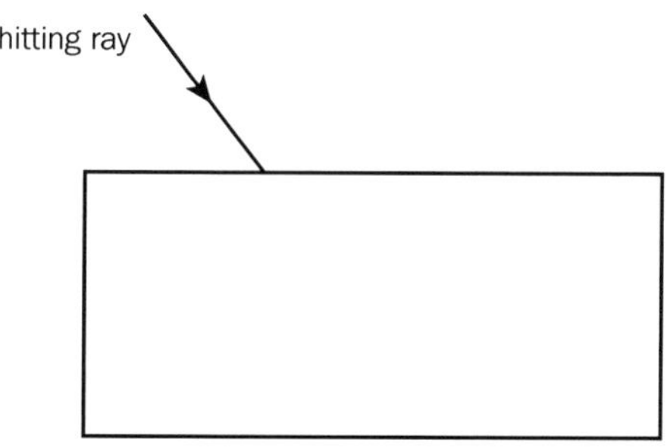

a Copy and complete the diagram showing the ray of light passing through and emerging from the block.

b Label an angle of refraction.

c What is special about the hitting ray and the emerging ray?

4 marks

19 A blue filter is put over the end of a torch.

a What colour will a red dress appear when the torch shines on it?

b What colour will a blue hat appear when the torch shines on it?

c What colour will appear on the screen when blue light passes through a glass prism like this?

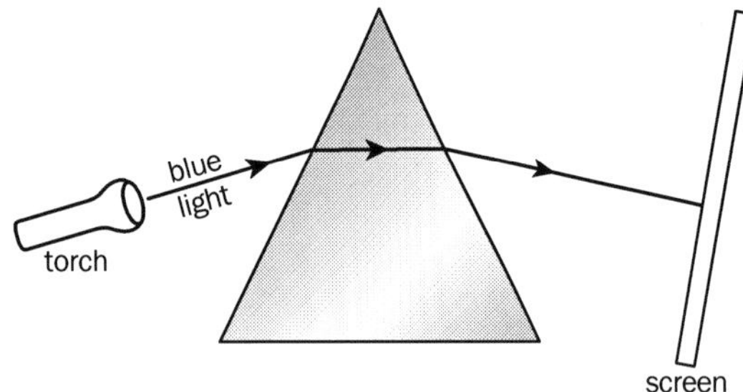

3 marks

20 The diagram shows an electric bell hanging inside a jar connected to a vacuum pump.

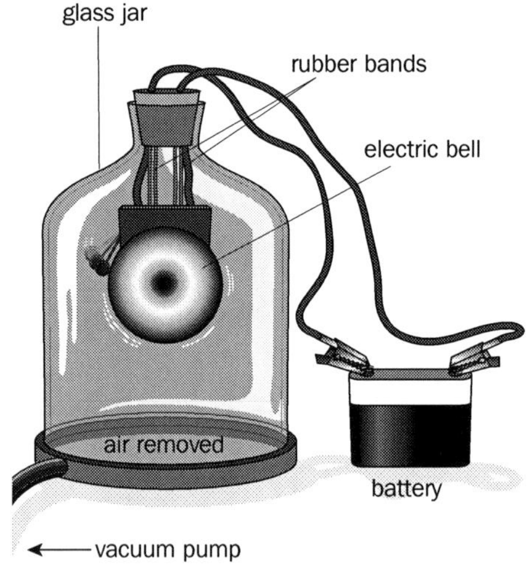

a Describe how the sound changes as air is pumped out of the jar.

b What does this experiment tell you about travelling sound?

c Sound waves can be shown as pictures on an oscilloscope. The diagram shows an example.

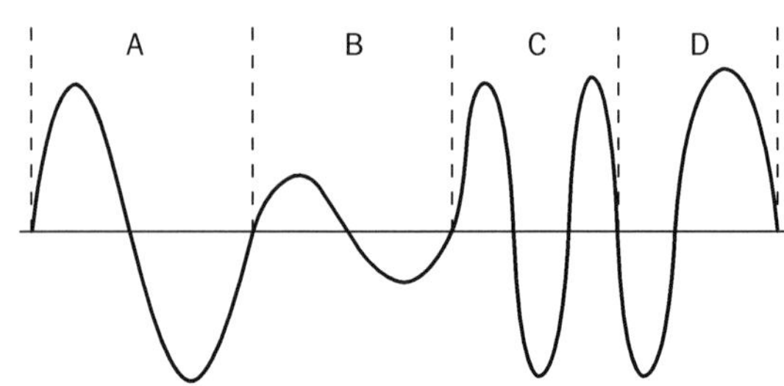

i Which part of the wave shows the greatest frequency?

ii Which part of the wave shows the lowest amplitude?

4 marks

Book 2 Examination Mark scheme

White

Question			Answer	Marks	Level
1	a	i	A	1	
		ii	B	1	
		iii	C	1	
		iv	D	1	D
	b		remove bacteria/food etc. (1) remove plaque (1)	2	D
				6	
2	a		carbohydrate	1	D
	b		potatoes	1	D
	c		fat	1	D
	d		energy	1	D
	e		fish	1	D
	f		building new cells	1	D
				6	
3	a		A	1	D
	b		C	1	D
	c		B	1	D
	d		D	1	D
				4	
4	a	i	strong beak/claws	1	
		ii	juicy stem/thick cuticle (skin)/ leaves are spines	1	
		iii	(long, thin) mouth parts act like drinking straw	1	
		iv	sharp teeth for tearing/ strong jaws for crushing	1	D
	b	i	no urine so no water loss (1) cool at night so no sweating (1)	2	
		ii	living in holes during day/ more difficult to be seen at night	1	D
				7	
5	a		three from zinc, chromium, nickel, and phosphates	3	D
	b		rapid growth of algae (1) reduce amount of oxygen in water (1)	2	D
	c		plankton at bottom of food chain/ without it all life dies	1	D
	d		treat waste before releasing into rivers, etc.	1	D
				7	
6	a		between 10 and – 20 °C	1	D
	b		0 °C and 40 °C	1	D
	c		more leaves for making food/ photosynthesis	1	D
				3	
7			Moon, Earth, Sun, galaxy, Universe (all in correct order = 4 marks, 4 in correct order = 3 marks etc.)	4	C
				4	
8	a		1 Earth year (accept anything between 0.25 and 12)	1	C
	b	i	more	1	
		ii	lower temperature means further from Sun	1	C
	c	i	A	1	
		ii	highest temperature	1	C
	d		B	1	C
	e		satellites, space probes, telescopes, etc. (any two)	2	D
				8	
9	a		gravity	1	D
	b	i	Venus	1	
		ii	closer to sun	1	D
	c		reflects light from Sun	1	C
				4	
10	a		move faster (1) further apart (1)	2	C
	b		move slower (1) closer together (1)	2	C
	c		move faster (1) further apart (1)	2	C
				6	
11	a		oxygen	1	D
	b		carbon dioxide is produced	1	D
	c	i	blue	1	
		ii	water produced during burning	1	D
	d		new things produced	1	D
	e		heat (energy)	1	D
				6	

Question			Answer	Marks	Level
12	a		ink dissolved in water	1	C
	b	i	B	1	
		ii	D	1	C
	c		D	1	C
	d		safety goggles, etc.	1	C
				5	
13	a		high	1	C
	b		high	1	C
	c		low	1	C
	d		low	1	C
	e		high	1	C
	f		low	1	C
				6	
14	a	i	between hands and handles/ feet and ground	1	
		ii	between sleigh and ice/air	1	C
	b		sit low in sleigh (1) wear streamlined helmets/smooth clothing (1)	2	D
	c	i	fastest speed	1	
		ii	brakes, etc. to stop	1	C
				6	
15	a		10 N	1	D
	b		2 kg	1	D
	c		6 times	1	D
	d		nothing/weightless	1	D
				4	
16	a		two from cling film, glass, or Perspex (both needed)	1	C
	b		two from brick, cotton wool, or wood (both needed)	1	C
	c		light travels in straight lines	1	C
				3	
17	a		reflected ray in correct position	1	C
	b		angle of incidence correct	1	C
	c		they are the same	1	C
	d		different direction/ i angle not same as r angle	1	C
				4	
18	a	i	bend/refract light	1	D
		ii	light bent/refracted in different directions	1	D
	b		convex lens	1	D
	c		any suitable equipment, e.g. microscope	1	D
				4	
19	a		sound gets louder	1	D
	b		vibrates more	1	C
	c		high pitch; vibrations fast/ more frequent (1) low pitch; vibrations slow/ less frequent (1)	2	D
	d	i	any correct instrument, e.g. guitar	1	
		ii	any correct instrument, e.g. trumpet	1	D
				6	
			TOTAL 99 marks		

Suggested grade/level boundaries
C = 37/99
D = 70/99

Book 2 Examination Mark scheme

Question	Answer	Marks	Level
1 a	very tiny living things	1	E
b i	tiny/no nucleus, etc.	1	
ii	virus smaller/not cells/ only reproduce inside other living things	1	E
c i	sewage works/compost/yoghurt/ biotechnology, etc.	1	
ii	disease	1	F
		5	
2 a	changing genetic make up of an organism	1	F
b	bacteria	1	F
c	breed/grow rapidly	1	F
d	they cut	1	F
e i	piece of a chromosome/DNA	1	
ii	kill weeds without harming soya	1	F
		6	
3 a	light coloured/yellow	1	F
b	some better suited to environment (1) so survive/breed (1)	2	F
c	breed them (1) and produce fertile offspring (1)	F	
d i	random change in chromosome/gene	1	
ii	x-rays/UV rays/cigarette smoke/ nuclear radiation (any one)	1	F
		7	
4 a i	any correct factor, e.g. other pupils	1	
ii	any correct factor, e.g. temperature	1	F
b	correct change, e.g. temperature	1	
	correct apparatus, e.g. thermometer	1	F
		4	
5 a	goes cloudy/milky	1	F
b	B	1	F
c	exhaled air contains more carbon dioxide	1	F
d	lime water forced out of test tube A	1	F
e i	respiration	1	
ii	in the blood	1	F
		6	
6 a	break up food into smaller molecules	1	
b	42 °C (+ or - 1 °C)	1	
c	enzymes do not work as well at low temperatures	1	F
d i	breaks up/digests protein	1	
ii	works best in acid/low pH conditions	1	F
		5	
7 a	position 4	1	E
b	position 4	1	E
c	position 4	1	E
d	position 2	1	E
		4	
8 a	365/365.25 days	1	E
b	one day	1	E
c	gravity of Sun and Moon combine in diagram 1 (1) gravitational pull is at right angles in diagram 2 (1)	2	E
		4	
9	balloon represents Universe (1) spots represent planets/galaxies (1) as balloon expands these move apart (1)	3	F
		3	
10 a	3	1	F
b	(3) protons and (4) neutrons	2	F
c	7	1	F
d	Dalton/Rutherford/Bohr/Thomson (any one)	1	F
		5	
11 a i	waste gases	1	
ii	iron ore	1	
iii	molten iron	1	
iv	hot air	1	E
b i	aluminium is very reactive/ high in reactivity series	1	
ii	electrolysis	1	E
		6	
12 a i	solution A	1	
ii	lowest pH	1	E
b	add alkali/neutralise it	1	E
c i	hydrogen	1	
ii	'pops' with lighted splint	1	E
		5	
13 a	40 cm^3	1	F
b	40 seconds	1	F
c	steeper/reach maximum faster	1	F
d	more particles/molecules to react	1	F
e	calcium carbonate + hydrochloric acid ⟶ calcium chloride + carbon dioxide	1	F
f	new substances formed	1	F
		6	
14 a	to stop it falling over	1	E
b	200 (1) Nm (1)	2	E
c	0.5 (1) m (1)	2	E
d	400 (1) Nm (1)	2	E
		7	
15 a	pressure = $\frac{force}{area}$ (1) = 120 (1) N/cm^2 (1)	3	F
b	$\frac{120}{0.1}$ = 1200 (N/cm^2)	1	F
		4	
16 a	wire (1) iron core (1) electricity (1)	3	F
b	increase current (1) more turns on coil (1)	2	F
		5	
17 a	LDR or thermistor	1	F
b	thermistor or LDR (depending on above)	1	F
c	NOT gate	1	F
d	NOT gate	1	F
e	AND gate	1	F
f	heater	1	F
		6	
18 a	correct refracted ray in block (1) correct refracted ray out of block (1)	2	E
b	label in correct place	1	E
c	they are parallel	1	E
		4	
19 a	black	1	F
b	blue	1	F
c	blue light only	1	F
		3	
20 a	loud to silent	1	
b	sound won't travel through vacuum/ sound needs something to travel through	1	E
c i	C	1	
ii	B	1	F
		4	
		TOTAL 99 marks	

Suggested grade/level boundaries
E = 34/99
F = 67/99

Book 1 ▶ Glossary

Chapter 2

Geologist	scientist who studies rocks
Igneous rock	rocks formed from liquid hot rock
Magma	hot liquid rock
Basalt	igneous rock formed when magma cools on the Earth's surface
Granite	igneous rock formed when magma cools in the Earth's crust
Seismology	the study of earthquakes
Seismic waves	vibrations in the Earth caused by earthquakes
Fault	lines of weakness in the Earth's crust
Weathering	break up of rocks
Transport	rock fragments carried by wind and/or water
Erosion	weathering and transport of rock fragments
Sedimentary rock	rock formed by sediment piling up and being squashed
Chalk/limestone	sedimentary rock formed from shells and skeletons of animals
Clay	weak sedimentary rock formed from tiny grains
Sandstone	sedimentary rock formed from small grains of sand
Metamorphic rock	rock made by changing another rock by heat/pressure
Slate	metamorphosed mudstone
Fossils	traces of prehistoric life found in sedimentary rock
Clay soil	badly drained soil with lots of tiny clay particles
Sandy soil	light soil with lots of sand particles
Loam	good soil with some sand, some clay and lots of humus
Humus	material in soil made from dead animals and plants

Chapter 3

Energy	things have energy if they can be used to do work
Energy conversion	change from on kind of energy to another
Sankey diagram	diagram showing what energy changes are taking place
Kinetic energy	moving energy
Potential energy	stored energy
Chemical potential energy	potential energy in fuel
Gravitational potential energy	potential energy in something high up
Hydroelectricity	electricity produced from the stored energy of water held back by a dam
Water/steam turbines	turn generators which produce electricity
Transmission	moved from one place to another
National grid	network which carries electrical energy around the country
Efficiency	the amount of energy available as useful energy
Step-up transformer	increases voltage
Step-down transformer	decreases voltage
Non-renewable	once gone, gone forever
Renewable	will not run out
Alternative energy source	energy sources that don't harm the environment by causing pollution
NIMBY	not in my back yard

Chapter 4

Organism	scientific name for a living thing
Life processes	things that animals and plants do, e.g. feed
Variation	differences between members of the same species
Continuous variation	features which are not easily separated; there are lots of possibilities, e.g. height
Discontinuous variation	distinct features; you either have them or you don't, e.g. ear lobes
Acquired variation	caused by what an organism does, e.g. dyes their hair
Classification	putting animals and plants into sets
Kingdoms	the biggest sets in classification
Vertebrate	animal with a backbone
Invertebrate	animal without a backbone
Monocotyledon	plant with long, narrow leaves and parallel veins
Dicotyledon	plant with broad leaves and a network of veins
Key	series of questions which help to identify an unknown organism
Sampling	small part of an area that is studied very closely
Transect line	piece of string or tape stretched across a survey area
Quadrat	0.5 m square grid used for sampling
Species	population of living things that can breed together to produce fertile offspring
Genus	group of organisms with common features, e.g. gulls

Chapter 5

State (of matter)	matter that is either solid, liquid or a gas
Solid	cannot move and keeps its shape and volume
Liquid	can flow, takes the shape of its container but volume stays the same
Gas	can flow, change its shape and volume
Changing state	changing from one state to another
Atom	smallest particles
Molecule	a group of atoms joined together
Diffusion	when one substance spreads through another
Air pressure	pressure caused by air molecules bouncing off a surface
Expand/expansion	take up more space
Contract/ contraction	get smaller
Density (of a substance)	the mass of 1 cm^3 of a substance

Chapter 6

Electrical conductor	allows electricity to pass
Electrical insulator	does not allow electricity to pass
Energy changer	device that changes electrical energy into another form of energy
Circuit	continuous path taken by electricity
Series circuit	circuit where each component follows the other on a single pathway
Parallel circuit	circuit where there is more than one pathway for electricity to follow
Ring main	a parallel circuit in the home
Current	flow of electricity in a circuit
Amperes/amps	what electric current is measured in
Ammeter	device used to measure current
Voltage	driving force of an electric current
Voltmeter	device used to measure voltage
Resistance	slowing/stopping the flow of electricity
Resistor	something which resists the flow of electricity

Chapter 7

Cell	'building block' from which living things are made
Cell membrane	thin skin surrounding the cell which allows substances to pass through
Cell wall	rigid coat round the outside of plant cells which gives the cell support
Chloroplasts	green structures inside the cytoplasm of plant cells
Chlorophyll	green chemical which traps sunlight energy needed for photosynthesis
Cytoplasm	jelly-like substance in a cell where chemical reactions happen
Nucleus	control centre of a cell
Vacuole	large space in the middle of plant cells filled with cell sap
Tissue	group of cells that do the same job
Organ	part of the organism, e.g. heart (made from different tissues)
Organism	scientific name for a living thing. Made up of different organs
Sexual reproduction	production of new individuals from sex cells
Secondary sexual characteristics	changes to the body that happen at puberty
Puberty	time when a boy or girl becomes sexually mature
Eggs	female sex cells
Sperm	male sex cells
Testes	male sex organs where sperm are made
Ovaries	female sex organs where eggs are made
Sexual intercourse	when the male moves his penis up and down in the female's vagina
Ejaculation	release of sperm from the testes through the erect penis
Fertilisation	when the nucleus of a sperm joins with the nucleus of an egg or the nucleus of a pollen grain joins with the nucleus of an ovule
Embryo	developing baby
Amnion	bag containing the embryo inside the womb
Amniotic fluid	liquid in the amnion which cushions/protects the embryo
Umbilical chord	connects the embryo to the placenta
Placenta	allows food/oxygen and waste to be exchanged with mother's blood in the womb wall
Afterbirth	placenta and remains of the umbilical chord
Menstruation (period)	monthly loss of blood from the womb lining which passes out of the vagina
Chromosome	thread-like structures in the nucleus of every cell

Genes	parts of chromosomes that determine the features of an organism
Gene alleles	different forms of the same gene
Dominant	stronger allele in a pair of alleles
Recessive	weaker allele in a pair of alleles
Pollination	transfer of pollen from anthers to stigmas
Cross pollination	transfer of pollen from the anther on one plant to the stigma on another plant
Self pollination	transfer of pollen from the anther to the stigma on the same plant
Fruit	developing ovary with the seeds inside
Germinate/ germination	when a seed begins to grow

Chapter 8

Element chemical substance that cannot be broken down into anything simpler because it is only made of one type of atom

Compound chemical substance that has more than one kind of atom in it

Mixture when two or more substances are mixed together without joining up

Formula chemical shorthand that tells you which atom and how many atoms are joined up

Chemical equation shorthand way of writing what happens in a chemical reaction

Chemical reaction production of new material by changing atoms or molecules

Burning chemical reaction caused by heating that leads to a permanent change

Flammable something that burns easily

Hydrocarbons substances made up of hydrogen and carbon

Air pollution harmful substances in the air

Global warming heat trapped near the Earth by 'greenhouse gases'

Greenhouse gas gas, e.g. carbon dioxide which causes global warming

Chapter 9

Kilojoule (kJ) what energy is measured in

Heat measure of energy

Temperature measure of hotness or coldness

Absolute zero (-273 °C) temperature at which a material has no heat energy

Kinetic theory the idea that everything is made of moving particles

Conductor (of heat) allows heat energy to travel through it

Insulator (of heat) does not allow heat energy to travel through it

Conduction heat energy being passed through a material

Convection heat energy carried upwards by a liquid or a gas

Convection current circulation of liquid or gas caused by convection

Radiation heat energy being 'thrown ' across open space as straight rays

Insulation bad heat conductor used to keep heat in

Solar energy heat (and light) energy from the Sun

Chapter 10

Photosynthesis	plants using light energy to make food
Carbohydrates	high energy foods
Veins	bundles of tiny tubes which carry food and water around
Midrib	large vein running along the middle of a leaf
Leaf stalk	connects the leaf to the stem
Limiting factor	something that slows down or stops something
Food chain	food passing from plant to animal to animal etc.
Food web	interconnecting food chains
Producers	green plants
Primary consumer	animal that eats green plants
Secondary consumer	animal that eats other animals
Predator	secondary consumer
Prey	animal eaten by a predator
Plankton	microscopic animals and plants that live in the sea
Habitat	place where an animal or plant lives
Pyramid of numbers	diagram showing the numbers of organisms in each link in a food chain
Pyramid of biomass	diagram showing the mass of organisms in each link in a food chain
Biomass	mass of living material

Chapter 11

Solar system	Sun, planets, comets, and asteroids
Year	length of time taken for a planet to orbit the Sun once
Day	time taken for the Earth to spin once
Orbit	the path followed by an object as it moves around a planet, moon, or the Sun
North Pole	northern point of the Earth's axis
South Pole	southern point of the Earth's axis
Axis	line between the N and S poles around which the Earth spins
Phases of the Moon	apparent changes in the shape of the Moon
Eclipse	when a planet or moon blocks light from another
Solar eclipse	when the Moon gets directly between the Sun and the Earth
Lunar eclipse	when the Earth gets directly in between the Sun and the Moon
Astronomer	scientist who studies stars and planets
Optical telescope	light is focussed by mirrors and lenses onto a detector
Radio telescope	radio waves are focussed onto a detector
Hubble space telescope	orbits the Earth using digital cameras to collect pictures and send electronic signals to Earth
Universe	all the galaxies; everything that exists
Light year	the distance that light travels in one year (9.46 million million km)
Galaxy	huge system of billions of stars
Black hole	what is left when a massive star collapses at the end of its life. A black hole's gravity is so strong that nothing can escape, not even light
Gravity	force of attraction between two large objects
Big bang theory	theory that the Universe was formed when a massive 'super atom' exploded and expanded outwards. Planets and stars were formed from condensing gases

Chapter 12

Force push or a pull. It is measured in Newtons (N)

Friction force produced when two surfaces rub against each other

Air resistance friction between the air and an object moving through it

Streamlining shaping something to reduce resistance

Upthrust upward pressure

Balanced forces the force on one side of an object equals the force on the other side, so the object stays still or moves at a constant speed

Unbalanced forces the force on one side of an object is different to the force on the other side, so the object starts to move or moves faster

Gravity force of attraction between two large objects

Weight the force of gravity pulling down on an object

Mass how much matter an object contains

Gravitational field the force of the Earth's gravity

Newton (N) the unit of force named after Sir Isaac Newton (1642–1727)

Kilogram (kg) what weight is measured in

Pressure result of a force pressing on a certain area. Measured in N/cm^2

Hydraulic system system filled with water or oil, e.g. a car braking system

Pneumatic system system filled with gas or air, e.g. pneumatic drill

Fluid something that flows, e.g. liquids and gases

Chapter 13

Carbohydrates	sugary and starchy foods that give energy
Fats	solid and liquid (oil) foods that give energy and insulate the body
Proteins	body-building foods
Minerals	substances that are needed for a healthy body
Roughage (fibre)	plant cell walls that pass through the gut without being digested
Vitamins	help control chemical reactions in the body
Plasma	the liquid part of the blood
Capillaries	very small blood vessels
Tissue fluid	mainly water which surrounds all the tissues in the body
Incisors	cutting teeth at the front of each jaw
Canine teeth	biting/tearing teeth
Pre molars and molars	back teeth used for grinding food
Dentine	soft layer beneath the enamel in teeth
Pulp cavity	living part of a tooth containing blood vessels and nerves
Enamel	hard, non-living surface of a tooth
Plaque	coating on the surface of teeth in which bacteria grow
Digestion	breaking down food into smaller, soluble pieces
Digestive system	the organ system where digestion happens
Enzymes	substances that break down (digest) food into tiny soluble pieces
Amylase	enzyme that digests starch
Breathing	getting oxygen into the body and carbon dioxide out
Diaphragm	sheet of muscle at the bottom of the rib cage
Cilia	moving hairs inside the nose, windpipe, and bronchi
Bronchi	tubes connecting the two lungs to the windpipe
Mucus	sticky fluid lining the nose, windpipe and bronchi
Alveoli	air sacs in the lungs where oxygen and carbon dioxide are exchanged between the air and the blood
Respiration	getting energy out of food using oxygen
Lock and key theory	idea for how enzymes work. Enzyme molecules and food molecules fit together like a key in a lock

Chapter 14

Malleable shape can be changed by hammering

Ductile metal drawn out into a thin wire

Reactivity series 'league table' with the fastest reacting metals at the top and the ones that don't react at the bottom

Displacement reaction reaction in which a metal higher in the reactivity series takes the place of (displaces) a metal lower in the reactivity series in a solution

Native metal pure metal found in the ground, e.g. gold

Metal ores metals found as part of compounds in the ground, e.g. iron oxide

Smelting heating a metal ore so the metal can be separated from it

Electrolysis using an electric current to break down a solution

Electroplating using an electric current to put a metal coating onto another a metal

Corrosion chemical (water, air, or acid) attack on the surface of a metal

Recycling using waste materials to make new ones

Metal alloy mixture of two or more metals

Proton part of an atom carrying a positive charge

Electron part of an atom carrying a negative charge

Neutron part of an atom which has no charge

Nucleus (of an atom) very dense centre of an atom where the protons and neutrons are found

Electron shells 'orbits' in which electrons spin around the nucleus of an atom

Atomic number number of protons in an atom

Atomic mass number of protons plus the number of neutrons in an atom

Periodic Table arrangement of elements in order of increasing atomic number. Families of elements are grouped together

Chapter 15

Shadow	area behind an object where light is missing
Transparent	lets light through
Opaque	does not let light through
Translucent	lets some light through
Plane mirror	mirror with a flat reflecting surface
Incident ray	ray of light hitting a mirror
Reflected ray	ray of light reflected from a mirror
Angle of incidence	angle between the incident ray and a line at right angles to the surface of the mirror
Angle of reflection	angle between the reflected ray and a line at right angles to the surface of the mirror
Concave mirror	mirror with a reflecting surface curved inwards
Convex mirror	mirror with a reflecting surface curved outwards
Concave lens	lens with two sides curved inwards
Convex lens	lens with two sides curved outwards
Refraction	bending of light as it passes from one transparent material to another
Spectrum	colours of the rainbow
Absorb/absorption	cutting out some or all light
Transmission/ transmitted	light passing through something
Primary colours for light	red, blue and green light
Secondary colours for light	cyan, yellow and magenta
Primary colours for paint	red, blue and yellow
Volume	make a sound louder or softer
Pitch	make a sound higher or lower
Vacuum	empty space
Noise pollution	harming the environment with too much noise
Soundproofing	reducing noise pollution by using sound absorbing materials
Decibel (dB)	what sound is measured in
Frequency	number of sound vibrations in one second. Measured in hertz (Hz)
Amplitude	size of the sound vibrations
Oscilloscope	instrument used to show sound vibrations as waves

Chapter 16

Microorganism very tiny living thing that can only be seen with a microscope

Microbe common name for a microorganism

Bacteria very small living cells with no nucleus

Virus simpler and much smaller than bacteria. They are not cells, just a protein shell with DNA inside. Can only reproduce inside living cells

Microscopic fungi yeasts and moulds which feed on dead and decaying material

Decomposers microbes which make things rot/decay

Biodegradable material which will decompose/rot

Biotechnology using microbes to make useful things

Genetic engineering taking genes from one organism and putting them into another

Insulin chemical (hormone) in the body that controls the level of sugar in the blood

Diabetic person who cannot make enough insulin of their own; they have diabetes

Disease caused by microbes getting into the body

Infection bacteria getting into the body

Immune system defence system of the body against microbe attack

Antibodies chemicals made by blood cells which kill microbes

Immune protected against a disease

Vaccine dead or weak microbes injected into the body to make the body produce antibodies against a disease

Chromosomes thread-like structures in the nucleus of every cell

Genes parts of chromosomes that determine the features of an organism

DNA deoxyribonucleic acid. What chromosomes and genes are made of

Gene alleles different forms of the same gene

Dominant stronger allele in a pair of alleles

Recessive weaker allele in a pair of alleles

Selective breeding another name for artificial selection

Artificial selection choosing and breeding the best animals or plants to produce a desired feature, e.g. cows that produce more milk

Clones genetically identical organisms

Asexual reproduction producing new individuals (natural clones) without using sex cells

Micropropagation growing new plants from very small (microscopic) pieces

Tissue culture new plants grown from a few cells (tissue)

Genetic variation differences between members of the same species

Gene pool all the genes in all the members of a species

Mutation a change in the instructions carried by a chromosome or gene

Chapter 17

Evaporate/ evaporation	when liquid particles get enough kinetic (moving) energy to escape the liquid as a gas
Condense/ condensation	when gas particles lose kinetic (moving) energy and become a liquid
Saturated air	can hold no more water vapour
Adiabatic cooling	when the pressure on a gas is reduced the gas expands and cools down
Solution	formed when two substances mix
Solute	the substance which dissolves in a solution
Solvent	the substance which does the dissolving in a solution
Soluble	will dissolve
Insoluble	will not dissolve
Saturated solution	can hold no more solid/solvent
Crystallisation	the formation of crystals when a saturated solution cools down or evaporates
pH scale	used to measure how acid or alkaline a substance is
Neutralisation	chemical reaction in which an acid or an alkali is cancelled out
Salts	one of the products when an alkali exactly neutralises an acid
Indicator	extracts from plants whose colours are affected by acids and alkalis
Catalyst	chemical which changes the speed of a chemical reaction without being changed in the reaction
Catalytic converter	helps convert harmful gases into harmless ones
Chemical change	chemical reaction in which one (or more) new substance is formed
Physical change	substance changes its form, e.g. ice to water, but no new substance is formed

Chapter 18

Electromagnet coil of wire around a soft iron core. It becomes magnetised when a current passes through the coil but loses its magnetism when the current is switched off

Magnetic field region around a magnet where forces act on any magnetic material

Electronic systems include an electronic circuit that processes information

Input converts information into electrical signals for a processor

Processor modifies information received from an input and makes a decision

Output receives electrical signals and reacts, e.g. by switching on a heater

Linear activator device that converts electrical signals into motion in a straight line

Diode only lets a current flow in one direction

Resistor reduces the current flowing through the components of an electronic system

Transistor can be used as an amplifier or an on-off switch

Analogue uses electrical signals to represent continuous physical quantities

Digital switching circuits that can either be on or off. Use binary logic

Amplifier increases voltages by multiplying them or adding them together

Power driver an amplifier chip which can handle very high power outputs

Binary logic uses only two numbers, 1 and 0. 1 means 'on' and '0' means off'

Logic gates the decision-makers of electronic processors

NOT gate has one input and one output. When input is 1 its output is 0 and visa versa

AND gate has two inputs and one output. Both inputs must be the same to give an output of 1

OR gate has two inputs and one output. The output is 1 when one or the other (or both) inputs are 1

Truth table table that shows the outputs for all the possible input combinations

Chapter 19

Environment scientific word for surroundings

Abiotic non-living parts of the environment

Biotic living parts of the environment

Ecosystem living things together with the abiotic parts of their environment

Acid rain caused when harmful gases, e.g. sulphur dioxide dissolve in rain making an acid

Global warming heat trapped near the Earth by 'greenhouse gases'

Pesticides chemicals used to kill pests, e.g. insecticides kill insects

Extinct animal or plant gone forever

Biodegradable material which will decompose/rot

Organic farming method of farming where no chemical fertilisers or pesticides are used

Fossil traces of prehistoric life found in sedimentary rock

Habitat where an animal or plant lives

Adapted the way animals and plants have developed special features to help them cope with their way of life

Colonisers first plants to grow on an area of bare earth

Evolution production of new species from existing ones by small changes over time

Archaeopteryx the first known bird. It evolved from a reptile

Pterodactyl flying reptile that lived millions of years ago

Natural selection animals and plants best suited to their surroundings survive and breed

Fossil record information from fossils about what animals and plants lived on Earth and when

Missing link gap in the fossil record

Mutation change in the instructions carried by a chromosome or gene (the DNA is altered)

DNA deoxyribonucleic acid. What chromosomes and genes are made of

Chromosomes thread-like structures in the nucleus of every cell

Genes parts of chromosomes that determine the features of an organism

Chapter 20

ET (Extra terrestrial)	intelligent life out in space
SETI project	the Search for Extra Terrestrial Intelligence
Pulsar	star which sends out signals in regular pulses
Trebuchet	ancient weapon similar to a huge lever
Biological warfare	using organisms or their poisonous products to kill people
Microorganism	very tiny living thing that can only be seen with a microscope
Microbe	common name for a microorganism
Microbial enzyme	enzymes produced and collected from microbes
Amylase	enzyme which breaks down starch into glucose
Lactase	enzyme which beaks down milk sugar (lactose) into glucose
Protease	enzyme which breaks down proteins into amino acids
Pectinase	enzyme which breaks down pectin (a carbohydrate) into soluble sugars
Americium	radioactive element used in smoke alarms
Laser	Light Amplification by Stimulated Emission of Radiation. A laser produces a very intense beam of light
GM	stands for genetic modification. It is what biotechnologists do when they genetically engineer living organisms
Fractional distillation	way of separating mixtures by evaporating at different temperatures
Haber process	industrial manufacture of ammonia from nitrogen and hydrogen
Input	converts information into electrical signals for a processor
Processor	modifies information received from an input and makes a decision
Output	receives electrical signals and reacts, e.g. by switching on a heater
Logic gates	the decision-makers of electronic processors